Public Undertakings of Nuclear Waste Storage

STUDIES IN POLITICS, SECURITY AND SOCIETY

Edited by Stanisław Sulowski
Faculty of Political Science and International Studies
University of Warsaw

VOLUME 50

Silvia Amato

Public Undertakings of Nuclear Waste Storage

The Role of the Government in the Dissemination of Public Knowledge

Bibliographic Information published by the Deutsche Nationalbibliothek
The Deutsche Nationalbibliothek lists this publication in the Deutsche Nationalbibliografie; detailed bibliographic data is available in the internet at http://dnb.d-nb.de.

Library of Congress Cataloging-in-Publication Data
A CIP catalog record for this book has been applied for at the Library of Congress.

Cover illustration: Courtesy of Benjamin Ben Chaim.

ISSN 2199-028X
ISBN 978-3-631-89018-9 (Print)
E-ISBN 978-3-631-89019-6 (E-PDF)
E-ISBN 978-3-631-89064-6 (EPUB)
DOI 10.3726/b20220

© Peter Lang GmbH
Internationaler Verlag der Wissenschaften
Berlin 2022
All rights reserved.

Peter Lang – Berlin · Bern · Bruxelles · New York · Oxford · Warszawa · Wien

All parts of this publication are protected by copyright. Any utilisation outside the strict limits of the copyright law, without the permission of the publisher, is forbidden and liable to prosecution. This applies in particular to reproductions, translations, microfilming, and storage and processing in electronic retrieval systems.

This publication has been peer reviewed.

www.peterlang.com

Preface

In this research literature about nuclear policy and the process of nuclear waste disposal, have been introduced some major arguments that can actually retrace a regulative profile on concurrent diversification processes explored in the case of renewable energy policies. A comprehensive focus put on local knowledge transfers in governance identifies some emerging factors that play a role in the regional state regulations and institutional industrial energy aims. In regional development policies, the current association of governmental knowledge transitions is analyzed in order to explain nuclear power policy and nuclear facilities management, responsive to a progressive regulative integration that takes into account science policy directions, local knowledge innovation, and transitional adaptive changes. More specifically, what comes into light in terms of multi-level energy strategies and public empowerment conditions has been the increasing complexity of practical and organizational dimensions of nuclear energy management systems that provide across international regions constructive elements for standardized innovation planning and public policy assessments, evolving in connection with environmental governance policies and local knowledge transfers. Therefore, the similitude of systemic approaches supporting national safety practices and structural regulative measures has reflected negotiated relational conditions that require a functional managerial complementarity but, at the same time, public-private transparency and cooperative mutual reliability.

TABLE OF CONTENTS

ACKNOWLEDGEMENTS 11

INTRODUCTION 13

CHAPTER ONE 15
  1.1 Cooperative Interactions 15
      1.1.1 Regional Environmental Aims 23
      1.1.2 Regulated Consensus in Governance 32
  1.2 Technology Risk Perspectives 35
      1.2.1 Incremental Group's Visibility 39

CHAPTER TWO 45
  2.1 Innovative Transition Systems 45
      2.1.1 Transitional Technology Cooperation 50
  2.2 Public Managerial Ties 53
      2.2.1 Public Networking 57
      2.2.2 Normative Transformation 65

CHAPTER THREE 69
  3.1 The Governance Networks 69
      3.1.1 Environmental Cooperation 75
      3.1.2 Collaborative Platforms 79

CHAPTER FOUR ... 85
4.1 Holistic Integration Approaches ... 85
    4.1.1 Clustering Provisions ... 88
    4.1.2 Territorial Reconstructions ... 93

CHAPTER FIVE ... 99
5.1 Public Territorial Approaches ... 99
    5.1.1 Provisional Requirements ... 102
    5.1.2 Community Attention ... 116
    5.1.3 Industrial Concentration ... 120

CHAPTER SIX ... 125
6.1 Socio-Technical Transitions ... 125
    6.1.1 Local Practice Projects ... 127
    6.1.2 Nuclear Waste Disposal ... 135

CHAPTER SEVEN ... 137
7.1 Preferential Ecosystems ... 137
    7.1.1 Local Adoption Agreements ... 142
    7.1.2 Formal Ethic Aggregations ... 144

CHAPTER EIGHT ... 149
8.1 Sectoral Participation ... 149
    8.1.1 Public Stratification ... 152

CHAPTER NINE ... 159
9.1 STS States Innovation Patterns ... 159
    9.1.1 Public Analytic Settings ... 162
    9.1.2 Processing Country Targets ... 165

TABLE OF CONTENTS 9

9.1.3 Supporting Technical Views 169
9.1.4 Public Sustenance Applications 175
9.1.5 Public Diversification Plans 179
9.1.6 Local Demonstrative Cases 182
9.1.7 Environmental Diffusion Optimization 186
9.1.8 Systemic Transfer Orientations 190

CHAPTER TEN 193
10.1 National Configurations 193
    10.1.1 Centralized Operations 197
    10.1.2 Mutual Communities and Critical Knowledge Intersections 201
    10.1.3 Material Dualist Interplay 204

CHAPTER ELEVEN 215
11.1 Territorial Governance Connectivity 215

CHAPTER TWELVE 223
12.1 Regional Specialized Networks (RIS) 223
    12.1.1 Local Ecological Policies 230
    12.2.2 Local PPP Indications 233

CHAPTER THIRTEEN 239
13.1 Local Territory and Local Identity 239
    13.1.1 Public Ecosystems Strategies 243

CHAPTER FOURTEEN 251
14.1 Transnationalism 251

CHAPTER FIFTEEN 257
  15.1 Local Mobilization Sectors 257

CHAPTER SIXTEEN 273
  16.1 Conclusion 273

BIBLIOGRAPHY 287

Index of Names 335

ACKNOWLEDGEMENTS

I would certainly acknowledge the support I have had during the Doctoral Research Study conducted at National Chengchi University, College of Social Sciences (CSS) (NCCU), (IDAS) International Doctoral Program in Asia-Pacific Studies, Taipei, Taiwan, where I obtained the PhD in 2017–18. This book is an earlier part of the doctoral dissertation that I submitted at the CSS/IDAS department. I have worked on the earlier draft of the dissertation and finally managed to produce this book. However, as of yet, there are no funding opportunities for this project.

Author: Silvia Amato

INTRODUCTION

The following chapters present a path for reviewing a body of literature that explains the international issues concerning nuclear policy, and the thematic reconstructions about nuclear waste disposal programs. Through a common view acquired about technology diffusion methodologies and responsive governmental models developing comparable science ecosystems and multi-level knowledge innovation transfers, it is important to scrutinize international nuclear safety domains and nuclear waste storage disposal activities of industrialized and semi-industrialized regions.

The application of the concept of 'knowledge innovation transfers' in governmental frameworks has been reviewed in the case of historic managerial alignments in public-private adaptative regional agreements. This study reveals the formal and informal coexistence of civilian participatory conditions and technology innovative productions in international renewable energy fields that have explored comparative dynamic collaborations of local governmental environments, particularly in specialized contexts.

At the preparation stage, there have been technical institutional orientations and managerial innovation programs which aimed to establish long-term development projects of nuclear energy facilities combined with nuclear waste disposal configurations. At a regional and local level, the implementation of systematically preferential programs and comprehensive decision-making regulations has resulted in drafting renewable energy policy transitions as well as comparable technology dynamic transfers.

Within the contextual regional environments, science and knowledge processes have committed to public integrated accesses and governmental adaptability issues. Essentially, this book discusses the connection between participatory governance and the emerging knowledge adaptation models which refer to multi-level nuclear power energy development and nuclear waste management processes that have been envisioned within public preferential dynamics tackling local environmental protection strategies.

For this reason, there is an analytical component to the science, technology, and society (STS) approach, identified in regional East Asian studies. By adopting a socio-political perspective on STS issues, the defining initiatives for more inclusive science and technology programs have been established in relation to societal changing processes and comparative knowledge innovation transfers. In this analysis, nuclear power energy and national renewable energy development concepts have been combined with public policy planning and managerial innovation systems that require participatory conditions as well as responsive facilitation mechanisms.

Local development and local integrated communities are particularly involved in public regulation plans that directly relate to technological material changes and prevention of local environmental risks. Fundamentally, as reported in this project, contextual public consensus for nuclear energy programs and nuclear waste siting process can determine different deliberative frameworks that promote participatory decision-making for environmental advocacy and public consultative arenas. Therefore, the discussion that follows based on nuclear science policy includes critical interpretative studies pointing at the relationship between multi-level governance interactions and public knowledge transfers. Taken in isolation the processing of industrial and technology innovation factors associated with renewable energy models has indirectly imported socio-economic differentiations favoring local knowledge transfers. At regional level, the energy development and governance adaptation plans have resulted in programmatic aims and defining scopes for nuclear waste disposal programs.

In summary, in this research study, we refer to local governmental regulations and to the type of relationship that has been implemented through private-public partnerships (PPPs) based on nuclear hazards management programs. Nuclear waste management policy (NWMP) has mirrored industrial political transformations that involve prevalent associations of participatory societies and political alliances, which remain difficult to fully integrate in knowledge innovation and technology transfers that pertain to versatile technical regulative dimensions.

CHAPTER ONE

1.1 Cooperative Interactions

Formally, major leading aspects of the identification process of local radioactive waste management and regional planning systems have been associated with the dynamic inter-operability of systematic production relations, evaluated in relation to policy programs designed at the local, national and international level. In this context, comparative institutional relations have been established through the adoption of participatory policy approaches.

Parallel to socio-technical initiatives, regional environmental planning, advanced with reference to progressive decision-making activities, has been reviewed according to different structural and technical interactions, especially when based on national energy security plans and nuclear waste hazard policies. In effect, the structural operating components linked to environmental approvals systems, while intertwined with decision-making transitions programs involving territorial development, have been important drivers for the common coordination of regional energy innovation paths.[1] Through a comparative literature review of nuclear science and policy studies that retrace the open partnerships for comparable countries within diversified energy areas, combined with local governance adaptation[2] models, it can be underlined that nuclear science policy has been formulated in relation to territorial integrative components involving, for instance, the nuclear cleanups management and nuclear remediation projects, or land restoration activities – which have been particularly tackled for parallel ecosystem dynamics, with the recognition of inclusion of major local stakeholders consulted for the disposition of large-scale nuclear waste disposal projects.[3]

Through the facilitation of public dialogue for possible societal and normative attributions, multi-level territorial experts have been engaged in non-profit organizations, governmental and management agencies, among other associations, in order to redraw common public and private socio-economic alliances (PPPs)

---

1 Joanna Burger, Michael Gochfeld, David S. Kosson, et al., "Science, Policy, and Stakeholders: Developing a Consensus for Amchitka Island, Aleutians, Alaska," *Environmental Management* 35. 5/2005, pp. 557–568.
2 See: Carmel Letourneau, (2010). "Adaptive Management of Complex Environmental Problems – Comparison of National Nuclear Waste Management Policies," PhD Dissertation at Graduate College University of Nevada, Las Vegas, United States.
3 Burger, Gochfeld, Kosson, et al., "Science, Policy, and Stakeholders," pp. 557–568.

that also refer to science policy partnerships.[4] At the same time, the evolution of participatory dialogue related to environmental political processes has been explored in this text through international academic studies focusing on national and regional public environmental confronted notions. For instance, the regional nuclear risk assessment procedures analyzed in comparative research studies[5] have been associated with local safety issues that have integrated the environmental impact analyses, regional shifts of renewable energy policy domains, and productive dimensions of responsible businesses.

Regarding the duality of science provisions reviewed through qualitative interpretations and empirical processes, the fundamental element of civilian participatory engagement has emerged in public governance cases, involving discussions on industrial transformation plans, while possibly retracing the level of protectionist interventions associated with environmental policy measures. According to environmental policy indications, the relationship established between mutual industrial parties in governance has involved a respective degree of understanding of public regulatory recognitions and local distributive powers that are also defined through consistent territorial analyses[6] targeting the evolution of international management relations.[7] From an environmental standpoint, the binding elements of the regional transition process influencing industrial power networks who operate through collaborative adaptations, have reflected participatory stages for civilian representative stakeholders. In particular, local collaborative stages have been determined in combination with organizational and interdependent accesses on transitional knowledge and information factors, which have been transferred in terms of ethical norms, strategic planning, and compatible business environments.[8]

---

4   Jonathan P. West, "Public-Private Policy Partnerships," *The American Political Science Review* 95.1/2001, pp. 219–220.
5   See: Yong-Jin Cha, (1997). "Environmental Risk Analysis: Factors Influencing Nuclear Risk Perception and Policy Implications," PhD Dissertation at Rockefeller College of Public Affairs and Policy, Department of Public Administration and Policy, State University of New York, U.S.
6   See: Laura Jeanne Russell, (2002). "Collective Reciprocal Causation: A Model of the Relationships Among Group Behavior, Group Environment, Group Cognition, and Group Effectiveness," PhD Dissertation at the Graduate Faculty of Auburn University, Alabama, United States.
7   Jon Burchell, Joanne Cook, "Sleeping with the Enemy? Strategic Transformations in Business-NGO Relationships Through Stakeholder Dialogue," *Journal of Business Ethics* 113/2013, pp. 505–518.
8   Burchell, Cook, "Sleeping with the Enemy?" pp. 505–518.

In comparative assessments, the transnational policy development referring to international business studies[9] has combined the exploration of regional energy market policies with public interest grounds put in relation to an emerging political activism, concerning non-governmental organizations (NGOs) among other institutional counterparts and local partners. Essentially, the systematic configuration of governance partnerships has included the interaction of multi-level agents for the adoption of industrial mutual arrangements established in line with technical managerial commitments.[10] Therefore, from theory to practice common governance notions that surface about collective environmental concerns and sustainable development conditions have been framed through national and international policy management guidelines drafted within respective administrative frameworks, which essentially can facilitate a consensus-type model of societal and business interactions, in order to generate potential feedbacks for long-term development impacts.[11]

Concurrently, there are affirmative diffusions of strategic business alliances for efficiency and responsiveness standards found in national implementation activities of industrial energy programs that have been modified in terms of public service formation, which has been assessed in view of common distribution factors, related to the propagation of legitimate norms and economic development concessions. But, it has to be specified that regional levels of collective representations for regulatory trading policies and states' mutual organizational interests have been introduced through concerted local accesses – delineated in terms of upgraded information and redefined regulative measures – including mass entrepreneurial alternative commitments for regional cases of environmental protection schemes and local conservation plans.[12]

In addition to this, the participatory implications regarding public-private managerial relations have been extremely contingent to national and local territorial entities and departmental institutions, which have experienced the aggregation of policy integrative practices directly affecting the development of local communities' welfare. On such critical accounts, the science technology field

---

9 See: Jason Hall McNichol, (2002). "Contesting Governance in the Global Marketplace: A Sociological Assessment of Business-NGO Partnership to Build Markets for Certified Wood," PhD Dissertation at Sociology Graduate Division of the University of California, Berkeley, U.S.
10 Burchell, Cook, "Sleeping with the Enemy?" pp. 505–518.
11 Kevin Gibson, "Stakeholders and Sustainability: An Evolving Theory," *Journal of Business Ethics* 109/2012, pp. 15–25.
12 Gibson, "Stakeholders and Sustainability," pp. 15–25.

of nuclear energy policy has been interrelated with nuclear waste management regulations which have been advanced according to the presence of industrial knowledge transfer capabilities, identified in connection with more inclusive environmental platforms and local consultation open forums, introduced in accordance with regional, national, or local deliberative practices.[13]

The international and national political economy[14] field experts[15] have pointed out the fact that participative and organizational contributions formulated within intraregional industrial societies have led to the affirmation of environmental justice criteria and public legitimacy practices. At the same time, the incrementation of environmental cooperation measures has been consolidated in parallel with the formalization of national managerial exchanges seeking in particular to actualize the diffusion of equal open accesses to transmissible information services.[16] The eventual elaboration of prospective national environmental goals intertwined with regional economic policies which can integrate the modified regional planning systems – through possible inclusive measures for local citizens' supporting activities – has been studied by comparing regional governmental analyses based on actualized political motivations and interdependency of socio-economic integrative spheres.[17]

Essentially, a multi-disciplinary comparative adaptation in governmental and knowledge management innovation areas can be performed through the consolidation of distinctive territorial approaches that have been examined for practical and theoretical PPPs frameworks. Through selective technology and knowledge areas, the common adoption of managerial innovation approaches looking into the establishment of participatory processes can be politically translated into binding norms, as well as, into non-binding practical formations,

---

13 Genevieve Fuji Johnson, "The Discourse of Democracy in Canadian Nuclear Waste Management Policy," *Policy Science* 40/2007, pp. 79–99.
14 See: Benjamin K. Sovacool, Scott Victor Valentine, "The Socio-Political Economy of Nuclear Energy in China and India," *Energy* 35.9/ 2010, pp. 3803–3813, https://doi.org/10.1016/j.energy.2010.05.033. (6 Aug.2022).
15 See: Lucie Laurian, "Public Participation in Environmental Decision Making: Findings from Communities Facing Toxic Waste Cleanup," *Journal of the American Planning Association* 70.1/2004.
16 Fuji Johnson, "The Discourse of Democracy," pp. 79–99.
17 See: 1) Raymond B. Ludwiszewski, "The Interdependence of Economic Development and Environmental Protection," *Economic Development Review* 10. 3/1992. 2) Seong-Jai Kim, "Korean Waste Management and Eco-Efficient Symbiosis – a Case Study of Kwangmyong City," *Clean Technologies Environmental Policy* 3/2002, pp. 371–382.

especially when referring to public regulative interactions connected to public stakeholders' positions and to industrial planners' production models.[18]

When we think in terms of national energy implementation policies from a governance perspective, we need to take into account the fact that governmental bureaus, as well as, networked ministerial agencies have performed corresponding multi-level functions according to public participation scenarios. In particular, for critical environmental decision-making activities the territorial formative meetings and convergent open discussions have been essential to the enhancement of public cooperation levels, through which local representative parties can formally address diversified environmental issues, also associated with national security aspects and territorial defense implications – both emerging from management activities run at military sites which in essence have been concurrently organized in view of territorial adaptive plans set for strategic land dispositions.[19]

Specifically, the case of nuclear waste management and its geological disposal regulations have been associated with the governmental distribution capacities defining major collective roles and local responsibilities for territorial management sites. It can be identified a common set of public assessment strategies and national policy directives, which have included the adoption of public participatory approaches that have been structured according to reciprocal affirmative conditions for the evaluation of environmental strategic expectations and local integrative contributions of comparative urban societies. For instance, international policy guidelines that have been highlighted about the application of regulatory decisional options concerning regional nuclear energy industrial designs, have also offered a comparative understanding of the multi-level structural processes conducted by public-private trading delegations, and committed national officials, in relation to civil nuclear energy projects. In this case, it can be applied this evaluative OECD/NEA scheme specifically pointing at major roles and responsibilities regarding managerial regulatory issues on nuclear waste safety that have been reported in OECD/NEA tabulation format described in Table 1.

---

18 Fuji Johnson, "The Discourse of Democracy," pp. 79–99.
19 Simon M. Clark, "Public Participation in Decisions Relating to the Environmental Management Ministry of Defence Sites," in: *Defence and the Environment: Effective Scientific Communication*, eds. K. Mahutova et al., (Kluwer Academic Publishers. Printed in the Netherlands, 2004), pp. 65–70.

Table 1: OECD International Regulatory Framework

"Stakeholders"	"Traditional expectations for roles and Responsibilities"	"Evolving expectations for roles and responsibilities"
"Policy-makers"	"Defining policy options, investigating their consequences under different assumptions, making policy choices."	"Informing and consulting stakeholders about policy options, assumptions, anticipated consequences, values and preferences. Setting the "ground rules" for the decision-making processes. Communicating the bases of policy decisions."
"Nuclear safety Regulators"	"Defining regulatory requirements and guidance. Defining a regulatory process, making choices regarding regulatory options. Reviewing the implementer's safety options and design and asking for possible complements or modifications. Making decision on the step forward. Reviewing and validating operational rules. Controlling the compliance of operation with operational rules. Communicating the bases of regulatory decisions."	"Maintaining open and impartial regulatory processes. Providing stakeholders with understandable explanations of the mechanisms of regulatory oversight and decision-making, including explanations of the opportunities available for stakeholder participation therein. Serving as a source of information and expert views for local communities."
"Scientific experts, consultants"	"Carrying out scientific/technical investigations with integrity and independence. Advising institutional bodies such as the nuclear safety regulator, other authorities and implementing agencies on technical issues in relation with safety concerns in view of providing balanced and qualified input for decision-making."	"Acting as technical intermediaries between the general public and the decision-makers within the limits of the mandate that they have received from the organization upon which they depend. Providing balanced and qualified input for all stakeholders and encouraging informed and comparative judgement."

Table 1: Continued

"Stakeholders"	"Traditional expectations for roles and Responsibilities"	"Evolving expectations for roles and responsibilities"
"Implementing agencies"	"Proposing safety options and designs for radioactive waste management solutions, investigating their consequences under different assumptions. Developing a chosen solution, implementing the solution."	"Co-operating with local communities in working through proposed options and designs in order to find an acceptable project for radioactive waste management. Co-operating with local communities in implementing the project. Interacting with policy-makers and regulators."
"Potential host communities"	"Accepting or rejecting the proposed facility."	"Negotiating with implementers to find locally acceptable solutions for radioactive waste management that help avoid or minimize potentially negative impacts and provide for local development, local control and partnership. Interacting with policy-makers and regulators."
"Elected local or regional representatives"	"Representing their constituencies in debates on radioactive waste management facilities."	"Mediating between several levels of governments, Institutions and local communities in seeking mutually acceptable solutions. Interacting with regulators and implementers."

Source: OECD/NEA Report. Radioactive Waste Management 2012. "The evolving role and image of the regulator in radioactive waste management: Trends over two decades." OECD Report 2012 (Organisation for Economic Co-operation and Development)/NEA (Nuclear Energy Agency). No.7083. (Page 14).

Fundamentally, one of the main reasons for the development of an inclusive institutional participation process with major societal stakeholders has emerged from the adopting the logic of deliberative approaches, leading to a comprehensive utilization of respective definitions and categorical understandings that can facilitate the regional knowledge determinants based on motivational characteristics and socio-economic drivers. On a more descriptive note, we can highlight another related aspect of the public's involvement in local governance analysis that is defined through a comparative search of applicable definitions. For instance, it has been pointed out that "public participation makes frequent reference to 'stakeholders' and it is important to understand what a stakeholder is and the different categories of stakeholder."[20]

In essence:

> a stakeholder is any person or body that has an interest in the problem being discussed. Interest can range from being directly affected by the project to those who have an interest in the particular subject area. Stakeholders can come as individuals or representing various groups ... All that is required is to ensure that the rest of the stakeholders are clear about which view is being presented. Other parts of government, the regulators and local government officials are often stakeholders representing their own organization's views.[21]

As observed by Delmas et al. (2004), the common participatory interactions enlarged at community level have been shaped across multiple organizational domains adapted for stakeholders' development interests, which have basically matured into the actual constitution of governmental decision-making directions, followed in the case of comparative environmental and managerial relations.[22] Moreover, the development of regional and local governmental relations has been reaffirmed through an interrelated innovation planning strategy applied, for instance, in the affirmation of international knowledge production systems positioned at industrial scale.[23]

However, we need to consider in particular that a resulting assimilation of public information systems has not been homogeneous for the concretization of interdependent aspects of deliberative supportive processes, that can also upgrade internal knowledge and local information contents, which have been catalyzed on public territorial issues by referring to intraregional environmental distribution conditions. Within the dimension of regional development, the use of common ICT information and knowledge organization standards applied in

---

20 Clark, "Public Participation in Decisions," pp. 65–70, p. 67.
21 Simon M. Clark, 2004. Page 67.
22 See: Magali Delmas, Michael W. Toffel, "Stakeholders and Environmental Management Practices: An Institutional Framework," *Business Strategy and the Environment* 13. 4/ 2004, pp. 209–222.
23 I. Linkov, A. Varghese, S. Jamil, T.P. Seager, G. Kiker, T. Bridges, "Multi-Criteria Decision Analysis: A Framework for Structuring Remedial Decisions at Contaminated Sites," in: *Comparative Risk Assessment and Environmental Decision Making*, eds. I. Linkov and A. Bakr Ramadan, (Kluwer Academic Publishers. Printed in the Netherlands, 2004), pp. 15–54.

governance has often involved the material distribution of overlapping corporate sources serializing technical information, at the national and local level, in addition to different relational coordination models and comparative participatory designs.[24]

1.1.1 Regional Environmental Aims

Regionally speaking, PPPs involving confrontation issues of industrial energy development and knowledge information management have been shifted according to public environmental aims tackled within transnational cooperation systems.

In consequence, the formulation of a governmental multi-criteria planning analysis employed in the decision-making process has been undertaken across separate industrially-based technology regions, which have been confronted with critical environmental knowledge issues such as the pollution of contaminated lands and the evaluation of territorial decommissioning plans and local rehabilitation programs. These programmatic activities have specifically reflected the regional environmental directions of policy implementation and evaluation plans that have been materially delivered through the possible diversification of convergent objectives, for instance, based on methodological approaches which have allowed multiple stakeholders' involvement, through systematic analyses for local scale information data collection and selection of flexible socio-economic distributive models, when operable.[25]

For the case of environmental nuclear waste disposal programs – international nuclear agencies and institutional regulators have offered a wide-oriented view about public information on nuclear waste management disposal proceedings; for instance the OECD/NEA international associations have favored a concerted information access through operable distribution channels organized for the preparation of public-private promotional dialogues, including regional and international initiatives, while reflecting multi-level collaborative trusting relations[26] such as: *the Forum on Stakeholder*

---

24 Linkov, Varghese, Jamil, Seager, Kiker, Bridges, "Multi-Criteria Decision Analysis," pp. 15–54.
25 Linkov, Varghese, Jamil, Seager, Kiker, Bridges, "Multi-Criteria Decision Analysis," pp. 15–54.
26 See: Tomoko Murakami, and Venkatachalam Anbumozhi, "Public Acceptance of Nuclear Power Plants in Hosting Communities: A Multilevel System Analysis," *Economic Research Institute for ASEAN and East Asia (ERIA)*, ERIA, Research Project

Confidence[27] *(FSC).*[28] In order to introduce each single country on respective technical specificities about nuclear energy production risks and comparative territorial management issues, public reforming agencies have focused on major thematic aspects concerning the national nuclear disposal activities, such as the national implementation of reversibility and retrievability processes[29] of hazardous waste disposal systems (OECD 2012),[30] where the communication protocols and common safety standards and societal trust measures have been pivotal. A comparative sample about national nuclear safety proceedings and information safety measures can be retraced through a brief summary in Table 2.

2018, No.18, published in December 2019. https://www.eria.org/uploads/media/Public-Acceptance-Nuclear-Power-Plants/RPR_FY2018_18.pdf

27 Note: OECD-NEA Organization: "The Forum on Stakeholder Confidence (FSC) was established by the NEA Radioactive Waste Management Committee (RWMC) in the year 2000. It fosters learning about stakeholder dialogue and ways to develop shared confidence, informed consent and acceptance of radioactive waste (RW) management solutions." FSC Representative Partners: (e.g.) Commonwealth Scientific and Industrial Research Organisation (CSIRO), Australia; Australian Radiation Protection and Nuclear Safety Agency (ARPANSA), Australia. NEA website available at:http://www.oecd-nea.org/jcms/pl_26865/forum-on-stakeholder-confidence-fsc.

28 See: Michael Scott Spranger, (1999). "Citizen Participation in Environmental Issues: A Comparative Study of Citizen Involvement Programs in Water Quality and Nuclear Waste Management." PhD Dissertation at Portland State University, Urban Studies, U.S.

29 Note: "Retrievability of nuclear waste is associated with the possibility to remove and recover waste packages after they have been emplaced in a repository." OECD, 2012, note 16, p. 19. In "Reversibility and Retrievability in Planning for Geological Disposal of Radioactive Waste."

30 OECD 2012. "Reversibility and Retrievability in Planning for Geological Disposal of Radioactive Waste." Proceedings of the "R&R" International Conference and Dialogue 14–17 December 2010 Reims, France. NEA No. 6993 (Nuclear Energy Agency) – Organization for Economic Co-operation and Development (OECD Secretary General), p. 67.

Table 2: National Disposal Proceedings

Nuclear Waste Management Retrievability	Reference/Full Quote	Doc/NEA No. 6993
Finland	"Retrievability is related In Finland to the potential availability of alternative technology in nuclear waste management and, especially, options in geological disposal. Retrievability has been used as a criterion when different methods of implementation were assessed and compared with direct geological disposal in the late 1990s. The options were hydraulic cage and deep holes. The comparison of benefits and disadvantages of alternatives is based on environmental impact assessment." (p. 28).	
Switzerland	"In 1999, the Federal Department of the Environment, Transport, Energy and Communications (DETEC) formed the 'Expert Group for Disposal Concepts for Radioactive Waste' (EKRA), which consisted of experts from a broad variety of fields. Its mandate was to formulate basic principles for a variety of waste management options, and its final report (DETEC), published in 2000, formed the basis for Switzerland's concept. The concept called 'monitored long-term geological storage' combines the isolation of radioactive waste in deep geological layers with technical and natural barriers, and the option of retrievability at society's request (being one feature of a reversible process)." (p. 30).	
United States	"'Retrievability' in the United States means the ability to remove high-level waste after it has been emplaced in a geologic repository; this typically implies permanent removal. … retrievability must be maintained for both economic and safety purposes. Nuclear Regulatory Commission (NRC) regulations further stipulate the time during which retrieval capability must be maintained for safety. Under the Nuclear Waste Policy Act of 1982, retrieval is maintained for safety, environmental or economic reasons. The Department of Energy (DOE) specifies the period of retrieval, subject to NRC approval. The NRC further requires retrievability throughout waste emplacement and performance confirmation programmes. … Maintaining retrievability after closure is not currently required in NRC regulations, though it is understood that the capability to retrieve could remain for some time beyond closure." (pp. 37–38).	

(Continued)

Table 2: Continued

Nuclear Waste Management Retrievability	Reference/Full Quote	Doc/NEA No. 6993
Sweden	"The issue of retrievability has so far not occupied a prominent position in the public discussion in Sweden. ... An important milestone in the approach to management of the Swedish nuclear waste was reached in the report of a governmental committee in 1976. The question of retrieval of spent nuclear fuel was not dealt with there, however; the report was completely focused on the question of a final disposal of the nuclear waste. ... In this context, the issue of retrievability became marginal. ...	
	The Swedish Nuclear Fuel and Waste Management Company (SKB) has in its tri-annual RD&D report several times claimed that the KBS-3 method neither presumes nor excludes retrieval. SKB has formulated its own requirement that the final repository for spent fuel must be designed in such a manner that it is possible to retrieve deposited canisters before closure. ... Single canisters may have to be retrieved from a deposition hole if something unforeseen happens during deposition. Retrieval of a large number of canisters in a later phase of operation of the repository must also be possible. If another method for disposing of or making use of the spent nuclear fuel is preferred in the future, technology for retrieving canisters will be needed then as well." (p. 46).	
Japan	"Here it is especially important to distinguish between reversibility of emplacement – recovery of waste packages in the event of operational problems – and retrieval after an emplacement module has been filled and sealed. In the context of the geological disposal programme in Japan, the relevant key words are understood to have the following meaning: – Radioactive waste ... spent fuel is not planned to be directly disposed of ... – Disposal ... Closure ... this includes the final engineering or other work required to bring the facility to a condition that will be safe in the long term ...	

Table 2: Continued

Nuclear Waste Management Retrievability	Reference/Full Quote	Doc/NEA No. 6993
	Reversibility ... retrievability denotes the possibility of reversing the action of waste emplacement ... The regulator has established the basic policy that geological disposal should proceed step by step, with decision making at each stage. ... In each stage, the implementer needs to obtain the consent of the local community. If consent is not received, the implementer is not allowed to proceed to the next stage. The NSC and the regulator require retrievability until the time of repository closure when the long-term safety is confirmed by safety assessment taking into account additional information obtained through repository construction and operation" (p. 50).	
United Kingdom	"Cmd 5552 (Managing the Nuclear Legacy: a Strategy for Action) and more recently the Energy Act 2004 charge the NDA with responsibility for ensuring appropriate research is carried out to underpin its short, medium and long-term decommissioning and clean-up activities. ... The Research Board on Decommissioning and Clean-up in the UK was formed in 2006. Since that time NDA has been given additional responsibilities by UK government including the development of a national strategy for LLW and implementation of a Geological Disposal Facility. ... – CoRWM recommends to Government that it ensures that there is strategic co-ordination of UK R&D for the management of higher activity wastes.	NDA EDRMS No. 24761455.
	Such co-ordination is required within the NDA, between the NDA and the rest of the nuclear industry, amongst the Research Councils and between the whole of the nuclear industry, its regulators and the Research Councils. – CoWRM recommends to Government that mechanisms are put in place to ensure that a wider range of stakeholders than to date will be involved in establishing R&D requirements for the long-term management of higher activity wastes and that accessible information will be made available to the public about R&D needs, plans and progress" (page 1).	

Source: 1. OECD 2012. "Reversibility and Retrievability in Planning for Geological Disposal of Radioactive Waste." *Proceedings of the "R&R" International Conference and Dialogue 14–17 December 2010 Reims, France*. NEA No. 6993 (Nuclear Energy Agency) – Organization for Economic Co-operation and Development (OECD Secretary-General). 2. UK's review: NDA Nuclear Decommissioning Authority, 2016. "Nuclear decommissioning authority independent research board terms of reference." NDA Research Board Terms of Reference, Issue 3, November 2016. EDRMS No. 24761455.

At the national level, the possible diversification of democratic and participatory approaches has meant the inclusion of public accountability campaigns and governance-oriented exercises. In fact, the combination of structural initiatives can increase regional public awareness about the management of high-technology energy systems incorporated with information knowledge measures that have been specifically related to technical arrangement mechanisms, influencing the potential impact on national human health and local levels of risk perception.[31] In effect, the exploration of national preparatory paths concerning stakeholders' environmental commitments has evolved according to contextual positions that have been developed together with integrative and participative knowledge factors. Participative knowledge approaches have been applied according to different relational stages of collaborative ground-level consultations while offering intra-regional constructivist platforms, also delivered in association with the energy and innovation transfer processes of individual countries.[32] According to the basic constructivist notions, the long-term common vision of science and technology development, and socio-economic evolution when based on democratic interactions has to be considered as well, which can lead to political legitimacy programs and regulative innovation processes, eventually undertaken in view of facilitating common regulatory recognitions, particularly when reflecting public deliberative choices about the systematic assimilation of national multi-level renewable energy projects.[33]

As a matter of fact, the structural combination of environmental government exercises[34] can be translated according to reciprocal management levels of multi-layered learning systems developed within organizational knowledge capacities that are either directly or indirectly able to respond to societal changing needs,

---

31 Matthijs Hisschemoller, "Participation As Knowledge Production and the Limits of Democracy," in: *Democratization of Expertise? Exploring Novel Forms of Scientific Advice in Political Decision-Making eds.* Sabine Maasen and Peter Weingart, – (Sociology of the Sciences 24/2005, Springer. Printed in the Netherlands), pp. 189–208.
32 See: Margot Hurlbert, "Evaluating Public Consultation in Nuclear Energy: The Importance of Problem Structuring and Scale," *International Journal of Energy Sector Management* 8.1/2014, pp. 56–75.
33 Hisschemoller, "Participation As Knowledge Production," pp. 189–208.
34 See: 1) Andy Gouldson, "Risk, Regulation and the Right to Know: Exploring the Impacts of Access to Information on the Governance of Environmental Risk," *Sustainable Development* 12.3/2004, pp. 136–149. 2) Sylvia I. Karlsson, "Institutionalized Knowledge Challenges in Pesticide Governance: The End of Knowledge and Beginning of Values in Governing Globalized Environmental Issues." *International Environmental Agreements: Politics, Law and Economics* 4/2004, pp. 195–213.

particularly regarding science policy innovation and direct public participation. Moreover, this process should be complemented by a methodological investigation, based on a multi-criteria theoretical dimension discussed at the governmental level and aligned with the additional technical adaptation of information capacity planning that can reaffirm material and societal aspects interconnected to regional institutional distribution systems.[35]

Therefore, when we take into account the relational mediation factors of influential distribution networks related to political approval systems and environmental decision-making conformations, we then need to emphasize the role of science-based technology communication applications for the public accessibility to technical information which has reflected concerted knowledge patterns indirectly associated to national education experiences, as well as, to societal learning exchanges. However, through progressive organizational initiatives, what needs more consideration is what happens at the level of governmental science technology programs, as that also depends on the actual implementation of environmental knowledge objectives that can facilitate a collaborative decision-making process, which has been based on the availability of monitoring information tools and complementary planning methods. Moreover, the configuration of environmental decisional patterns has to, in turn, rely on national innovation cycles involving renewable energy production systems, which can affect the local administrative criteria and decentralized distributive proceedings for the selection of policy regulatory options.[36]

Essentially, the socio-economic patterns responsive to national technical innovation productions and intensive knowledge transferring modes – embedded within mass energy consumption models – have suggested that the wide range of local environmental knowledge approaches that have characterized the type of integration of environmental decision-making activities, has been the result of public collaboration and public identified targets, similarly evolving through the possible regulation options that are commonly set according to local distribution

---

35 Matthijs Hisschemoller, Jan Eberg, Anita Engels, and Konrad von Moltke, "Environmental Institutions and Learning: Perspectives from the Policy Sciences," in: *Principles of Environmental Sciences*, eds. Jan J. Boersema and Lucas Reijinders (Springer Media B.V, 2009), pp. 281–303.
36 Maryke van Staden, Francesco Musco, "Three Streams of Local Action: Strategy and Policy; Technology and Measures; People and Lifestyle," in: *Local Governments and Climate Change*, eds. M. van Staden and F. Musco, (Advances in Global Change Research 39/2010, Springer Media B.V), pp. 111–172.

goals projected in the long-term period.[37] Nonetheless, because of the fact that national and local ecology-based programs have to address the specific level of sustainability of related local resources, the governmental regulatory groups have necessarily pointed their attention at functional leading actions still within common knowledge networking areas of respective local and provincial communities, including regional and national public policy representatives, particularistic interest groups, and industrial international producers.

This type of mutual recognition of multi-level strategic operating groups has been in essence specified according to the different progress that has been made in science innovation technologies through a specific set of national production determinants, which have been interrelated to local ecosystems' preservation measures and environmental protection policies.[38] As such, a comparative research knowledge orientation has emerged concerning multi-level strategic operating mediators that have become responsible on environmental investigation matters through the direct establishment of regional cooperation mechanisms and development adaptive practices, which have been solidified in time for local development and specific environmental conservation plans.[39] As reference, for clarification purposes of major conceptual assumptions, Table 3 provides a summarized overview referring to technical energy related issues and environmental definitions reported by OECD/NEA/EPA/EEA agencies.

---

37 See: Shu-Fen Kao, (2002). "Risk Perceptions and Environmental Mobilization. Tracking the Transformation of Collective Actions in a Radiation Contamination Incident in Taiwan." PhD Dissertation at Michigan State University, Department of Sociology, U.S.
38 James Aronson, Florian Claeys, Vanja Westerberg, Philippe Picon, Guillaume Bernard, Jean-Michel Bocognano, and Rudolf de Groot, "Steps Towards Sustainability and Tools for Restoring Natural Capital: Etang de Berre (Southern France) Case Study," in: *Sustainability Science: The Emerging Paradigm and the Urban Environment*, eds. M.P. Weinstein and R.E. Turner, (Springer Media, LLC, 2012), pp. 111–138.
39 See 1) Shan-Shan Chung, Lo Carlos Wing-Hung, "The Roles of Grassroots Local Government in Sustainable Waste Management in China," *International Journal of Sustainable Development and World Ecology* 14.2/2007, pp. 133–144. 2) Peter Jones, "Local Environment Agency Plans: New Environmental Management nitiatives in England and Wales," *Management Research News* 22.10/1999, pp. 12–16.

Cooperative Interactions 31

Table 3: Technical Concept Overview by OECD/NEA/EPA/EEA Agencies

About Communities
The Activity of Uranium Mining: "By the 1970s, escalating impacts from the operations of the health workers, the environment and the communities located nearby became increasingly evident. Societal pressure typically driven by unions representing miners, led to a number of investigation boards, commissions of inquiry and numerous health studies that clearly identified the extent and far-reaching impact of historic mining operations that lacked proper operational and waste management practices" (page 18)
Source: OECD/NEA Nuclear Development Report 2014. "Managing Environmental and Health Impacts of Uranium Mining." Nuclear Energy Agency NEA No. 7062 & Organisation for Economic Co-operation and Development OECD.
About Ecosystems
"Ecosystem impacts can be expressed at the level of the individual organism or at the system level. Examples of effects on individual organisms include death, reduction of health or vitality, accumulation of toxic substances, and alteration of reproductive success. Examples of ecological system effects include changes in birth or death rates; changes of toxic element concentrations throughout entire food webs; and changes in population size, habitat, or community structure" (NRC 1977, Section B)."
Source: EPA 2008. "309 Reviewers Guidance for New Nuclear Power Plant Environmental Impact Statements." EPA Publication 315-X-08-001 (U.S. Environmental Protection Agency, pp. 5–13).
About Energy Sectors
"Energy gives personal comfort and mobility to people, and is essential for the generation of industrial, commercial and societal wealth. On the other hand, energy production and consumption (including heat and electricity production, oil refining and final uses in households, services, industry and transport) place considerable pressures on the environment. These pressures include the emission of greenhouse gases and air pollutants, land use, waste generation and oil spills. They contribute to climate change, damage natural ecosystems and the man-made environment, and cause adverse effects to human health" (page 5)
"Attitudes to nuclear power are mixed. Although nuclear power produces little pollution under normal operations, highly radioactive wastes are accumulating for which no generally acceptable disposal route has yet been established or implemented at large scale. There is still a risk of accidental radioactive releases for existing new nuclear power plants" (page 9).
Source: EEA Report 2006. "Energy and Environment in the European Union. Tracking progress towards integration." European Environment Agency EEA Report No.8/2006.

(Continued)

Table 3: Continued

About Green Economy
"The concept of a green economy recognizes that ecosystems, the economy and human well-being are intrinsically linked. At the core of a green economy are twin challenges; a) ensuring *ecological resilience* of natural systems and b) improving *resource efficiency*. Natural capital is not only of interest from an ethical point of view, but vital for the long-term sustainability of our economic activities. In turn, increasing resource efficiency is not only desirable from an economic point of view; it is also a necessity to reduce the pressure on the natural ecosystems that sustain us" (page 3).   **Source:** EEA 2012. "Agriculture and the Green Economy." European Environment Agency TH-31-12-196-EN-N.

1.1.2 Regulated Consensus in Governance

The regional public mandates put into existence for independent ecosystems have been the result of decisional functions that have encouraged consensus rules on renewable energy programs and industrial diversification plans delivered across Europe, East Asia and the U.S., for instance. If we decide to shift our theoretical focus on the consensus-based regulatory instrumental governance, there have been regional implementation policies which have directly or indirectly been confronted with multiple science innovation facets of complex development environmental issues pertaining to stability of natural ecosystems. On such inferring aspects, the maturation of environmental negotiation processes that involve multi-stakeholder domains, has become publicly crucial because of complementary knowledge dynamics, in fact it is highlighted that "most governments have come to acknowledge that an environmental policy based solely on unilateral regulation is no longer adequate. Thus, there is a need for a new way of making environmental policy making. The new approach must be based on a mix of traditional and a novel instrument of environmental policy... The negotiated agreement is one of these 'novel' instruments."[40]

Nonetheless, throughout the recent decades from the 1990s up until the 2000s, the process of cooperation on national and local environmental sustainable programs[41] has had serious limitations due to a set of institutionalized

---

40 Akim Seyad, Steven Baeke, Marc de Clercq, "Success Determining Factors for Negotiated Agreements," in: *Co-operative Environmental Governance*, eds. P. Glasbergen, (Kluwer Academic Publishers, 1998), p. 111. pp. 111–132.

41 See: Matthew Potoski, Aseem Prakash, "The Regulation Dilemma: Cooperation and Conflict in Environmental Governance," *Public Administration Review* 64.2/2004, pp. 152–163.

structural components, which have not yet fully merged into the comparative knowledge adoption of so-called ecological modernization (EM)[42] approaches.[43] The actual implementation of ecological standards in association with nuclear waste disposal programs has specifically included uncertain risk factors, which have been related to societal cooperation and instrumental level of public knowledge interactions.[44] For example, in Figure 1, we have drafted a basic matrix concerning the nuclear waste disposal (NWD) implementation and framework conditions, in order to specify the conditions for the negotiation dimension to apply to nuclear waste disposal agreements.

---

42 See: Wai-Hang Yee, Carlos Wing-Hung Lo, Shui-Yan Tang, "Assessing Ecological Modernization in China: Stakeholder Demands and Corporate Environmental Management Practices in Guangdong Province." *The China Quarterly* 213.3/2013, pp. 101–129.
43 Andrew Blowers, "Power, Participation and Partnership. The Limits of Co-operative Environmental Management," in: *Co-operative Environmental Governance*, eds. P. Glasbergen, (Kluwer Academic Publishers, 1998). pp. 229–249.
44 See: Jonathan L. Katz, "A Web of Interests: Stalemate on the Disposal of Spent Nuclear Fuel," *Policy Studies Journal* 29.3/2001, pp. 456–477.

```
┌─────────────────────────┐
│ Nuclear Waste Disposal  │
│ Geological Implementation│
└─────────────────────────┘
```

What should be considered:	What key factors should be present:	What needs to be managed:
- The decision-making process country to country difference for stakeholders' involvement - Understanding key concerns of involved stakeholders - Development of effective interactions - Lessons learnt from national – international efforts - Nuclear agencies mediation for adaptation of radioactive waste disposal to stakeholders' feedback	- Complementarity of needs with constitutive members and civil society - Nuclear regulatory bodies' accountability and public trust - Clarity about the legacy of past nuclear activities and future disposal of radioactive waste for society - Periodic long-time frameworks of national arrangements and sites project development - Recognition of organizational disposal implementation approaches with stakeholders' acceptance	- A strategic identification of public perception and confidence - The reaching of approval to develop nuclear waste repositories at specific locations - Information reviews of memorial records and theoretical perspectives - An opening-up of waste management dialogues through local representatives having role as intermediaries - Supporting agreements to quality programs with technical and transparent socio-economic components

Source: Model Adapted from: OECD/NEA 2000. "Stakeholder Confidence and Radioactive Waste Disposal." *Inauguration, First Workshop and Meeting of the NEA Forum on Stakeholder Confidence in the Area of Radioactive Waste Management.* Paris, France. Nuclear Energy Agency and Organization for Economic Co-operation and Development.

1.2 Technology Risk Perspectives

From a comparative perspective, we can analyze an available association of public international technology studies[45] (STS)[46] made in combination with national industry and managerial studies[47] which have included energy technical forecasts based on single country's historical backgrounds, including the functional interactions operated with specialized managerial agencies that have produced annual energy knowledge reports.[48] Furthermore, it becomes possible to identify an interchangeable understanding of regional socio-economic dynamics in a more comprehensive manner.

In particular, contemporary literature commonly refers to policy and technology risk analyses,[49] nuclear waste reduction[50] and nuclear fuel energy programs,[51] which have then incorporated theoretical demonstrative issues. Local risk management theories, for example, have been elaborated for the identification of multi-level functional relations dependent on societal organizational components involving public environmental dialogues, public risk perceptions

---

45 Ryuma Shineha and Masaki Nakamura, "Diversity in STS Communities: A Comparative Analysis of Topics," *East Asian Science, Technology and Society: An International Journal* 7.1/2013, pp. 145–158, DOI: 10.1215/18752160-2075813.
46 See: Nick Pidgeon, "Public Understanding of, and Attitudes to, Climate Change: UK and International Perspectives and Policy," *Climate Policy* 12/2012, pp. S85–S106.
47 See: 1) Charles Seabrook, "Nuclear Power Industry Tries to Improve Image with Public," *Journal Record [Oklahoma City, Okla]* 04/1991; 2) Michael Gochfeld, Sandra Mohr, "Protecting Contract Workers: Case Study of the US Department of Energy's Nuclear and Chemical Waste Management," *American Journal of Public Health (AJPH)* 97. 9/ 2007, pp. 1607–1613.
48 See: Yi-Chong Xu, "IAEA: Facilitating Multilateral Cooperation on Nuclear Fuel Services," *Social Alternatives* 28. 2/2009, pp. 36–41.
49 See: Maria C. Powell, Martin P.A. Griffin, Stephanie Tai, "Bottom-Up Risk Regulation? How Nanotechnology Risk Knowledge Gaps Challenge Federal and State Environmental Agencies" *Environmental Management* 42/2008, pp. 426–443.
50 See: Gillian Chi-Lun Huang, Tim Gray, Derek Bell, "Environmental Justice of Nuclear Waste Policy in Taiwan: Taipower, Government, and Local Community," *Environment, Development, and Sustainability* 15/2013, pp. 1555–1571.
51 See: 1) Interfax: Ukraine Business Weekly, "Economic Policy; Energy Ministry Developing Nuclear Fuel of Ukraine State Program for 2014–2019," *Interfax-America, Inc.* 11/2011; 2) Jay Donohue, (2010). "Resolving Past Liabilities for Future Reduction in Greenhouse Gases; Nuclear Energy and the Outstanding Federal Liability of Spent Nuclear Fuel," Master thesis at the George Washington University, Law School, United States.

and public deliberative participation expressed at the local level.[52] In addition to this, local participatory evaluations, local reporting status, and local monitoring assessments have been progressively established, and correspond to the public affirmation of the *de facto* interrelated managerial cooperative spheres that have already emerged in the regional civil nuclear industry, which has been observed in conjunction with complementary knowledge criteria, critically adopted for policy legitimation and transparency of the normative process.[53] For instance, in the case of European Union changing regulations, the presence of respective countries' regulatory guidelines tackling renewable energy industrial goals has been interrelated to local environmental issues, including the management of nuclear waste disposal programs. Moreover, the collaborative formation of civil community networks expanded through multi-level adaptation knowledge platforms – for maintaining regional debating opportunities – has potentially granted favorable access to national and local environmental policy deliberations, calling on the multi-level domestic parties for practical planning[54] and learning involvement.[55]

This emergent collaborative path has meant that science and technology innovation organizations have strictly come to terms with local participatory designs in governance regulatory activities, through which the multi-level stakeholders have been able to engage in regional decision-making performances that concern deliberative processes. These, in turn, relate to energy management policies, focused on nuclear waste storage facilities and nuclear functional directives.[56] In comparative territorial cases described about science policy research initiatives and knowledge innovation dynamics intertwined with structural regulatory

---

52 Goran Sundqvist, and Mark Elam, "Public Involvement Designed to Circumvent Public Concern? The 'Participatory Turn' in European Nuclear Activities," *Risk, Hazards & Crisis in Public Policy* 1. 4. 8/2010, pp. 203–229.
53 See: Annika Beelitz, Doris M. Merkl-Davies, "Using Discourse to Restore Organizational Legitimacy: 'CEO-Speak' After an Incident in a German Nuclear Power Plant," *Journal of Business Ethics* 108/2012, pp. 101–120.
54 Roopali Phadke, Christie Manning, Samantha Burlager, "Making It Personal: Diversity and Deliberation in Climate Adaptation Planning," *Climate Risk Management* 9/2015, pp. 62–76. https://doi.org/10.1016/j.crm.2015.06.005.
55 Sundqvist, and Elam, "Public Involvement Designed to Circumvent Public Concern?" pp. 203–229.
56 Annelise Poetz, "What's Your 'Position' on Nuclear Power? An Exploration of Conflict in Stakeholders Participation for Decision-Making About Risky Technologies," *Risk, Hazards & Crisis in Public Policy* 2.2.2/2011, pp. 1–38.

matters – as was the case with Canadian Nuclear Power Plants[57] – the practical safety and security information have remained as a central policy domain. Moreover, environmental knowledge that has been centered on nuclear PPP safety risk cases has specifically involved the critical contribution provided by public deliberative forums, non-profit sectors, and local cooperative partners – who have been confronted with regional material implications of local managerial conflicts and public participation conditions.

Similarly, public strategic polices emanating in the form of regulatory proceedings and material environmental measures have in addition been defined according to national organizational programs adopted for targeted production areas, where the local nuclear energy facilities have been secured and directly managed.[58] However, we can consider the fact that the interrelated science and technology processes *per se* have been mainly focused on the procedural and technical understanding of complex managerial systems. Technical territory accounts have been framed according to the need of putting information first in reference to the science and energy development route, re-establishing delivery mechanisms while involving comparative protection perspectives.[59]

Nonetheless, when we take into account corresponding systemic needs of international, national, and local energy management frameworks consolidated into public-private configurations through PPPs, according to the emerging flow of industrial changing methodologies and governance territorial dynamics,[60] the deliberative environmental factors play a critical role as well. It can be noted that local planning measures and regional development programs that favor the implementation of nuclear energy and industrial technology designs, have become particularly responsive to a combination of multiple programmatic schemes taken up in complex production networks, which in turn have involved

---

57 See: 1) Dave Tod, "Ontario Power Generation to Assume Armed Security Duties at Nuclear Plants," *SNL Energy Power Week Canada* 08/2007; 2) R. Lane, E. Dagher, J. Burtt, P.A. Thompson, "Radiation Exposure and Cancer Incidence (1990 to 2008) Around Nuclear Power Plants in Ontario, Canada," *Journal of Environmental Protection* 4.9/2013, pp. 888–913.
58 Annelise Poetz, "What's Your 'Position' on Nuclear Power?" pp. 1–38.
59 See: Steven M. Albrecht, "Forging New Directions in Science and Environmental Politics and Policy: How Can Co-operation, Deliberation and Decision Be Brought Together?" *Environment, Development and Sustainability,* 3/2001, pp. 323–341.
60 See: Jungah Bae, (2012). "Green Governance Innovation: The Institutional Political Market for Energy Sustainable Communities," PhD Dissertation at the Florida State University, College of Social Sciences and Public Policy, United States.

multi-level stakeholders, and civil society such as: Nuclear Safety Commissions; Institutional Agents; Non-Profit Organizations; Public Forums; and Scientific Experts. In essence, the national formation of aligned administrative entities and regional markets' knowledge trading bodies has provided diversity and affirmation of different regulatory positions when the case.[61]

For such reasons, the comparable technical factors that have referred to national environmental risk's perception studies and local understanding of multi-party stakeholders, while resuming designated local exchange analyses, have been scrutinized according to specific territorial findings, which have been based on investigative results reaffirming a tendency of either positive or negative integrative accounts.[62] In theory, comparative environmental positions have been considered in relation to the elaboration of territorial risk perception analyses that have included determinant individual factors; as such composite science knowledge orientations have therefore provided mutual information exchanges regarding the process of public risk attitudes and public trust formation, and, particularly in the case of nuclear waste disposal policies, incrementally redefined for appropriate regulatory responsive models.[63]

As a matter of fact, periodic country-based surveys about nuclear energy power systems and nuclear waste disposal issues related to public attitudes' formations have been in place.[64] Institutional and academic research projects have been collected and analyzed in the form of national opinion polls and scientific institutional reports[65] that have underlined how and why public national orientations have resulted in dramatic shifts, particularly after the nuclear post-disaster events occurring during the Fukushima NPP crisis of Japan in 2011.[66]

---

61 Poetz, "What's Your 'Position' on Nuclear Power?" pp. 1–38.
62 See: Darko Pavo Kulas, (2004). "Contemporary Political Risk Analysis and Management: Civil Society and Sustainable Development," Master's Thesis at the University of Calgary, Department of Management, Alberta, Canada.
63 Lennart Sjoberg, "Explaining Individual Risk Perception: The Case of Nuclear Waste," *Risk Management: An International Journal* 6.1/2004, pp. 51–64.
64 See: Ann S. Bisconti, J. Scott Peterson, "Congressional Actions on Nuclear Energy: Do They Follow Public Opinion?" *Natural Gas & Electricity* 20.12/2004, pp. 11–17.
65 See: 1) Sung Roe Lee, (1997). "A Policy Deadlock in South Korea: The Case Study of Siting Nuclear Facilities," PhD Dissertation at the University of Tennessee, Knoxville, United States.; 2) Harold Paul Rubinson, (2008). "Containing Science: The U.S. National Security State and Scientists' Challenge to Nuclear Weapons During the Cold War," PhD Dissertation at the University of Texas, Austin, United States.
66 M.V. Ramana, "Nuclear Power and the Public," *Bulletin of the Atomic Scientists* 67.4/2011, pp. 43–51.

Successively, scientific national institutes and international advisory committees and public civil bodies[67] have been commonly engaged in the amelioration of information gathered and locally processed within regional knowledge programs operated through public mobilization platforms and environmental grassroots campaigns, and supported in corresponding local areas.

Some important aspects of the nuclear energy industry and the regional local assurances given on security and safety promotion have basically been highlighted in order to bring closer local, national, and international communities[68] facing specific socio-economic implications after the nuclear environmental crisis taking place in Japan after March 2011. In line with comparative environmental directions, civil society operating groups have essentially reaffirmed regional and national commitments toward the maintenance of common ethical values that have favored concurrent national opinions, which have also been underlined by local civic opposition parties – who have criticized the use and diffusion of alternative renewable energy production sources – when solely based on the structural distribution capacities developed at nuclear power plants and local organizational facilities.[69] At the same time, the affirmation of science and technology perspectives which have been articulated in association with public local risks and management evaluation criteria[70] that refer to the implementation of nuclear waste disposal programs, has determined a specific contrasting ground for contingent institutional choices involving the national construction plans of nuclear waste burial repositories.

1.2.1 Incremental Group's Visibility

In terms of performing managerial organizations and global risk scenarios, the nuclear energy industry has maintained a modulated degree of international raising expectations that have been modified across nuclear states' technology programs operated in contextual intraregional sectors.

---

67 See: US Fed News Service, "NRC Advisory Committee on Nuclear Waste and Materials Issues Report on Engagement with International Commission on Radiological Protection," *US State News [Washington, D.C.]* 9/2007, Publisher HT Media Ltd. U.S.
68 See: Jeff Kingston, "Nuclear Power Politics in Japan, 2011–2013," *Asian Perspective* 37.4/2013, pp. 501–521.
69 Ramana, "Nuclear Power and the Public," pp. 43–51.
70 See: Rodney P. Carlisle, "Probabilistic Risk Assessment in Nuclear Reactors: Engineering Success, Public Relations Failure," *Technology and Culture* 38.4/1997, pp. 920–941.

At research level, in relation to the identification of public risk perception analyses[71] there is a reference case about the long-term disposal policies that have been based on the scenario of the U.S. Nevada nuclear repository plans[72] at Yucca Mountain,[73] which has been negotiated over the last three or more decades.[74] Comparative studies on international resource management issues and regional business literature[75] have highlighted the importance of contextual regional directions and national socio-economic interactions that are taken for local environmental sustainability and industrial planning schemes both connected to public-private partnerships PPPs in multinational adaptive energy systems.[76] This integral duality has been inevitably related to social distribution elements and common ethical characteristics that have concerned expansive learning aspects of nuclear waste disposal programs, favoring the knowledge sharing norms and common supportive responsibilities.[77]

In fact, according to specific regional governmental and managerial organization relations based on the territorial dimensions and regulatory distribution spheres, the effective aggregation of multi-level production functions and policy adaptation rules has involved respective institutional roles and responsibilities of nuclear industrial consortia[78] and public interest groups, which have been integrated through progressive programmatic policies. Basically, the multi-level affirmation of respective environmental governance criteria and concurrent local

---

71 See: Lennart Sjoberg, "Risk Perception, Emotion and Policy: The Case of Nuclear Technology," *European Review* 11.1/2003, pp. 109–128.
72 See: EPA, "What Is the Yucca Mountain Repository," *Unites States Environmental Protection Agency (EPA)*/2021, Radiation Protection Topic. Available at: https://www.epa.gov/radiation/what-yucca-mountain-repository
73 See: Jeanne Nelson Ratliff, "The Politics of Nuclear Waste: An Analysis of a Public Hearing on the Proposed Yucca Mountain Nuclear Waste Repository," *Communication Studies* 48.4/1997, pp. 359–380.
74 Howard Kunreuther, Douglas Easterling, William Desvousges, and Paul Slovic, "Public Attitudes Toward Siting a High-Level Nuclear Waste Repository in Nevada," *Risk Analysis* 10.4/1990, pp. 469–484.
75 See: Claus Dierksmeier, "The Freedom-Responsibility Nexus in Management Philosophy and Business Ethics," *Journal of Business Ethics* 101/2011, pp. 263–283.
76 See: Catherine Dwyer, "The Relationship Between Energy Literacy and Environmental Sustainability," *Low Carbon Economy* 2/2011, pp. 123–137.
77 Marshall Alan, "The Social and Ethical Aspects of Nuclear Waste," *Electronic Green Journal* 1.21. 4/2005, p. 1–28.
78 See: Marvin Fertel, "Consortia Confirm Nuclear Resurgence," *Electric Perspectives* 29.4/2004, pp. 52–53.

organizational plans has also been established through the incremental visibility of societal networks and engaged environmental movements that have been aligned according to public agency's implementation and distribution plans[79] for the enhancement of common legitimacy conditions and local justice elements emerging in recent ideological acceptances.

Moreover, the environmental impact assessment plans have required methodological approaches based on case-by-case national documentation processes, scrutinized in reference to the management of nuclear power plants and parallel local environmental impacts, as pointed out in the case of the U.S. Uranium Mills Contamination in Canon City,[80] Colorado.[81] At the same time, we can observe that an actual polarization of sociological and political narratives about the presence of political hegemonic issues that refer to nuclear power institutional relations[82] has been systematically substantiated through international comparative theoretical and empirical analyses.

Specifically, the environmental knowledge area assessed in connection with nuclear policy, regulatory planning, and industrial energy blueprints, has been scaled-down to the socio-economic conditions, which have converged into local conflicting factors that have been interrelated with public policy motivations for either acceptance or rejection of aggregative models, based on the case of radioactive waste management and local disposal facilities activities.[83] In contrast, national and regional policy standards identified for the affirmation of collective acceptance criteria about nuclear policy designs, have been combined according

---

79 See: Magali A. Delmas, "The Diffusion of Environmental Management Standards in Europe and in the United States: An Institutional Perspective," *Policy Sciences* 35.1/ 2002, pp. 91–119.
80 See: WISE Uranium Project, "Issues at Canon City Uranium Mill (Colorado)." *World Information Service on Energy Uranium Project (WISE)*/2014, Available at: https://www.wise-uranium.org/umopcc.html
81 Chris M. Messer, Alison E. Adams, and Thomas E. Shriver, "Corporate Frame Failure and the Erosion of Elite Legitimacy," *The Sociological Quarterly* 53/2012, pp. 475–499. (Midwest Sociological Society).
82 See: 1) David H. Sacko, (2003). "Hegemonic Governance and the Process of Conflict," PhD Dissertation at the Pennsylvania State University, Department of Political Science, United States.; 2) V. Edward Sachse, (1989). "Hegemonic Stability Theory: An Examination," PhD Dissertation at the Louisiana State University and Agricultural and Mechanical College, United States.
83 Ji Bum Chung, Hong-Kew Kim, and Sam Kew Rho, "Analysis of Local Acceptance of a Radioactive Waste Disposal Facility," *Risk Analysis* 28.4/2008, pp. 1021–1032.

to local emerging levels of societal trust and respective risk impacts measured through environmental assessment analyses.

For instance, the management and local risk impact policies and public trust issues have been put in conjunction with the application of public adaptability models involving dynamic communication, and innovation practice frameworks, which have been essentially characterized by the harmonization capacity of governmental planning tools, when also addressing the evolutive divide of regional development programs implemented across eastern and western countries.[84] For example, in South Korea's case, the local community and local residents' orientations about nuclear risk perceptions and related institutional cooperation issues have supported public acceptance models based on tolerance orientations, which have been expressed at public forums and national information debates.[85] As of 2021, in South Korea, the World Nuclear Association stated that "wide support for nuclear power has been maintained. In a September 2021 poll of 1000 adults by EmBrain Public ... 72 % of respondents supported the use of nuclear power, while 24.3 % opposed it. ... 72.3 % of respondents also believe nuclear power is safe, and 81.5 % support resuming construction of Shin Hanul 3 & 4, which had been halted for more than four years due to the government's phase-out policy."[86] Therefore, research and development organizations in South Korea's case have tackled nuclear energy programs as valuable regional energy assets, requiring major financial investments on knowledge innovation capacities and safety maintenance plans that are periodically reviewed for public consensus levels and local environment agreements.

In addition to this, when looking at science and policy innovation drivers introduced at an international level, the dependent regulatory dynamics of nuclear energy research projects[87] have been diversified in countries such as the United States, or in comparative nuclear power member states. The possible diversification of institutional policy mechanisms supporting the specificities of industrial territorial locations – for nuclear energy facilities and waste disposal

---

84 See: Charles Hostovsky, (2002). "Integrating Planning Theory and Waste Management: A Critical Analysis of Current EIA Practice in Ontario," PhD Dissertation, at the University of Waterloo, Ontario, Canada.
85 Chung, Kim, and Kew Rho, "Analysis of Local Acceptance," pp. 1021–1032.
86 World Nuclear Association, 2021. "Nuclear Power in South Korea." WNA Country Profiles, updated October, 2021. Web-page available at: https://world-nuclear.org/information-library/country-profiles/countries-o-s/south-korea.aspx (01 August 2022).
87 See: Peter DeLeon, "Comparative Technology and Public Policy: The Development of the Nuclear Power Reactor in Six Nations," *Policy Sciences* 11/1980, pp. 285–307.

programs[88] – has concurrently been implemented according to societal organizational components analyzed in view of public local opposition factors that are framed within progressive and transitional governance networks. For example, about the adoption of common international governmental definitions such as: *Not in my backyard syndrome* (NIMBY)[89] and *Concentrating locations at major plants* (CLAMP),[90] the national policy awareness on environmental planning activities has been a resulting integrative reflection about local regulatory adaptive mechanisms.[91] Essentially, the analytical connection between *Local Nuclear Sites Radiation Risks*[92] in developed and in newly developed nations has emerged through an enhanced interest on local environmental risk studies, and on the reporting status of public risk analyses.[93] In fact, the analytical information areas and comparative knowledge innovation backgrounds that emerge at public level have been essentially developed in order to establish and try to redefine similar material thematics about local environmental governance determinants, integrating local community perceptions about science technology and nuclear waste polluting risks. In essence, current public opposition exposures have formed, e.g., as in Taiwan's case, about the controversial debate on nuclear waste disposal and local siting policies, operated under managerial innovation models that have been affirmed within the regulatory organization

---

88 See: Joanna Burger, James Clarke, Michael Gochfeld, "Information Needs for Siting New, and Evaluating Current, Nuclear Facilities: Ecology, Fate, and Transport, and Human Health," *Environmental Monitoring and Assessment* 172.1.4/2011, pp. 121–134.
89 See: 1) Carol Mansfield, Georgie Van Houtven, Joel Huber, "The Efficiency of Political Mechanisms for Siting Nuisance Facilities: Are Opponents More Likely to Participate Than Supporters?" *Journal of Real Estate Finance and Economics* 22.2.3/2001, pp. 141–161; 2) Douglas Easterling, Howard Kunreuther, *The Dilemma of Siting a High Level Nuclear Waste Repository* (Kluwer Academic Publishers, Norwell, Mass., 1995), vii+ 286 pp.
90 See: Gillan Chi-Lun Huang, Tim Gray, Derek Bell, "Environmental Justice of Nuclear Waste Policy in Taiwan," pp. 1555–1571.
91 Michael R. Greenberg, "NIMBY, CLAMP, and the Location of New Nuclear-Related Facilities: U.S. National and 11 Site-Specific Surveys," *Risk Analysis* 29. 9/2009, pp. 1242–1254.
92 See: Nathan Edward Busch, (2001). "Assessing the Optimism-Pessimism Debate: Nuclear Proliferation, Nuclear Risks, and Theories of State Action," PhD Dissertation at the Graduate Department of Political Science, University of Toronto, Canada.
93 See: Doug Mercer, Thomas Leschine, Christina H. Drew, William Griffith, Timothy Nyerges, "Public Agencies and Environmental Risk: Organizing Knowledge in a Democratic Context," *Journal of Knowledge Management* 9. 2/2005, pp. 129–147.

fields of nuclear power plants and nuclear disposal facilities, still coordinated on contrasting political grounds.[94]

Overall, the formal interpretations provided according to the provisional conditions of local environmental conflicts associated with the public risk perceptions' analyses, referring to nuclear waste management programs,[95] have reflected the permanency of structural imbalances associated with public cognitive orientations that permeate differently in the case of critical local risk acceptances. Moreover, there are additional intrinsic causal factors that have been intertwined with science innovation programs and regional knowledge integrative mechanisms, which have been connected to a set of important socio-economic issues concerning in particular: land policy, demographic shifts, proximity to locations, collectivism, elitism, and generational passage of time.

In many ways, when putting together concurrent practice elements directly or indirectly favoring a critical core of mutual decisional alternatives in nuclear policy, interlinked with the national interchange of administrative conditions, they cause reflective evolving approaches, structuring the societal environmental paths through local adaptive changes of industrial knowledge-based components and public acceptability factors; which overall have also determined comparative different strategies for national territorial policies designed for the nuclear waste management development.[96]

---

[94] Hung-Chih Hung, and Tzu-Wen Wang, "Determinants and Mapping of Collective Perceptions of Technological Risk: The Case of the Second Nuclear Power Plant in Taiwan," *Risk Analysis* 31.4/2011, pp. 668–683.

[95] See: Tom Vander Beken, Nicholas Dorn, Stijn Van Daele, "Security Risks in Nuclear Waste Management: Exceptionalism, Opaqueness, and Vulnerability," *Journal of Environmental Management* 91/2010, pp. 940–948.

[96] Hank C. Jenkins-Smith, Carol L. Silva, Matthew C. Nowlin, and Grant de Lozier, "Reversing Nuclear Opposition: Evolving Public Acceptance of a Permanent Nuclear Waste Disposal Facility," *Risk Analysis* 31.4/2011, pp. 629–644.

CHAPTER TWO

2.1 Innovative Transition Systems

In the case of managerial political frameworks[1] analyzed in association with multi-level actors' technical knowledge for collaborative system strategies, it is essential to more directly address the functional responsive delivery of institutional arrangements that rely on national energy service provisions.[2] Such joint strategies directed at multi-level governance portfolios in different regulative areas have particularly been questioned in relationship to the application of supportive distribution alignments that refer to public transitional aims, and democratic environmental acceptances, which have corresponded to the structural managerial sustainability of productive resources.[3] On the one hand, we can consider that the combination of managerial transfers in national production systems has comprised both the availability of distributive energy resources as well as the implementation of regional production networks, which have been connected to public-private entities operating within technical normative domains for specified territorial boundaries.[4]

But on the other hand, recent regional studies scholars such as Makard,[5] Raven, and Truffer[6] 2012,[7] have highlighted some comparative reflections about

---

1 See: Hideaki Shiroyama, Masaru Yarime, Makiko Matsuo, Heike Schroeder, Roland Scholz, Andrea E. Ulrich, "Governance for Sustainability: Knowledge Integration and Multi-Actor Dimensions in Risk Management." *Sustainability Science, supplement* 7.1/ 2012, pp. 45–55.
2 See: Lars Hogberg, "Root Causes and Impacts of Severe Accidents at Large Nuclear Power Plants," *Ambio* 42. 3/2013, pp. 267–284.
3 See: Carolyn M. Hendriks, "Policy Design Without Democracy? Making Democratic Sense of Transition Management," *Policy Sciences* 42/2009, pp. 341–368.
4 See: Jane Holder, Antonia Layard, "Drawing Out the Elements of Territorial Cohesion: Re-scaling EU Spatial Governance," *Yearbook of European Law* 30.1/2011, pp. 358–380.
5 See: Filippo Celata, Liz Dinnie, and Anne Holsten, "Sustainability Transitions to Low-Carbon Societies: Insights from European Community-Based Initiatives," *Regional Environmental Change* 19/2019, pp. 909–912. https://doi.org/10.1007/s10 113-019-01488-6.
6 See: Bernhard Truffer and Lars Coenen, "Environmental Innovation and Sustainability Transitions in Regional Studies," *Regional Studies* 46.1/2012, https://doi.org/10.1080/ 00343404.2012.646164.
7 Jochen Markard, Rob Raven, Bernhard Truffer, "Sustainability Transitions: An Emerging Field of Research and Its Prospects," *Research Policy* 41/2012, pp. 955–967.

the regional comprehensive integration of emerging societal dynamics concerning local transformation and knowledge innovation elements, which have involved political, economic, and social organizational transitions. The intrinsic determinism of social production and local innovative spheres identified through the constitution of distributed organizational factors, has essentially been perceived in reference to the sustainable development of interrelated knowledge management processes, which have been adapted in view of multi-level territorial modifications, eventually established across the overlapping institutional dimensions.[8]

From a technological production perspective, the public organizational models have been sustained through research integration on changing aspects of societal vs. technical interactions,[9] for the purpose of establishing formal national directions and local cohesive plans, targeting common needs concerning the approach to modern technological transitions in local governance. Moreover, the states' similarities about local regulatory performances of complementary multilateral production functions have been related to the capacity of enhancing a comparative regional understanding that has been adapted in due course according to multi-level operating networks and innovative managerial distribution systems.[10]

Overall, multi-purpose global distribution systems can simultaneously involve the national multi-sectoral l dimension that can refer to the composite nature of multilateral managerial environments including among others: the water distribution services; the local energy consumption patterns, and the national distribution of municipal territorial practices.[11] Moreover, the inclusion of different

---

8 See: 1) Andre Lecours, Daniel Beland, "The Institutional Politics of Territorial Redistribution: Federalism and Equalization Policy in Australia and Canada," *Canadian Journal of Political Science* 46.1/2013, pp. 93–113; 2) Robert McCorquodale, Raul Pangalangan, "Pushing Back the Limitations of Territorial Boundaries," *European Journal of International Law (EJIL)* 12.5/2001, pp. 867–888.

9 See: 1) Luis F. Luna-Reyes, Jing Zhang, J. Ramon Gil-Garcia, Anthony M. Cresswell, "Information Systems Development As Emergent Socio-Technical Change: A Practice Approach," *European Journal of Information Systems* 14/2005, pp. 93–105; 2) Stuart Anderson, Massimo Felici, "Classes of Socio-Technical Hazards: Microscopic and Macroscopic Scales of Risk Analysis," *Risk Management* 11.3–4/2009, pp. 208–240.

10 See: Jorn Birkmann, Matthias Garschagen, Frauke Kraas, Nguyen Quang, "Adaptive Urban Governance: New Challenges for the Second Generation of Urban Adaptation Strategies to Climate Change," *Sustainability Science* 5. 2/2010, pp. 185–206.

11 See: Ryan Bowie, "Indigenous Self-Governance and the Deployment of Knowledge in Collaborative Environmental Management in Canada," *Journal of Canadian Studies* 47.1/2013, pp. 91–121.

types of local organizational development operated through institutional collaboration and territorial public assessments also needs to be considered. Eventually, the bilateral diffusion of institutional PPPs has drawn together economic and governmental adaptative phases associated with territorial regulatory policies, which have been the result of regional distributive choices – incrementally introducing comparative local practices – connected with environmental partnering strategies, as well as, with commercial and socio-technological networking activities that are responsive to public diffusion patterns.[12]

From a regional institutional perspective, it is key to assess the comprehensive understanding that has been formed concerning territorial and ecological transitional innovation activities linked to cooperative proximity interactions that have critically progressed through the actual integration of comparative operational management associations, which have required regional technical identifications as well.[13] In Cooks' view,[14] regional political adaptation approaches evolving through strategic regulative patterns that have emerged according to national identification processes, have fundamentally involved the systematic combination of local technical knowledge, also interrelated to eco-innovation plans and regional governance distributive measures.[15]

In addition to this, we can underline that the value of instrumental measures solidified in terms of knowledge innovation policies has allowed the systematic regulation of local innovation process activities put in relation with comparable operating clusters attaining developmental aims based on internal – external production characteristics.[16] In fact, the regional knowledge formations and instrumental deliveries of national energy productions have corresponded to the level of application of functional industrial cohorts, which have grown through sectoral diffusion cycles, including the incremental innovation dynamics. In the

---

12 Markard, Raven, Truffer, "Sustainability Transitions," pp. 955–967.
13 See: Bjorn Hassler, "Accidental Versus Operational Oil Spills from Shipping in the Baltic Sea: Risk Governance and Management Strategies," *Ambio* 40.2/2011, pp. 170–178.
14 Philip Cooke, "Transitions Regions: Regional-National Eco-Innovation Systems and Strategies," *Progress in Planning* 76/2011, pp. 105–146.
15 See: Rudiger K.W. Wurzel, Anthony Zito, Andrew Jordan, "From Government Towards Governance? Exploring the Role of Soft Policy Instruments," *German Policy Studies* 9.2/2013, pp. 21–48.
16 See: Jungwon Yoon, (2011). "Exploring Regional Innovation Capacities of PR. China: Toward the Study of Knowledge Divide." PhD Dissertation at the Georgia Institute of Technology, United States.

meantime, the correspondence of local regulatory dispositions on public environmental measures has remained in place for the regional territorial collectivities.[17]

Presumably, the evolutive expression of collective territorial interests elaborated in terms of knowledge innovation management and environmental protection policies has been theorized according to the different transition of common internal determinants for multinational innovation markets.[18] Essentially, the integration of national regulatory conditions, for instance, regarding local energy diffusion market activities than can result in respective industrial outcomes, has been formally recognized through mutual institutional agreements, depending on comparative PPP transitions, and local organizational changes. Therefore, this composite PPP dimension overall has been operated according to the level of organizational integration of national and regional public-private stakeholders, through strategic local activity and continuous public involvement.[19] Across the broader spectrum of knowledge production and innovation functions, multi-level actors have been involved through the devolution of decisional distribution powers developed in competitive marketing systems. Comparative regulatory integrated systems have been concurrently composed according to the different degree of local knowledge transfers and specialized innovating dynamics, which have accompanied national and regional energy ecological plans with field-related implementing programs.[20]

Therefore, what can become achievable in terms of regional coordination process for regional innovation energy markets designed through the necessary infrastructural development and respective economic transition, it is constructively dependent upon the participation of implementing PPP members – supporting the programmatic objectives for concerted institutional changes – which need to be addressed through the evolution capacity of technological and ecological

---

17 See: Suzanne Gladys Tilleman, (2009). "Aligning Institutional Logics to Enhance Regional Cluster Emergence: Evidence from the Wind and Solar Energy." PhD Dissertation at the Department of Management, the Graduate School of the University of Oregon, United States.
18 See: Yaqun Yi, Yi Liu, Hong He, Yuan Li, "Environment, Governance, Controls, and Radical Innovation During Institutional Transitions," *Asia Pacific Journal of Management* 29.3/2012, pp. 689–708.
19 See: I Ketut Putra Erawan, (2003). "Why Do Regional Actors Comply? Subnational Structure and Collective Action in Indonesia, 1990–2001." PhD Dissertation at Northern Illinois University, Department of Political Science, United States.
20 Cooke, "Transitions Regions," pp. 105–146.

transfer models planned across time. Through a prospective view, the regional planning strategies[21] have been associated with the different confrontation of human affirmation activities that independently rely on public technological diffusion and sectoral technical expansions. This is because of the corresponding mutuality of comparative societal converging interests, where the affected groups have favored a structural differentiation in public and private developmental spheres, which have been consequentially regulated according to national and local productive innovation dimensions.

To these elements, the actual perception of necessary material identifications regarding the industrial planning process has been shifted to the managerial innovation issues for the operational dimension of local production systems, with national and local impacts that have been aggregated according to the public allocation of available resources.[22] In a way, this has resulted in a form of social dialogues about public-private strategic confrontations, based on the systematization of either centralized or decentralized territorial resources, which have been influenced by the presence of multilateral renewable energy markets with competitive distributed operations indirectly affecting the number of territorial organizations and federal administrative systems.[23]

In combination to complex national levels of sectoral structural adjustments, the multiple centralized bodies and decentralized organizational entities have facilitated the knowledge integrative process for the social adaptation by including respective operable conditions applied according to distinct geographical zones, also with the goal of reaching major economic stability and an evolutive cooperation on environment exchange programs. Therefore, the regional integration of common institutional changing elements, which in Cooke's view[24] can be applied to multi-dimensional adaptive transitions, has been based on deliberative collective dynamics that essentially require the direct involvement

---

21 See: Bravishwar Mallavarapu, (2013). "Regional Pathways to Technological Upgrading: The Impact of Agglomeration Economies and Its Regional Covariates on Upgrading in Post-Reforms India's Manufacturing Sector." PhD Dissertation at the University of California, Los Angeles, Department of Urban Planning, United States.
22 See: Yosseph Leibovitz, (2001). "Associative Governance? The Political Economy of Institutional Change in Two Ontario City-Regions." PhD Dissertation at the Graduate Department of Geography, University of Toronto, Canada.
23 See: Nicole C. McDonald, Joshua M. Pearce, "Renewable Energy Policies and Programs in Nunavut: Perspectives From the Federal and Territorial Governments," *Artic* 65.4/ 2012, pp. 465–475.
24 Cooke, "Transitions Regions," pp. 105–146.

of progressive regulatory organizations putting forward technical collaborative plans, tied up to regional intra trade exchanges.[25]

However, part of the international political economy IPE literature[26] draws attention to public policy aspects which have basically indicated additional groundwork on theoretical aspects that have involved the regional economic dynamics[27] compared with environmental sustainability issues. Regional and local interdependency processes have been framed in view of understanding the production requirements and systemic innovation needs, that are adapted *for modern societies shown in transitions.*[28] In fact, the discursive integration of technological innovation elements reviewed for IPE academic debates has contributed to bring into focus additional explanatory factors that can refer to the direct or indirect material composition of socio-economic conditions co-existing in multi-layered innovation systems for commercial trading assessments, catalogued according to different territorial economic dimensions.

2.1.1 Transitional Technology Cooperation

Epistemic communities founded on public trust and mutual protection principles have tended to reaffirm multiple collaborative knowledge domains in science and technology, among others, which have required social cultural identifications and local approval systems – too often isolated in terms of intrasectoral mediation dialogues and cooperative adaptive designs. Essentially, the multi-layered provisions of managerial territorial innovations and environmental development dynamics have been put in conjunction with fundamental

---

25 See: Chen-Yu Wang, (2006). "The Impact of Regional Economic Integration under the GATT/WTO Regime toward the Peace Process: the Case of Conflict Resolution between Taiwan and Mainland China." PhD Dissertation at the Faculty of the Washington College of Law, American University, United States.
26 See: Milton L. Holloway, "Innovation Dynamics and Policy in the Energy Sector," *Academic Press* Copyright @ 2021 Elsevier Inc. All rights reserved. https://doi.org/10.1016/C2020-0-01005-4.
27 See: Thomas Pellerin-Carlin, Jean-Arnold Vinois, Eulalia Rubio, and Sofia Fernandes, "Making the Energy Transition a European Success," *Notre Europe*, Jacques Delors Institute, Studies & Reports 114/2017. Webpage available at: https://www.sipotra.it/wp-content/uploads/2017/06/MAKING-THE-ENERGY-TRANSITION-A-EUROPEAN-SUCCESS-TACKLING-THE-DEMOCRATIC-INNOVATION-FINANCING-AND-SOCIAL-CHALLENGES-OF-THE-ENERGY-UNION.pdf.
28 See: Giovanna Dore, Tanvi Nagpal, "Urban Transition in Mongolia: Pursuing Sustainability in a Unique Environment," *Environment* 48.6/2006, pp. 10–24.

concurrent aspects that have been considered for economic performing practices able to retrace for instance:

a. the specification of societal actors involved with local structural identifications;
b. the networking provisions for collaborative organizational changes;
c. the configuration of multiple know-how technical components of regional energy infrastructures;
d. the independence of food system sustainability with protectionist designs concerning the international capacity distribution;
e. the national and local affirmation of clustering urban planning strategies

Source: Farla, Makard, Raven, Coenen, 2012; Williams, 1999; Coenen, Bennworth, Truffer, 2012; Jorgensen, 2012.

For example, in terms of urban governance designed within comparative institutional frameworks[29] we can consider the regional policy studies that have revealed different degrees of urban adaptive environments modified in relation with structural changing conditions. Moreover, other conditional elements have also been in view, such as ones that refer to the constitution of large metropolitan cities aligned across industrial integrated territories like China[30] and Australia, which have also emerged in connection with local reconstructing policies addressing the correlated decisional domains of previous local urban assets, identified through comparative knowledge transfers.

In terms of socio-economic industrial changes, the specific evolution of 'modernizing' cities has acquired territorial versatile configurations that reflect a comparative degree of cumulative impact related to governance planning policies, which have been associated with the distribution of adaptive managerial criteria that have been functional to the geographical rural-urban landscapes.[31] Interestingly, we have found in Williams's study that urban metropolitan designs explored in governance have been affected by the existing levels of public urban interactions, which have already been developed according to comparative

---

29 Gwyndaf Williams, "Metropolitan Governance and Strategic Planning: A Review of Experience in Manchester, Melbourne and Toronto," *Progress in Planning* 52/1999, pp. 1–100.
30 Jichuan Sheng, and Xiao Han, "State Rescaling, Power Reconfiguration and Path Dependence: China's Xin'an River Basin Eco-Compensation Pilot (XRBEP)," *Regional Studies* 2022, DOI:10.1080/00343404.2021.2009454.
31 See: Colleen Crystal Hiner, (2012). "Changing Landscapes, Shifting Values: A Political Ecology of the Rural-Urban Interface." PhD Dissertation at the Geography Department, Office of Graduate Studies, University of California Davis, United States.

spatial transitions, including structural urban changes based on technological and commercial identifications.

In addition, the formal interaction route advanced by public and private planning officials has been interrelated with the modified nature of national and local territorial distribution processes, which have been dependent on internationally based commercial trading exchanges, associated with comparable operating innovation practices applied to global goods provisions and global services delivery. As a significant remark, the growing interrelation of urban developmental factors has been defined according to the structural intensification of changing cities that decide to follow progressive cohesive plans, which have been formalized through functional regulatory transfers and regional inclusions of distributed marketing capacities. Through which, both internal and external knowledge innovation spheres of multi-level trading provisions have reflected local adaptation programs with an active mobilization of industrial resources, which have been allocated according to the diffusion of transnational foreign trades.[32]

Basically, a need has emerged to combine widespread regional innovation distribution patterns with territorial production areas affirmed e.g., in East Asia, while supported according to the present intensification of harmonized regulatory requirements.[33] At the same time, there have been quantifiable changes in terms of urban-rural development determinants, influencing the regional marketing transitions with interrelated strategic processes based on local industrial productions, which have been regulated in terms of commercial trading networks, while formalized through the economic liberalization of international negotiation measures.[34] Under such conditions, finding respective intra-regional contemporary trends in technological emerging productions within harmonized regulative provisions has also meant planning local environmental implications,

---

32 See: Matt Hern, (1997). "Making Space: Radical Democracy in the Megalopolis." PhD Dissertation at the Union Institute Graduate School, Urban Studies, United States.
33 See: Robert White, "Implementing a Strategic Approach for Energy Efficiency Regulations," *Report on WTO Workshop on Nontariff Barriers Trade in Information Technology Products*, 2015. Web-page available at: https://www.wto.org/english/trato p_e/inftec_e/workshopmay15_e/white.pdf
34 See: 1) Siri Terjesen, Jolanda Hessels, "Varieties of Export-Oriented Entrepreneurship in Asia," *Asia Pacific Journal of Management* 26/2009, pp. 537–561; 2) Laszlo Bruszt, Bela Greskovits, "Transnationalization, Social Integration, and Capitalist Diversity in the East and the South," *Studies in Comparative International Development* 44/2009, pp. 411–434.

which have essentially included a certain level of technological coordination and collaborative embedded diffusion. For these reasons, the industrial multilateralism that has been arranged through regional urban-rural systematic processes has entailed the intra-mechanical uses of industrial ecosystems performing within the urban-rural industrial areas, while also involving the urban zones and regional districts located globally.[35]

2.2 Public Managerial Ties

As result, we have assisted on a regional basis at the introduction of 'strategic production alliances' already structured within the complexity of urban developed locations, which have directly or indirectly favored the growing divide created between cities' agglomerated systems, and the actual flow of human environmental interactions.[36] Substantially, there has been a respective direction marked in governmental sectors engaging in rules that delve into the diffusion of industrial innovation networks, which have become intertwined with the national and regional spheres of socio-economic exchanges, reflecting common participatory levels of local adaptation plans.

At the same time, in view of coping with the PPP organizational consolidation and transitional knowledge diversification, operated throughout countries' common innovation trading areas, the actual formation of innovative linking ties aggregated with the changing distribution of information networking capacities, has been monitored according to traditional performing PPP interactions. In fact, there has been a specific interest in trying to better redefine the PPP knowledge affirmation in local capacities interlinked with organizational managerial ties, which have particularly been developed through a number of different inclusive factors such as: 1. local spatial identities; 2. technological access to clustered machineries; 3. national growth plans with performing financial investments; 4. concentration of labor in aggregated sectors; and 5. multi-level cooperative interactions within multi-sectoral industries, among others.[37] In

---

35 See: Jonathan Quentin Morgan, (2004). "The Role of Regional Industry Clusters in Urban Economic Development: An Analysis of Process and Performance." PhD Dissertation at the Graduate Faculty of North Carolina State University, Public Administration, United States.
36 See: David Wachsmuth, (2014). "Post-City Politics: US Urban Governance and Competitive Multi-City Regionalism." PhD Dissertation at the New York University, Department of Sociology, UMI Dissertation Publishing, United States.
37 See: 1) Nicolai J. Foss, Bo. B. Nielsen, "Researching Multilevel Phenomena: The Case of Collaborative Advantage in Strategic Management," *Journal of CENTRUM*

essence, the establishment of global knowledge information capacities, and regional innovation production objectives, has involved the implementation of regional industrial knowledge networks – spreading across multiple institutional associations of different national societies – which have been connected to the facilitation of local changing processes for the elaboration of more tangible outcomes, when based on comparative local knowledge capacities and local innovation accesses.

For instance, one of the prevailing factors that can put into evidence the orientation of multi-level decision-making activities that can address respective degrees of power centralization and decentralized performing roles related to diversified environmental issues – which are connected to national and regional industrial innovation productions – has been identified in terms of local environmental sustainability, as it is fundamentally required for resource management capacity supporting regional energy plans. Moreover, comparable and integrated energy managerial initiatives have corresponded to the technological development of concentric unitary systems.[38] In comparison, the current knowledge-based competition between regional local economies has been maintained in terms of mutuality pursued through the national institutional differentiation, which has also remained a central aspect for determining a specific core of comprehensive knowledge transfer mechanisms and strategic innovation objectives.

The managerial innovation knowledge determinants when associated with standardized regulative aggregations as well as with national implementation processes, have introduced regional technical requirements including the local trading implications. In addition, the objective regulatory diversification that leads to the establishment of national trading control practices and local risk assessments for the amelioration of development strategies associated to regional innovation goals have remained in place. Therefore, the formal responsive environment extrapolated in conjunction with socio-economic relations through updated policy configurations, has resulted into an attainable adaptive

---

Cathedra 5.1/2012, pp. 11–23; 2) Susan L. Reynolds Fisher, (2000). "A Multilevel Theory of Organizational Performance." PhD Dissertation at the Graduate College of the Oklahoma State University, Degree of Doctor of Education, United States.

38 See: The-ENPI.ORG, "The Four Concentric Cycles of a Circular Economy," *The European Network of Policy Incubator*, Policy Paper 2/2016, Web-page available at: https://the-enpi.org/2016/08/26/project-sustainabilty-paper-1-the-four-concent ric-cycles-of-the-circular-economy/. Jacco Farla, Jochen Markard, Rob Raven, Lars Coenen, "Sustainability Transitions in the Making: A Closer Look at Actors, Strategies and Resources," *Technological Forecasting & Social Change* 79/2012, pp. 991–998.

framework for the provision of legislative guidelines and organizational implementation measures that have been put into place when needed.[39]

Essentially, the presence of PPPs with regional innovation stakeholders aligned with the institutionalization process has been harmonized through local governance agencies, which have been able to determine temporary adaptation conditions for comparative managerial shifts that have involved multi-layered commercial mechanisms, as well as, regional development interests, put forward in association with public environmental norms and municipal public acceptances.[40] However, we may ask until which point can we determine what specific types of innovative production issues have been at stake? Moreover, which public policy's trajectory has been followed up for local deliberation practices, which have been established under conditional distribution factors, referring to regional innovation systems and local knowledge capacities? And how have private firms' managers operated within managerial embedded processes that have directly or indirectly affected the comparative transitional innovation paths for the application of sustainable technologies? Moreover, which type of societal deliberation profile inserted into specific capacity integration domains has been associated with established technology innovation industries? To add to this, which type of public interaction favors a common knowledge ground for public dialogues revisiting the major issue of remaining locally sustainable?[41]

When considering such open-ended questions, a comparative type of ideological divide between dualistic transitional domains can be described, as it invests in the search for a public regulatory consensus while following the global production changes of dynamic trading interactions.[42] According to these authors: Farla, Markard, Raven, Coenen, 2012,[43] the regional intensification of

---

39   Farla, Markard, Raven, Coenen, "Sustainability Transitions in the Making," pp. 991–998.
40   See: Judith A. Layzer, "Citizen Participation and Government Choice in Local Environmental Controversies," *Policy Studies Journal* 30.2/2002, pp. 193–207.
41   See: Arnim Wiek, Barry Ness, Petra Schweizer-Ries, Fridolin S. Brand, Francesca Farioli, "From Complex Systems Analysis to Transformational Change: A Comparative Appraisal of Sustainability Science Projects," *Sustainability Science*, 7/ 2012, (Supplement 1), pp. 5–24.
42   See: Carl Grafton, Anne Permaloff, "Public Policy for Business and the Economy: Ideological Dissensus, Change and Consensus," *Policy Sciences* 34.3–34/ 2001, pp. 403–434.
43   Farla, Markard, Raven, Coenen, "Sustainability Transitions in the Making," pp. 991–998.

networked knowledge exchanges in both social and productive dimensions has engaged the systemic diffusion of changing trading environments – depending on the very nature of dynamic production transitions that will reflect the day-to-day operational maintenance activities and local environmental sustainability plans.[44] Therefore, through a specific type of sectoral identification process and mature technological development, basically, what was offered here is the attributable integration of multi-level organizational shared practices adopted for differentiated energy production systems that can facilitate the search of cyclical interchanges and material objective arrangements for structural technology innovation assets.

In such way, the territorial exploration approaches attempted by respective institutional organizations and public interacting associations have allowed the progressive cognitive assimilation of local distributive conditions which concretize societal orientations that tend to emerge particularly within the industrial spatial dimension providing support on local managerial regulative measures. At the same time, the combination of socio-political determinants linked to the structural intensification of industrial knowledge assets has been influenced by national and regional regulatory practices following a mutuality model[45] of international trading exchanges. Similarly, the competitive factors of emerging energy innovation markets with technological managerial affirmations have been explored through respective economic transitional reviews mostly focused on regional and local innovation performing capacities.[46]

From an environmental policy perspective, local environmental sustainability has been considered in terms of geographical concentration zones operated through the consolidation of respective socio-economic productive innovation

---

44   See: Stephen M. McCauley, Jennie C. Stephens, "Green Energy Clusters and Socio-Technical Transitions: Analysis of a Sustainable Energy Cluster for Regional Economic Development in Central Massachusetts, USA," *Sustainability Science* 7.2/2012, pp. 213–225.

45   See: Alistar Colin-Jones, and Sudhir Rama Murthy, "Mutuality and Concepts of Responsible Business," in: *Putting Purpose into Practice: The Economics of Mutuality*, (Oxford University Press, 2021), chapter 6. Web-page available at: https://purposeintopractice.org/mutuality-and-concepts-of-responsible-business

46   See: 1) Kimberly Bates, E. James Flynn, "Innovation History and Competitive Advantage: A Resource-Based View Analysis of Manufacturing Technology Innovations," *Academy of Management Journal* 1995, pp. 235–239; 2) Jon-Arild Johannessen, Bjorn Olsen, "Systemic Knowledge Processes, Innovation and Sustainable Competitive Advantages," *Kybernetes* 38. 3–4/2009, pp. 559–580. Emerald Group Publishing Limited.

systems. This integration process of multi-purpose organizational adaptive systems has been contextually experienced according to different knowledge transfers and scale technologies emerging in national and regional industrial configurations, assisted in selective periods of time through comparative PPP for the sharing of allocated resources.[47] Nonetheless, by taking into account the regional development designs of processed industrial innovation systems, it may be noted that the essential delivery of local regulatory mechanisms has basically differed in terms of the local affirmation of cyclical production changes, which have facilitated the mutuality of legitimate adaptive measures for international technology assets and innovation energy markets. While at the same time common territorial participants have been faced with a formal aggregation of recurrent organizational tendencies that can influence socio-economic conditions due to the imposition of multiple local production policies fully investing in regional renewable energy trading targets.[48]

However, there are arguments claiming that local sustainability involving socio-economic resources through a mutual degree of public interactions are conducive to a complexity of compatible trading innovation systems, increasing a comparative different understanding of technological diffusion practices operated in specific territorial manufacturing areas. Due to participative inductive processes of public deliberations, it is important to consider the co-evolution of common innovation development patterns can retrace the diversification of political and societal embedded networks, while reaffirming the regional interdependency of management knowledge productions shifting toward continental contexts.[49]

2.2.1 Public Networking

The actual specification of knowledge information networks and intra-regional trading flows has led to characteristic PPP formations in view of sub-territory configurations promoting modern national assets' innovation and assimilation at local production level. At different organizational stages, public environmental interactions have reflected comparative responsive priorities determining mutual understanding and progressive local acceptances.

---

47 Lars Coenen, Paul Benneworth, Bernhard Truffer, "Toward a Spatial Perspective on Sustainability Transitions," *Research Policy* 41/2012, pp. 968–979.
48 See: Neil Strachan, Stephen Pye, Nicholas Hughes, "The Role of International Drivers on UK Scenarios of a Low-Carbon Society," *Climate Policy* 8/2008, pp. 125–139.
49 Coenen, Benneworth, Truffer, "Toward a Spatial Perspective," pp. 968–979.

Through descriptive accounts, the common elements identified in the case of public environmental concerns associated with variations of industrial local development have been polarized on basic socio-economic needs that characterize public actors' participatory relations, converging on nodal aspects of the centralization of distributive dynamics and technical decisional demands, which entail social environmental interactions.[50] But again, the knowledge diffusion of comparative levels of common understanding of the confluence of national regulative proceedings that can shift to local power distributive entities has involved respective degrees of local programmatic affirmations. For regional firms' innovation capacities, this systematic affirmation has been affected by regulatory assessment practices and by public agencies implementing requirements that may trace down the technical monitoring dissemination, which needs to avail the industrial processing changes evolving under uncertain territorial conditions.[51]

Moreover, there have been interrelated public development issues that have surfaced about the critical stabilization of local environmental cases that refer to domestic and international technology innovation supported capacities, which have been assimilated according to the actual exploration of national adaptations based on local ruling conditions of common normative systems, favoring the active integration of multi-layered organizational trading components. While trying to redefine the upcoming alternative policies and local development perspectives, the public decisional approaches have addressed the inclusion of national collaborative visibility for incremental societal memberships. This empowerment process has basically involved an enhanced availability of deliberative cooperation platforms, through which a higher presence of participatory decisions[52] about international trading strategies has emerged in connection with mutual learning agendas and local innovation practices, establishing respective

---

50 See: Carlos A. Bana e Costa, Joao Carlos Lourenco, Monica Duarte Oliveira, Joao C. Bana e Costa, "A Socio-Technical Approach For Group Decision Support in Public Strategic Planning: The Pernambuco PPA Case," *Group Decision and Negotiation* 23.1/ 2014, pp. 5–29.
51 See: Sofka Wolfgang, "Globalizing Domestic Absorptive Capacities," *Management International Review* 48. 6/2008, pp. 769–792.
52 See: SEI report, 2021. "The Local 2030 Coalition for the Decade of Action." Stockholm Environment Institute, DOI: 10.51414/sei2021.035. Web-page available at: https://cdn.sei.org/wp-content/uploads/2021/11/local2030-coalition-for-the-decade-of-action.pdf.

intraregional areas for mutual local exchanges based on environmental trading conditions.[53]

In particular, the consistency of structural organizational rules at a societal level has been referred through participatory societal needs in order to verify the institutional coexistence of multi-level policies' directions. Also noteworthy is the fact that representative decision-making routes have characterized the national and local prevalence of common knowledge activities for mutual collaboration and cooperation, similarly leading to a comparative applicable identification of sectoral economic transitions.[54] In this sense, public governance scholars[55] have conceptually explained that the societal knowledge diversification associated with common trading innovation areas can be better understood by adopting a multi-level perspective, especially regarding the innovation process *per se* put in conjunction with an institutional formation of regional alliances, comparatively targeting a selection of innovation industry environments.[56]

In essence, the governmental roles and shared responsibilities connected to performance criteria and protection mechanisms have been applied according to different science knowledge cases which have involved the decentralization of managerial resources, while including regional energy systems and technological development plans. In turn, the sectoral alliances have been implemented in comparative management environments which have adapted by proposing relational strategies, while taking into account the compartmentalization of international industrial niche economies (e.g., electronic semi-conductor industry).[57]

Substantially, the sectoral participation of national specialized economies has been followed through different public strategies with the adoption of PPP organizational approaches, adapted according to the multi-level perspectives of

---

53 Ulrik Jorgensen, "Mapping and Navigating Transitions – The Multi-Level Perspective Compared with Arenas of Development," *Research Policy* 41/2012, pp. 996–1010.
54 See: Yosseph Leibovitz, (2001). "Associative Governance?".
55 See: 1) Bing-Sheng Teng, T.K. Das, "Governance Structure Choice in Strategic Alliances. The Roles of Alliance Objectives, Alliance Management Experience, and International Partners," *Management Decision* 46.5/2008, pp. 725–742; 2) Carl Marcus Wallenburg, Thorsten Schaffler, "The Interplay of Relational Governance and Formal Control in Horizontal Alliances: A Social Contract Perspective," *Journal of Supply Chain Management* 50.2/2014, pp. 41–58; 3) Dan Li, Lorraine Eden, Michael A. Hitt, R. Duane Ireland, Robert P. Garrett, "Governance in Multilateral R&D Alliances" *Organization Science* 23.4/2012, pp. 1191–1210.
56 Ulrik Jorgensen, "Mapping and Navigating Transitions," pp. 996–1010.
57 Adrian Smith, Rob Raven, "What Is Protective Space? Reconsidering Niches in Transitions to Sustainability," *Research Policy* 4/2012, pp. 1025–1036.

institutionalization processes. In such type of associative agreements, comparative regional societies have been operating within the comparable socio-economic knowledge areas promoted through international production programs, incorporating the local innovation factors assessed by proximate industrial networks, which have also shifted leading positions beyond the conventional set of related national expectations.[58] Hence, what happens to public local deliberations and to public networks specifying environmental advocacy campaigns has been perceived in terms of critical improvement through which to address the specific inductive context of technical innovation platforms. Critical improvement has been also gradually introduced for more direct or indirect environmental cognitive assessments, elaborated about public-private PPP organizational involvement within the innovation knowledge process *per se*.

Nonetheless, the sectoral differentiation of adaptive knowledge information domains related with economic and social innovation transfers has created a formal disaggregation of transitional management systems in terms of technological blueprints that tend to reflect the actual projection of national sectoral interests. Therefore, managerial progressive undertakings[59] have relied on the PPP knowledge transfer practices for the configuration of national decision-making strategies that in time have emerged through a formal representation of combined regulatory management powers which indirectly affect the competing production factors.

Essentially, the consequential normative adoption of regional sustainable policies that are able to reaffirm local economic transitions operated within comparative ruling systems, has included an identification of corroborative paths retracing public modelled spaces.[60] Common governance spaces have been articulated according to the resulting local decisional frameworks which have been aggregated through highly fragmented and composite directions – including the technological knowledge development implying transitioning processes

---

58 See: Richard Patrick Bixler, (2014). "Is There an Heir Apparent to the Crown? A More Informed Understanding of Connectivity and Networked Environmental Governance in the Crown of the Continent." PhD Dissertation at the Department of Sociology, Colorado State University, Fort Collins, Colorado, United States.

59 See: Gwendolyn E. Lock (2010). "Who Shares? Managerial Knowledge Transfer Practices in British Columbia's Ministry of Health Services." PhD Dissertation at the College of Management and Technology, Walden University, United States.

60 See: Breena Holland (2005). "Environment and Capability: A New Normative Framework for Environmental Policy Analysis." PhD Dissertation at the Department of Political Science, University of Chicago, United States.

that have been expanded across larger societies. For such reasons, comparable adherence strategies about regional innovation niche economies[61] have practically advanced the terms of compatible organizational conditions, which have been structured through multiple interrelated local innovation activities still pertaining to the knowledge transfer processes.

In effect, public integral access on local governance through a sustained degree of programmed agendas set for public participation conditions and regional innovation alliances are facilitated through the actual deployment of a broader understanding.[62] At the same time, there have been comparative learning objectives reaffirmed across regional contexts about the sectoral variations of economic distributive resourcing bodies, which have been arranged in accordance with national and local innovation networks, configured around respective learning programs that have been operated through practical designs and expansive production systems.[63]

In reality, comparative political analysis which addresses the problem of sectoral distribution of dynamic innovation models incorporating technological industrial changes, has resulted into a composite descriptive exercise based on the actual levels of public interactions, where public actors' integration has impacted on governance profiles and concurrent interrelations. In addition to this, contextual learning systems associated with knowledge transfer factors have indirectly assumed a national distinction about sectoral types of practical industrial achievements for the material translation of regional and local development patterns.

In reference to the literature formulated in connection to multi-dimensional theoretical and empirical descriptive approaches[64] for the case of technological performances and knowledge innovation conditions, PPP reviews can be better observed under a common matured environment. The actual display of the negotiating table set for modern interpretative propositions presupposes the possible

---

61 See: Michele Mastroeni, (2008). "How Small Economies Finance High-Tech Industries: Seeking Variance in Innovation Commercialization." PhD Dissertation at the Department of Political Science, University of Toronto, Canada.
62 See: Celeste Richard F., "Strategic Alliances for Innovation: Emerging Models of Technology-Based Twenty-First Century Economic Development," *Economic Development Review* 14.1/1996, pp. 4–8.
63 Smith, Raven, "What Is Protective Space?" pp. 1025–1036.
64 See: Jeffrey Roy, (1999). "Government and Governance in High-Technology Localities: Ottawa-Carleton and Canada's Technology Triangle." PhD Dissertation at Carleton University, Ottawa, Ontario, Canada.

expansion of cross-fields of thought, which have been focused on the capitalized transition of socio-technical systems that have been carried forward in view of respective PPP managerial interests, which essentially have been constructively and economically interdependent.[65] Through the evolution of interdependent perspectives, the authors Safarzynska,[66] Frenken, and Bergh have stressed out the fact that the reconfiguration of PPP roles for knowledge trading economies has required the open adaptation on local resource provisions and local governance transitions, which can reflect different levels of development processes.

Fundamentally, regional technology clusters that form through inter-management areas of industrial innovation assets have been systematically interdependent while interrelated to applicative organizational conditions which belong to different socio-economic environments. As a matter of fact, structural changing conditions can either favor or disfavor the multiple technical adaptation scenarios and constructive guidelines that refer to the recent technology transformation models in which development regulatory plans have entailed distinctive practical choices about national managerial strategies as well as local inclusive approaches.[67]

In addition to this, societal adaptive powers resulting in connection with knowledge innovation transitions have emerged through cyclical periods of time but on occasion with material limitations for the constitution of contextual productive and learning adaptive stages, which fundamentally have been very different in nature. Thus, the innovation process itself has been the result of possible societal identifications involving cooperation and integration within complex productive dynamics. This type of progressive material evolution has been characterized by specific managerial adaptation features applied through the knowledge transfers occurring in societal changing processes.[68]

Similarly, when combining regulatory orientations of public distributive regimes and regional industrial innovation plans,[69] the corresponding levels of

---

65 Karolina Safarzynska, Koen Frenken, Jeroen C.J.M. van den Bergh, "Evolutionary Theorizing and Modelling of Sustainability Transitions," *Research Policy* 41/2012, pp. 1011–1024.
66 Tommaso Ciarli and Karolina Safarzynska, "Sustainability and Industrial Challenge: The Hindering Role of Complexity," *SPRU Working Paper Series*, SPRU – Science Policy Research Unit, University of Sussex Business School, 2020.
67 Safarzynska, Frenken, van den Bergh, "Evolutionary Theorizing," pp. 1011–1024.
68 See: Mari Jose Aranguren, Miren Larrea, "Regional Innovation Policy Processes: Linking Learning to Action," *Journal of the Knowledge Economy* 2.4/2011, pp. 569–585.
69 See: Dinah A. Koehler, "The Effectiveness of Voluntary Environmental Programs – A Policy at a Crossroads?" *Policy Studies Journal* 35.4/2007, pp. 689–722.

international and national cooperation have directly reflected strategic decisional possibilities adapted by multi-level agencies, progressively signaling the need for development of integral technological diffusion practices, and local innovation capacities (e.g. the renewable energy fields). In fact, the technology innovation process has been expanded in association with major socio-economic determinants relating to formal and informal national complementary criteria.

Comprehensive industrial restructured models have been favored through public accesses and continuous interaction paths at different development stages in order to support multi-level functioning systems turning up according to the regional organizational dynamics.[70] The dynamic distribution of regional innovation systems has also corresponded to functional technical interpretations that have been provided regarding the actual transformation of local economic planning, which involves comparative PPP knowledge transfers and changing innovation paradigms, particularly assessed in view of favoring mutual levels of public transparency and local accountability.[71]

As already emphasized, the quantitative and qualitative transformation of knowledge transfer systems has been coordinated through the preparation of development strategies and technical innovation policies, which have been tested in comparative industrial exchange zones based, for instance, in European countries and/or in the newly emerging economies of East-South-West Asian zones,[72] among other territorial contexts. The pursuance of national regulatory plans supporting transitioning processes has been followed through the progressive stages of PPP regional productions for integration as well as validation of concurrent practical aspects that relate to local regulatory activities, which can specify the path for transitional innovation measures foreseen within national management systems.

---

70   See: Marlete Beatriz Macaneiro, Sieglinde Kindl da Cunha, "Theoretical Analysis Model of the Adoption of Reactive and Proactive Eco-Innovation Strategies: The Influence of Contextual Factors Internal and External to Organizations," *Brazilian Business Review* 11.5/2014, pp. 1–23. ISSN 1808-2386.
71   K. Matthias Weber, Harald Rohracher, "Legitimizing Research, Technology and Innovation Policies for Transformative Change. Combining Insights From Innovation Systems and Multi-Level Perspective in a Comprehensive 'Failures' Framework," *Research Policy* 41/2012, pp. 1037–1047.
72   See: Morris Teubal, "What Is the Systems Perspective to Innovation and Technology Policy (ITP) and How Can We Apply It to Developing and Newly Industrialized Economies?" *Journal of Evolutionary Economics* 12.1–2/2002, pp. 233–257.

However, it can be taken into account that similar PPP coordination aspects have been polarized with resulting implications. For example, the critical political oppositions and shared ideological views able to frame the societal interpretations of systematic environment events, have required coherent legitimacy with a possible local correspondence. In sum, the overall national management directives servicing the structural innovation plans and regulatory changing policies, have been directly faced with material implications of local integrated environments, interlinked with local industrial production lines which follow on comparative operating policies. Procedural operable contexts have been reaffirmed at the local level in view of decentralized diffusion practices targeting competitive multi-layered innovation assets and technology marketing processes.[73]

Essentially, the public identified recognitions of public-private technological innovation systems have relied upon the institutionalization of complementary inferential factors referred to comparative innovation market conditions.[74] Under complex regional industrial environments, the governmental stakeholders have undertaken management adaptation choices according to mutual levels of understanding of progressive exchanges designed according to technical industrial stages. At the same time, multi-level representative administrations operating in cooperation with national and local stakeholders have been able to determine the deliberative processes for the actual implementation of environmental targeted measures, which have been assessed in parallel with the establishment of adaptive legislative plans and concerted industrial practices.

In fact, international scholars[75] have pointed at the common ground that has been established for the structural regulatory conditions developed in view of selective science industrial programs associated with knowledge information activities. In some cases, the national decision-makers have opted to start public exploratory reviews for the possible adoption of technological innovation approaches, which have been motivated by mutual levels of formal or informal involvement with corresponding societal empowerment directions.[76]

---

73 Weber, Rohracher, "Legitimizing Research, Technology and Innovation Policies," pp. 1037–1047.
74 See: Joseph Feller, Patrick Finnegan, Jeremy Hayes, and Philip O'Reilly, "Institutionalising Information Asymmetry: Governance Structures for Open Innovation," *Information Technology & People* 22.4/2009, pp. 297–316.
75 See: Kevin Zhu, Shutao Dong, Sean Xin Xu, Kenneth L. Kraemer, "Innovation Diffusion in Global Contexts: Determinants of Post-Adoption Digital Transformation of European Companies," *European Journal of Information System* 15/2006, pp. 601–616.
76 Evita Paraskevopolou, "Non-Technological Regulatory Effects: Implications for Innovation and Innovation Policy," *Research Policy* 41/2012, pp. 1058–1071.

In essence, a common aspect has emerged, referring to the concretization of public-private integrated knowledge approaches related to the fact that PPP innovation transfers and knowledge identification processes have been combined with sectoral local development. Creative programming initiatives have been mediated through the actions undertaken by respective innovation policy implementing agents, which have been focused on the long-term horizon. Moreover, the competitive regional productions and national adaptation plans have also offered concrete regulative and operational objectives, favoring the convergence of multi-level institutional performances that can address the transnational and local maintenance of cooperation management programs, which have involved concurrent organizing configurations arranged according to different political landscapes.[77]

2.2.2 Normative Transformation

The multiple PPP adherences to prescribed normative rules have led to the diffusion of institutional initiatives for research and development (R&D) that entrust the public accountable partners with concrete objectives which refer to the transforming distribution and service programs of national local economies.

This means that the new formulation of respective organizational boundaries involving the industrial firms' technological networks has included flexible innovation routes for the local adaptation process – stimulating progressive common directions that have also had policy implications – connected to national environmental orientations and participatory domains of local constituencies.[78] In essence, the possibility of translating the industrial formation of innovation networks into national collaborative paths has moved beyond the point of simple fragmentation which incorporates the technical regulatory rules, because for both public local organizations and private innovation hubs have been included modified policy measures and local participatory factors. The combination of

---

77 See: 1) Adam D. Sheingate, "Political Entrepreneurship, Institutional Change, and American Political Development," *Studies in American Political Development* 17.2/2003, pp. 185–203; 2) Gregory Jackson, Richard Deeg, "Comparing Capitalisms: Understanding Institutional Diversity and Its Implications for International Business," *Journal of International Business Studies*, suppl. Part Special Issue: Institutions and International Business 39.4/2008, pp. 540–561.

78 See: Ion Bogdan Vasi, Brayden G. King, "Social Movements, Risk Perceptions, and Economic Outcomes: The Effect of Primary and Secondary Stakeholder Activism on Firms' Perceived Environmental Risk and Financial Performance," *American Sociological Review* 77.4/2012, pp. 573–596.

national or local adaptability linked with transitioning aggregated procedures has been carried out through complementary functional conditions and resulting evolutive plans, which have been layered out according to the prioritization of PPP through private managerial requirements, and regulative decisional options undertaken for regional economic assets.[79]

With regard to these aspects, it is important to assume that sectoral development with complementary exercising entities that can involve local regulatory innovation mechanisms, has included concurrent technology propagation aspects directed toward common organizational targets and local practical aims. This operable integration has conducted to the affirmation of diverse structural components involving the social interacting domains, which have been particularly tested on territorial conditional implications, affecting the local regulatory cases on environmental conservation plans and multilateral preservation policies, among other strategic determinants.[80]

There have been recent managerial development studies which have looked at comparative regional perspectives explored by Buuren, Buijis, Teisman, 2010[81] and Achillas et al., 2011[82] highlighting the actual presence of an ideological divide forming between knowledge innovation firms that deal with regional operational practices, and PPP managerial implementing actors. In this case, multi-level technology and innovation systems have been faced with environmental development cases which have affected the application of theoretical models as well as the practical programmatic patterns identified according to local productions and technology transfer frameworks.[83]

As consequence, the environmental managerial approaches have been formulated in view of achieving coherent public aims while clarifying the dependency of inferring aspects for the quality assessment of the standardization process compared with local integrative managerial formations. However, the industrial

---

79 Paraskevopolou, "Non-Technological Regulatory Effects," pp. 1058–1071.
80 Paraskevopolou, "Non-Technological Regulatory Effects," pp. 1058–1071.
81 Arwin van Buuren, Jean-Marie Buijs, Geert Teisman, "Program Management and the Creative Art of Cooperation: Dealing With Potential Tensions and Synergies Between Spatial Development Projects," *International Journal of Project Management* 28/2010, pp. 672–682.
82 Ch. Achillas, Ch. Vlachokostas, N. Moussiopoulos, G. Banias, G. Kafetzopoulos, A. Karagiannidis, "Social Acceptance for the Development of a Waste-to-Energy Plant in an Urban Area," *Resources, Conservation and Recycling* 55/2011, pp. 857–863.
83 See: Davorin Kralj, "Dialectal System Approach Supporting Environmental Innovation for Sustainable Development," *Kybernetes* 37.9–10/2008, pp. 1542–1560.

innovation process at system level has been combined with separate coordination spheres therefore resulting in PPP interactions for local participatory plans and for public-private adaptive mechanisms, which need to be put in place for the programmatic conjunction of the regulatory implementation schemes applied through distribution planning programs.[84]

Moreover, there is an inner complexity in line concerning comparative verification stages of local innovation projects, which can cause further structural imbalances due to the systematic organizational and technical fragmentation of industrial conditions, affected by the increasing levels of global production assets and transnational integrated innovation zones.[85] Additionally, the sharing of similar dimensional development approaches with respect to the functional programming reliability – based on the local technology upgrading and the innovation assessment programs – has enhanced the concurrent aggregation of national organizational capacities. In essence, the institutional adaptive process of changing areas of PPP put under regulative transitioning aspects, can create additional technical measured requirements, including internal and external transitional steps of local productive dimensions.

Therefore, the managerial project aims formulated in terms of local structural identifications for technology and innovation development have involved a concurrent set of multi-layered factors, which have been defined in terms of functionalities regarding e.g.: the level of mutuality of public policies; the administrative contouring procedures; the international scientific research backgrounds; and the agency-based models of knowledge information diffusion. Such intertwined specific factors have become quite fundamental for the analysis of local organizational configurations, where the private industrial groups and codependent governmental agencies have been evaluated according to their respective different degrees of responsive operational abilities able to fulfil local adaptive governmental management plans.

In order to identify the contemporary progressive stages of environmental societal interactions, in this text is considered the interrelated affirmation of institutional protection mechanisms that can address local participative orientations and public ecology discourses which have involved the administered

---

84 See: Rene Kemp, Jan Rotmans, "Transitioning Policy: Co-Production of a New Strategic Framework for Energy Innovation Policy in the Netherlands," *Policy Sciences* 42.4/2009, pp. 303–322.
85 van Buuren, Buijs, Teisman, "Program Management and the Creative Art of Cooperation," pp. 672–682.

populations.[86] Similarly, consideration constructive ethical aspects highlighting that societal territory integration and participative modifying factors have both acquired respective significance in view of PPP socio-economic knowledge transitions and technology asset implementations associated with local planning activities.

---

86   See: Sabina Scarpellini, Alfonso Aranda, Juan Aranda, Eva Llera, Miguel Marco, "R&D and Eco-Innovation: Opportunities for Closer Collaboration Between Universities and Companies Through Technology Centers," *Clean Technologies and Environmental Policy* 14.6/2012, pp. 1047–1058.

CHAPTER THREE

3.1 The Governance Networks

The actual maintenance of regional innovation networks and local production strategies has led to the practical formulation of programmed scenarios based on the societal adaptability and legitimacy mechanisms, which have been redefined through constructive applicable agreements that have been aggregated into respective environmental assimilation processes, following governmental intervening approaches.[1] In the public policy literature have been identified societal cognitive determinants of participatory approaches which continue to remain quite key in reference to transnational innovation networks that reflect the industrial and semi-industrial models of newly formed social democracies.

The social democratic states have actively introduced concurrent questions about the development of national science innovation policies. Essentially, in order to specify the critical levels of local public acceptances in connection to science knowledge domains, have been enlisted common environmental principles and territorial acting roles which can define what an industrial policy adaptation really is.[2] In addition, there have been institutional attempts to include other fundamental exploratory questions about the combination of possible motivation elements that concern the legitimacy of action, local adaptation models, and transitioning steps intertwined with public-private partnerships PPPs. Such theoretical explorations can therefore lead to the prioritization of some basic participatory requirements still dependent on the ways that constituent bodies, centric decision-makers and public stakeholders can effectively decide on how to adopt

---

[1] See: Gilbert L. Rochon, Dev Niyogi, Souleymane Fall, Joseph E. Quansah, Larry Biehl, Bereket Araya, Chetan Maringanti, Angel Torres Valcarcel, Lova Rakotomalala, Hildred S. Rochon, Bertin Hilaire Mbongo, Thierno Thiam, "Best Management Practices for Corporate, Academic and Governmental Transfer of Sustainable Technologies to Developing Countries," *Clean Technologies and Environmental Policy* 12.1/2009, pp. 19–30.

[2] See: 1) Thomas Hickmann, "Science-Policy Interaction in International Environmental Politics: An Analysis of the Ozone Regime and the Climate Regime," *Environmental Economics and Policy Studies* 16.1/2014, pp. 21–44; 2) Eva Heiskanen, Sirkku Kivisaari, Raimo Lovio, Per Mickwitz, "Designed to Travel? Transition Management Encounters Environmental and Innovation Policy Histories in Finland," *Policy Sciences* 42.4/2009, pp. 409–427.

decisional strategies fostering the regional planning partnerships. Consequently, sectoral development objectives have spread through transnational innovation knowledge networks geographically clustered.[3]

There have been similar analytical indications regarding the comparative establishment of political adaptation norms and environmental networking transfer measures which have been put together for local management evaluations that have involved the ecological and climate environmental strategies.[4] Through multi-level governmental approaches, environmental governance and functional adaptive distributions that are available at societal level, have recalled attention upon the sectoral cooperative transformation – where regulation powers provide the legal connecting ties for comparative implementing options – also in view of different judicial emerging perspectives.

In such cases, the characteristic argumentations about legal environmental issues – based on climate change[5] factors and local collaborative designs of comparative regulatory measures, also adopted through instrumental geo location-based policies – have been related to contextual scales of public acceptances and legitimating policies. In fact, the pursuance of environmental sustainability policies has reflected the application of practical guidelines defined by national and regional regulatory bodies having fundamentally acted in connection with formal governance alignments and legal distribution of territorial concerted powers.[6]

At the same time, the reasons of environmental uncertainty have been influential in a complex way, because meaningful local adaptation strategies codified by public regulative entities and identified through widespread official administrations have been designed according to the evolution of national

---

3   See: Marcelo Bucheli, Min-Young Kim, "Political Institutional Change, Obsolescing Legitimacy, and Multinational Corporations. The Case of the Central American Banana Industry," *Management International Review* 52.6/2012, pp. 847–877.

4   Arwin van Buuren, Peter Driessen, Geert Teisman, Marleen van Rijswick, "Toward Legitimate Governance Strategies for Climate Adaptation in the Netherlands: Combining Insights From a Legal, Planning, and Network Perspective," *Regional Environmental Change* 2013. DOI 10.1007/s10113-013-0448-0. (Springer-Verlag Berlin Heidelberg).

5   See: Reuben Mondejar, Hongxin Zhao, "Antecedents to Government Relationship Building and the Institutional Contingencies in a Transition Economy," *Management International Review* 53.4/2013, pp. 579–605.

6   See: Valentina Mele, Donald H. Schepers, "E Pluribus Unum? Legitimacy Issues and Multi-Stakeholder Codes of Conduct," *Journal of Business Ethics* 118.3/2013, pp. 561–576.

legal orientations, addressing multi-level environmental dynamics. In fact, public regulatory decisions have included the legitimate participation of societal stakeholders for local consultation forums and public networking meetings which can directly confront the PPPs agreements for mutual understanding, especially regarding the application of environmental adaptability criteria that have been elaborated in association with knowledge information systems, and territorial diffused practices promoted within respective empowered regions.[7]

Therefore, the multilateral adaptive frameworks have emerged in combination with collaborative dynamics focused on the search of an equitable balance between the public and private legitimate counterparts that eventually can offer associated opportunities, drafted according to the national and local governmental evolutive outcomes re-framed in terms of environmental regulation procedures. Relating to these aspects are some national implementing stages of PPP and collaborative environmental schemes which have specifically assimilated corresponding transnational intentions for the promotion of possible industrial innovating plans that incorporate the compatible environmental managerial experiences, which can support the structural avoidance of terminal power abuses.[8]

Under similar tones, the practical organizational aspects regarding compatibility of innovation and technology management programs have been followed up in terms of transnational cooperating patterns involving the national policy representatives, and local citizens' environment networks who have been either directly or indirectly involved through local production processes. Moreover, the actual collaborative interactions have been translated into constructive management transition domains for the common legitimate environmental undertakings, which have emerged from local adaptation protection policies and governmental distribution capacities.[9]

In essence, comparative environmental institutional undertakings have reached up to the public environment coalitions in terms of collective local

---

7   Buuren, Driessen, Teisman, Rijswick, "Toward Legitimate Governance Strategies," DOI 10.1007/s10113-013-0448-0.
8   See: Soonsil Lee, (2003). "Successful Implementation Strategies for Environmental Management Systems in Public Organizations." PhD Dissertation at the Department of Work Environment, University of Massachusetts Lowell, United States.
9   See: Chang Bum Ju, (2008). "Institutional Contestation, Network Legitimacy and Organizational Heterogeneity: Interactions Between Government and Environmental Nonprofits in South Korea." PhD Dissertation at the Graduate School University of Southern California, Public Administration, United States.

requirements that incorporate the need of maintaining a performing exposition of common safety interests, which have become an integral part of multi-level transnational regulative processes. In addition, the systematic innovation processes have been based on the inclusion of essential determinants that refer to local regulative interactions and technical monitoring advancements made possible through complementary converging aspects related to public decision-making, and to managerial distribution practices, when feasible. Such dual character of public interaction models – functional to common knowledge environments – has been applied through a contextual structural association seen in terms of national environmental provisions, while framed in connection with local implementation and managerial delivery methods; to consider that the PPP public-private participation levels have been established through a broader institutional spectrum,[10] including local development objectives and systemic practical accountabilities.[11]

Moreover, there have been considerations regarding long-term plans on democratic governance and sustainable development in view of multi-level adaptation activities which have resulted from the networking interactions of complex local trading organizations, which characterize socio-economic growth patterns reviewed on a yearly basis. As highlighted, the process of legal normative representation and concerted PPP approaches touching upon the political legitimacy issues and institutional progressive capacities[12] has increased the tendency of offering a local capacity distribution of multi-level sectoral domains, also for decentralized municipal agencies. Similarly, the adaptation to prescribed rules has been put in relationship with the societal learning convergence shifted toward local diffusion practices, while considering the national or local technological consolidation of innovative tools requirements.[13]

---

10   See: Kenneth W. Abbott, and Duncan J. Snidal, "The Spectrum of International Institutions. An Interdisciplinary Collaboration on Global Governance," *Publisher Routledge*, 2021. Available on Taylor & Francis eBooks. ISBN 9780367629731.
11   Buuren, Driessen, Teisman, Rijswick, "Toward Legitimate Governance Strategies," DOI 10.1007/s10113-013-0448-0.
12   See: Alan Paul Diduck, (2001). "Learning Through Public Involvement in Environmental Assessment: A Transformative Perspective." PhD Dissertation at the University of Waterloo, Department of Geography, Ontario, Canada.
13   See: Kristina Lundberg, "A Systems Thinking Approach to Environmental Follow Up in a Swedish Central Public Authority: Hindrances and Possibilities for Learning From Experience," *Environmental Management* 48.1/2011, pp. 123–133.

Regarding the case of social democratic states assimilating the regulative environmental aggregations intertwined, for instance, with: transnational economic interests, public collaborative participation, local legitimacy factors, and systematic structural changes, the formalized multi-states innovation programs have been run in conjunction with multilateral comparable application domains. Overall, this experience has resulted into regional exploratory exercises in which such interrelated territorial factors have operated according to normative facilitated conditions that involve a concurrent liberalization of regional innovation production systems.

Moreover, it can be stressed out that the actual affirmation of industrial innovating clusters has been expanded through different types of structural local designs based on progressive PPP distributed plans involving the local entities, such as in Taiwan's case with independent territorial and technological transnational production hubs.[14] In particular, the transnational clustering of integrated regional innovation hubs has manifested itself with different types of knowledge innovation models and technology identification mechanisms, still dependent on core production functions; where the socio-economic needs have been addressed according to territorial merged production dimensions. In which case, the governmental planning objectives projected through different local territorial policies have been undertaken within open learning systems of knowledge-based Asian economies, while international organized transactions have evolved in connection with transnational collective distribution modes of mutual modelling organizations, performing in interdependent adaptative environments.[15]

For instance, the transnational industrialization process in the case of Taiwan's technological localized assets has been established in connection with competitive international and regional trade changing dynamics.[16] Through which, there have been contextually-associated factors interlinked with territorial developments and corresponding IT economic aggregations, which have emerged within separate organizational spheres of sectoral interest groups, which can propose a type of dual relationship that has been created between the trading innovation firms and the regulating central authorities. Such political distributed public-private relations have been mutually modified according to

---

14 See: Christopher Howe, "Taiwan in the 20th Century: Model or Victim? Development Problems in a Small Asian Economy," *The China Quarterly* 165/2001, pp. 37–60.
15 See: Cristiano Antonelli, "Models of Knowledge and Systems of Governance," *Journal of Institutional Economics* 1.1/2005, pp. 51–73.
16 See: Christopher M. Dent, "Taiwan and the New Regional Political Economy of East Asia," *The China Quarterly* 182/2005, pp. 385–406.

local affirmative expressions and critical insights. In fact, in reference to mutual organizing PPP plans, the concentration of newly developed production assets has corresponded to the specific territorial and local partnering dimensions.[17]

For simplification, it can be highlighted that in Taiwan's case the process of IT industrial clustering of production and local innovation networks has been combined with governmental political strategies, which have been characterized by the traditional and interdependent conditions of socio-economic cultural systems. The mutual presence of dynamic identifications for collaborative relationships can provide material conditions for the expansion of industry that is related to scientific operable models and local IT knowledge programs with embedded technology production networks.

Fundamentally, the so-called *regional embedded knowledge networks* have been formalized through emerging managerial strategies with transnational and regional trading measures, targeting the current availability of local innovation capacities, which have reflected practical levels of PPP public-private marketing interrelations. At the same time, the local integration path of programmatic guidelines for the establishment of common marketing innovation relations has been assessed through an actual availability of national and local institutional capacities.

Basically, the economic shifting focus of maturing logics regarding the PPP multi-level institutional and managerial convergence has been monitored according to the differences of present industry cases, especially at the level of local production capabilities that have enabled step-by-step local innovation stages put in conjunction with political interacting reforms which stimulate the common learning perspectives.[18]

The specific institutional progress that has been made regarding the local managerial and innovation reforms has been renewed according to territorial concerted accesses to the specialized areas of industrialization. The clustering process of industrial development areas has also been arranged in terms of spatial and temporal scales during at least the last three decades, and according to an emerging affirmation of global transactional building dynamics.

---

17   Liang-Chih Chen, "The Governance and the Evolution of Local Production Networks in a Cluster: the Case of Taiwan's Machine Tool Industry," *GeoJournal* 76/2011, pp. 605–622. DOI 10.1007/s10708-009-9317-2.
18   See: Woojin Yoon, Eunjung Hyun, "Economic, Social and Institutional Conditions of Network Governance," *Management Decision* 48.8/2010, pp. 1212–1229.

In addition to this, the formalized provision of functional distribution capacities, that run according to the governmental progressive regulations and environmental responsive measures, has been materialized in order to ensure the comparative marketing exchanges of supply and demand needs. Market-based innovation trading expansions can be balanced in view of specific PPP public-private decentralized converging conditions, facilitating the skilled labor allocation and the local contractual arrangements. Such types of decentralized comparative conditions have contributed to better define the terms of industrial local innovation and technological systemic inclusions, while also favoring public environmental assessment conditions.[19]

3.1.1 Environmental Cooperation

Historically speaking, the connected empowerment interpretations of PPPs for corporate and public organizational assets have usually recalled the necessity to adjust during national transition periods. From an environmental standpoint, the public resolutions regarding local empowerment accesses set across distinctive societies have failed to confront comparative sides of similar stories regarding local environmental degradation and the changing terms of local collective solutions.

In general, the states' compensating factors regarding transnational innovation and managerial cooperation have brought to the combined facilitation of public-private PPP interests represented by involved stakeholders, which have operated within respective national or local institutional boundaries. In comparative view, there has been a critical observation made in part of the literature about the multilateral innovation of ICTs related issues, which have been explored here as additional thematic notions. As such, the problematic delivery of IT local governance through specified adaptation plans, faced with confrontation determinants of civil society groups and international production associates, has led toward the temporary affirmation of structural production relations, which have been particularly focused on the recurrent integration patterns applied to the local contexts.

In substantive terms, the possible assimilation of disaggregated environmental socio-economic determinants can be explained according to the different transitions experienced in democratic, national, local and regional territories, which have maintained conditional local exchanges and regional participative orientations regarding public regulative conditions. Changing regulatory

---

19   Chen, "The Governance and the Evolution," pp. 605–622.

provisions have been adapted through the characterization of public monitoring schemes involving, for instance – international academic actors, international business groups, societal group alliances, and public representation platforms[20] which can prepare the field of common exchanging interfaces.[21]

On a similar ground, the regional differentiation regarding collaborative ruling designs formulated in association with decision-making propositions has offered additional knowledge areas on multi-level interaction dynamics, which have reflected the availability of responsive organizational environments. This set of regional collective platforms has also indicated the particular divide still existing on the implementation of structural innovation plans in governance, due to complementary adaptation procedures based on local public exchanges that have involved civilian participative groups.

What has come as firm criticism is the inherent emphasis on the possible integration of multiple coexisting PPP spheres which incorporate both open civilian partnerships and local governmental agreements for the affirmation of collaborative environmental engagements, mutually arranged in terms of societal and institutional applicable connections.[22]

Of particular importance is the fact that the affirmation of a common understanding of operable policy interactions for local participation and policy dialogues, has included the formation of public and private PPP environment networks that can determine the level of corresponding outcomes of the national incrementation of knowledge information models. The PPPs networking adoptions have been undertaken with the assistance of public and private national cooperative entities which can draw attention on the local environmental implemented processes for common governance configurations, as well as, technical distribution policies.[23]

Moreover, there have been regional economic links able to re-establish informal and formal planning rules integrating into public regulatory assets (including social interactions and knowledge diffusion patterns), which have

---

20   Jurian Edelenbos, Nienke van Schie, Lasse Gerrits, "Organizing Interfaces Between Government Institutions and Interactive Governance," *Policy Science* 43/2010, pp. 73–94.
21   See: Feller, Finnegan, Hayes, O'Reilly, "Institutionalising Information Asymmetry," pp. 297–316.
22   Edelenbos, Schie, Gerrits, "Organizing Interfaces," pp. 73–94.
23   See: Tyler Mccreary, Vanessa Lamb, "A Political Ecology of Sovereignty in Practice and on the Map: The Technicalities of Law, Participatory Mapping, and Environmental Governance," *Leiden Journal of International Law* 27.3/2014, pp. 595–619.

resulted into multi-level local exchanges, assimilating the formal operating practices. Nevertheless, through common knowledge interfaces,[24] the local administrative bodies have been faced with the problematic differentiation of governmental multi-level adaptation criteria resulting in targeted strategies for public accessibility conditions.[25] When in place, the formalization of governmental knowledge arrangements conducted through the involvement of respective national and local organizations, has been encountered with a series of organizational challenges, which have been dependent on regional and local integration rules associated with political powers' representative conditions.

In fact, the public transitional rules supported by decision-making activities have been drafted for the collective sensitization on knowledge-based PPP industrial regulatory programs, particularly aligned through the formal acceptance of national programmatic time-frames. For instance, in terms of contextual innovation factors pointing at the direction of multi-level development aggregations, there have been common occurrences regarding (e.g.) the formation of environmental measures facilitating the mutual deliberative interactions, which have included the elaboration of PPP harmonized mechanisms that can set closer the localities with participant communities.[26]

Fundamentally, the actual consolidation of constructive environmental objectives has favored the sustainability of local communities and, at the same time, the harmonization of common governance exercises pursued through public participatory aims, which may determine additional local practical implications. For the local practical implications, there have been corresponding PPP interactions with the political centralized expertise bodies, which have been oriented on the practical classification of environmental organizational adjustments that can be subscribed through compatible decision-making activities.[27]

Essentially, the presence of a wide range of interrelated management trends in the local environmental domains and the PPP science innovation policies,

---

24 See: Beverly Clarke et al., "Enhancing the Knowledge-Governance Interface: Coasts, Climate and Collaboration," *Ocean & Coastal Management* 86.12/ 2013, pp. 88–99. https://doi.org/10.1016/j.ocecoaman.2013.02.009.
25 See: Neil Bradford, "Public-Private Partnerships? Shifting Paradigms of Economic Governance in Ontario," *Canadian Journal of Political Science* 36.5/2003, pp. 1005–1033.
26 Edelenbos, Schie, Gerrits, "Organizing Interfaces," p. 73–94.
27 See: Robyn Keast, Myrna Mandell, Kerry Brown, "Mixing State, Market and Network Governance Modes: The Role of Government in 'Crowded' Policy Domains," *International Journal of Organization Theory and Behavior* 9.1/2006, pp. 27–50.

has provided transferable assumptions linking the international, national and local level of governance to the constitution of public environmental ownerships. In which, the inclusion of respective multilateral stakeholders and civilian participants has required the application of diversified local processing stages. Also, various complementary ruling domains are in place, which interact with jurisdictional decisions about the public inclusion of environmental practice agreements.

In fact, the precise alignment of inter-operating partners including representative structural bodies, has led to resulting configurations in public mediation activities performed within national and transnational boundaries of distributed industrial territorial systems.[28] For a more comprehensive analysis of industrial remodeling strategies in uncertain environments and institutional adaptive performances, there is the influence played by sub-regional systems interrelated with the internationalization of hybrid mechanisms and governance attributions, specifically for the interplay of national and local environmental dynamics that integrate mutual practice standards and information knowledge transfers.[29]

In particular, the possible combination attributed to public and private PPPs with production innovation programs referring to sub-regional regulation activities has been performed in cooperation with institutional local actors intertwined with the local controlling bodies constituted by national or local technical counterparts. For accountability reasons, the cooperative delivery of industrial trading mechanisms for multilateral commerce has provided major structural differences mostly perceived in terms of responsiveness criteria, local capacity adaptation, and organizational context-related practices. In fact, the actual correlation of several political factors regarding the adaptive regulatory dimension that has been established according to territorial industrial areas has created the basic conditions for promoting local partnering assets and responsive implementing programs, with referential collective assessments, including the management of local environmental resources.[30]

---

28 See: Naomi Roht-Arriaza, "Private Voluntary Standard-Setting, the International Organization for Standardization, and International Environmental Lawmaking," *Yearbook of International Environmental Law* 6.1/1996, pp. 107–163.
29 See: Arun Agrawal, Maria Carmen Lemos, "A Greener Revolution in the Making? Environmental Governance in the 21st Century," *Environment* 49.5/2007, pp. 36,38–45.
30 Ian Thynne, "Governance and Organizational Eclecticism in the Public Arena: Introductory Perspectives," *Public Organiz. Rev.* 13/2013, pp. 107–116. DOI 10.1007/s11115-013-0235-x.

Nonetheless, there have also been related difficulties regarding favoring constitutive regulatory frameworks based on the rotational applicative functions of governmental programs, involving concurrent national local openings. In fact, the national technical applicability of industrial energy capacities performing in accordance with local implementation guidelines which associate the evaluation and monitoring schemes, has been directed at the management of material distributive formations, interrelated with collective available resources. Such types of institutional efforts have been essentially regimented through the rotational applicability of international and regional arrangements, dependent on the mutual regulative agreements, which have been undertaken in view of competitive transnational provisional policies.[31]

3.1.2 Collaborative Platforms

The opening up of territorial consultations bringing together multilateral industrial partners which can respond according to transnational organizational capacities, has been deepened in time because of the need to coordinate logistic and instrumental technical plans that reflect as well the public direct involvement. In effect, in the case of mutual organizational dimensions interlinked with the public and private PPP innovation and technological ownerships, there have been intertwined management conditions regarding different levels of applicability in governance for local responsive provisions. As a matter of fact, public-private PPP partnering agreements targeting the industrial innovation development process have been aligned, across time, with international selective stages which have been confronted with different definitions of regulatory government models, because of historical affirmations and administrative orientations that foster collective designs for the territorial management adjustments.[32]

Basically, there has been a comparable prevalence of institutional developmental characteristics which refer to socio-democratic political formations. Such participative formations have involved the transmigration of systematic administrative conditions put in connection with local participative practices, that also emerge through the progressive understanding of national regulators and control management associations. For instance, the separation of decisional

---

31 See: Anna Helena Valberg, "Brazil's Role in Environmental Governance," *Report for the Norwegian Ministry of the Environment*, FNI Reports 8/2011, pp. 1–49.
32 See: Shiroyama, Yarime, Matsuo, Schroeder, Scholz, Ulrich, "Governance for Sustainability," pp. 45–55.

management domains regarding collective distribution choices has been operated in view of favoring the decentralization process regarding the industrial energy production networks.[33]

In addition to this, there has been the actual presence of interdependent multilateral regional specialists connected with specific territory identifications set across multi-state regions, contributing to structural development issues pointing at e.g.: the evidence of mutual political leaderships, the systematic hierarchical lines of powers in comparative regimes, and the managerial collaborative legitimation for local stakeholders. Regional managerial leaderships in governance have also involuntarily led to stratified levels of specific understanding of territorial institutional interactions, which have been mostly valued through the central involvement of collaborative production processes that can facilitate the actual mobilization of policy internal and external dialogues followed within national territorial organizations.[34]

On these accounts, there have been comparative contrasting views on the introduction of public evaluation criteria which have also emerged. Because, the formalization of governmental approaches which have explored the simplified aspects of regional *rapprochement* criteria of self-governance models, have been reaffirmed under a divisive prevalence of comparative organizational perspectives formulated according to the specification of territorial inner boundaries.[35]

In fact, the political coexistence of either external or internal industrial innovation development reaffirmed in terms of constructive linkages has especially been impacted by the classification of domestic relations based on correspondent technical information levels. As result, the governmental performance and local evaluative activities have been regularly adapted for the possible aggregation of centralized but also decentralized sustainable programs, involving targeted

---

33 See: 1) Edward Weber, "A Wish list for 21st Century Environmental Policy: Decentralization, Integration, Cooperation, Flexibility, and Enhanced Participation by Citizens and Local Governments," Policy Studies Journal 26.1/1998, pp. 185–195. Yooil Bae, "Decentralized Urban Governance and Environmental Collaboration in South Korea: The Case of Hyundai City," Pacific Affairs 86.4/2013, pp. 759–783.
34 Thynne, "Governance and Organizational Eclecticism," pp. 107–116.
35 See: Hannah Wittman, Charles Geisler, "Negotiating Locality: Decentralization and Communal Forest Management in the Guatemalan Highlands," Human Organization 64.1/2005, pp. 62–74.

managerial domains for the economic transnational practices which allow for multiple knowledge innovation conjugations.[36]

For instance, in the case of East Asian economies with comparative transnational business distributive models already adopting the self-governing path as in China's case, through territorial administrative rules applied in: Hong Kong, Taiwan, and Singapore, there has been a pervasive level of domestic cohesive independence which has created practical policy implications. Regarding the business entrepreneurial organizations in East Asia, the industry implementing agencies have operated in terms of aggregative adaptation models, which can reflect changing regulatory relations locally structured within the social and political interconnected spheres.[37]

This scholar H. Yeung, 2004 has produced a regional policy report in which he has observed the structural implications deriving from liberal progressive orientations of knowledge-based transnational economies such as: the Singaporean international commerce trading hubs. In fact, the institutional actualization of mutual roles and functional powers emanating in common administered localities, he noted, has brought into evidence some cognitive elements, which have been related to the translation of modernity with a value-based transfusion of national local economies. It has also reaffirmed the possibility of undertaking common proclivity steps that refer to regional institutional conducts, which have resulted into national cooperating environments, assessing common local management guidelines for the social affirmations and mutual configurations of development patterns.

Overall, societal organizational patterns have essentially reflected the integration of territorial political arrangements put into place through socioeconomic innovative configurations.[38] On this aspect, the prioritization of societal inclusive patterns for the formation of common ethical relations established in regional local governance has been intertwined with entrepreneurial

---

36 See: Dan Miodownik, Britt Cartrite, "Does Political Decentralization Exacerbate or Ameliorate Ethno-Political Mobilization? A Test of Contesting Propositions," *Political Research Quarterly* 63.4/2010, pp. 731–746.
37 Henry Wai-chung Yeung, "Strategic Governance and Economic Diplomacy in China: The Political Economy of Government-Linked Companies From Singapore," *East Asia* 21.1/2004, pp. 40–64.
38 See: Ibrahim Gamal, Galt Vaughan, "Ethnic Business Development: Toward a Theoretical Synthesis and Policy Framework," *Journal of Economic Issues* 37.4/2003, pp. 1107–1119.

organizational models, which have been organized through societal mediation contributions with corresponding local facilitation procedures.

Through this type of understanding, the institutional local agencies have tentatively restructured the regulatory conformation of significant political dichotomies. In terms of dualistic models which characterize local governance relations, the socio-economic factors have been explored according to the primary destination routes of capitalized systems. Thus the states internal investments have been transferred to the cooperative regional dimension, which has been associated with structural local decisions implemented in typified regions such as: the East Asian regional economies facilitating technological innovation in global intermittent clusters.[39] Moreover, the systematic incorporation of decision-making processes has been promoted along the lines of vertical and horizontal strategic coordination systems, which have been designed in view of allowing essential local business and local integration practices adjusting within comparative socio-economic conditions spread at various organizational levels.

Within classified East Asian knowledge economic transitions, the structural regulatory path of technology innovation campaigns promoting local networking programs has driven main established representatives to initiate local supportive actions adopted through the national configured institutions. The progressive infusion of some distinctive elements materializing from corresponding leading technical operations, reaffirmed through the transnational managerial systems, has also increased common levels of effectiveness in the provision of an operable implementation focused on technology production hubs, which have been spread across different regional local states promoting convergent social trajectories.[40]

In historical terms, the East Asian governance manifesting regional organizational trends has inherited the post-colonial fragmentation of local governance practices through the affirmation of development administrations, which have been able to rearrange institutional relations through common cognitive patterns referring to societal urban-rural linkages, that have been expanding under the regional economic liberalization of competitive trading dynamics focused on comparative living conditions, which have involved material local investments and transitional production logics.

Such complex development performances have integrated across time comparative organizational notions about the inter-operability of local standardization accesses designed between multinational companies for innovative

---

39 Yeung, "Strategic Governance and Economic Diplomacy," pp. 40–64.
40 Yeung, "Strategic Governance and Economic Diplomacy," pp. 40–64.

operating performances.[41] At the same time, the public and private PPP with local clustering patterns of comparative knowledge-based innovation assets have relied on PPP cross-bordering relations, which deal with permanent regulatory frameworks for the marketing flow of integral delivery aspects, surrounding newly developed economies such as: China's state trading organizations (STOs) acting within institutionalized collective formations of regional trading systems.[42]

Therefore in this discussion, we can take into consideration a comparative theoretical understanding of more inclusive aspects of regional socio-economic patterns that have recently been industrially configured in terms of multilateral governance tools and local innovation approaches – within public or private organizations – which have been interlinked with normative mutual designs and coexisting tendencies coming into place. Basically, the rational delivery of interrelated performing sub-systems involving multi-level political segments, has put forward the need of favoring the sectoral comparable innovation areas characterized by local sustainability issues. So that multi-areas innovation performances have been redefined in relation to competitive local innovation strategies that should correspond to the industrial managerial assessment plans.[43]

---

41 See: Aura Caramizaru, and Andreas Uihlein, "Energy Communities: An Overview of Energy and Social Innovation," *JRC Science For Policy Report* 2020, (European Commission, Belgium). https://publications.jrc.ec.europa.eu.
42 See: Sayeeda Bano, Sriya Kumarasinghe, Yih Pin Tang, "Comparative Economic Performance and Stock Market Performance: Some Evidence From the Asia-Pacific Region," *Asian Journal of Finance & Accounting* 3.1/2011, pp. 1–22.
43 See: Chih-Pin Lin, Hsin-Mei Lin, "Maker-Buyer Strategic Alliances: An Integrated Framework," *Journal of Business & Industrial Marketing* 25.1/2010, pp. 43–56.

CHAPTER FOUR

4.1 Holistic Integration Approaches

The institutional reviews made on local learning economies nowadays have resulted in complex governmental visioning exercises where compatible local exchanged solutions have been considered through a differentiation of regional changing processes that lead to consistent mutual affirmations.

In similar ways, we can emphasize that the formation of theoretical and interpretative assumptions on strategic local governance has included a multi-disciplinary public management analysis that has been drafted according to the arrangement levels of economic liberalization processes established for common regional trading areas. Through the possible inclusion of regional knowledge and innovation standards for local learning economies, the adoption of more holistic notions about the corresponding political adaptive approaches is more favorable.

In order to understand the type of relationship that has been created between the private enterprises focused on knowledge-transfer developments and the public agencies operating through local constructive interrelations, the author Frey, 2005[1] has pointed out that this discussion converges on the characterization of multiple levels of coexisting organizational domains. Multiple knowledge integrated domains have been managed by public actors through public or private PPP institutional exchanges, which have reflected public local strategies emerging within democratic governance systems, when the case.

In particular, public agencies have maintained the prevalence of instructive practices made available for the specific incorporation of states' programmatic relations. Where, major PPP stakeholders have been particularly influenced by competitive dynamics of regional production systems. Such emerging trade dynamics have involved the governmental adoption of common local alliances approached through flexible and effective practices oriented at local organizational processes.[2]

---

1 B.S. Frey, "Public Governance and Private Governance: Exchanging Ideas," in: *Multidisciplinary Economics*, eds. P. de Gijsel and H. Schenk (Springer, printed in Netherlands, 2005), pp. 167–186.
2 See 1) Antonin Wagner, "Good Governance: A Radical and Normative Approach to Nonprofit Management," *Voluntas: International Journal of Voluntary and Nonprofit Organizations* 25.3/2014, pp. 797–817; 2) Peter Vincent-Jones, "Contractual

In fact, there have been recurrent themes regarding public decision-making based on the principle of mutuality regarding decentralized information business assets, associated with the methodology stages that are required for interpreting local visible implications, which can lead eventually toward the integration of democratic and trustworthy relations. In essence, the regional managerial organizations and public local agencies have been focused on cooperation for emerging integration processes that can facilitate common leading cognitive standards, involving the local affirmation of innovation and technology systems. For specific territorial management sites, the comparative outlook provided about the regulatory actualization process has been referred to the constitution of participatory local interests forming across distinctive national governmental sectors.

The supportive exchange process regarding the facilitation of multiple organizational attitudes has been elaborated through the practical distribution of functional administrative roles and local responsibilities, which have constantly emerged within nationally-based institutional components, involving the assimilation of practical changing technologies and local innovation schemes.[3] Basically, the changing organizational criteria of national institutional settings have adhered to a distinctive composition of legislative, judicial, and administrative functions, which have become integrated in the form of primary delivery interfaces that allow for public accountability of local transferring proceedings.

In addition, the PPPs in local modified settings have formally established multi-level responsive linkages through national organizing counterparts verifying the local adaptability of public and private regulative performances. At the same time, environmental corresponding policy performances have included monitoring and evaluation M&E procedures which can address the diversification of local controlling interests managed in parallel with formal technical requirements that support the collective societal inputs provided by local citizens engaged in participative programming initiatives.[4]

Similarly, in view of introducing motivated management indications regarding the regional and national strategic planning of industrial innovation activities have been taken into account the dynamic mechanisms of local collaborative

---

Governance: Institutional and Organizational Analysis," *Oxford Journal of Legal Studies* 20.3/2000, pp. 317–351.

3   See: James Aaron, "Distributive Justice Without Sovereign Rule: The Case of Trade," *Social Theory and Practice* 31.4/2005, pp. 533–559.

4   See: Agni Kalfagianni, "Addressing the Global Sustainability Challenge: The Potential and Pitfalls of Private Governance From the Perspective of Human Capabilities," *Journal of Business Ethics* 122. 2/2014, pp. 307–320.

relationships, which have consolidated multi-lateral territorial governance systems. In essence, institutional innovation changes have been intertwined with compensatory distribution schemes combined in turn with structural territorial shifts incubating the diffusion of local innovation strategies. The technology innovation strategies have been operated in terms of mutuality and coherence principles, assumed through different levels of regulatory integration for each territorial dimension.

As a matter of importance, the sectoral innovation capabilities have been developed in line with the common promotion of territorial industrial dimensions in which the PPPs multilateral emerging inputs have already been delineated according to the type of legal and normative applications foreseen in the case of new business innovation models, which suggest the prior establishment of governmental local trading policies. Moreover, the association of corresponding governance and economic sustainable measures has been pursued by allowing regional stratified designs of industrial territorial programs to identify the economic integrative drivers e.g. the case of European member states and the regional member's local economies.[5]

Moreover, there are other determinant growth factors at play which have contributed to the spatial identification of competitive global production hubs that have been spread through the regional institutional networks supporting the adaptation character of public local agencies. The networking activity of public local bodies has been in place for the provision of regulatory conditions which reaffirm common integration factors for the local territorial development. Through different temporal stages, the functional assimilation of socio-economic local approaches has emerged according to case-by-case supportive rural-urban developments, for instance, with newly-formed adaptation plans regarding the construction of national maritime ports, or otherwise for the local expansion of innovative building infrastructures.[6]

In addition to this, there is the need to identify common regulative elements regarding the formation of socio-economic innovation policies that refer to official environmental sustainability requirements. In the exploratory field of scientific knowledge and comparative scientific groups' expertise with public-private

---

5   See: Burkard Eberlein, "The Making of the European Energy Market: The interplay of Governance and Government," *Journal of Public Policy* 28.1/2008, pp. 73–92.
6   Arnoud Lagendijk, Frans Boekema, "The Territoriality of Spatial-Economic Governance in Historical Perspective: The Case of the Netherlands," in: *The Disoriented State: Shifts in Governmentality, Territoriality and Governance*, eds. B. Arts et al., (Springer Science-Business Media B.V, 2009), pp. 121–140.

regional affiliations, the reinforcement of environmental networking interactions with local regulatory agencies has been maintained through supportive multi-level participant stakeholders; which however have been dependent on the mutual level of internalization and externalization of multilateral regional trading innovation assets related e.g. to: the knowledge adaptation measures; the decentralization of productions for the innovation transfer process; and the combination of an increased spatial autonomy.[7]

Certainly, we should consider that there are comparative historical implications about, for instance, the European economies with comparative integrated finance institutional assets arranged from previous centuries. There have been specific developmental characteristics connected with local administrative regions within larger countries such as: France, Italy, and Spain, which have consistently emerged across time, especially in terms of public sharing powers intertwined with political provincial leaderships.[8]

4.1.1 Clustering Provisions

The programmatic orientations expressed about compatible institutional adaptations in STS environments have implied progressive knowledge transfer options that have been determined for the aggregation of figurative modern practices involved. For instance, a recurrent divisive line regarding regional governance can be clearly observed, which addresses the functional centralization of national power relations conducted in view of fulfilling local regulative aims.

Through regional states' planning policies, the actual identification of knowledge transitions with sectoral distribution networks spread across internal provinces and local municipalities has been commonly acquired through the propagation of corresponding norms and common rights associated with common duties, which have been addressed to meet local needs of both administrators and the administered.[9] In fact, the role of state institutions has been modified according to historical normative backgrounds materializing through progressive social and economic relations.

---

7   Lagendijk, Boekema, "The Territoriality of Spatial-Economic Governance," pp. 121–140.
8   See: Marc Lazar, "Testing Italian Democracy," *Comparative European Politics* 11.3/2013, pp. 317–336.
9   See: Peter John, Alistair Cole, "When Do Institutions, Policy Sectors, and Cities Matter? Comparing Networks of Local Policy Makers in Britain and France," *Comparative Political Studies* 33.2/2000, pp. 248–268.

This type of a common historical ground has fundamentally determined a functional organization of the regional and local parties, who have been involved on different national fronts for the material flow of economic and political interactions. In addition, the transition of civic local parties which have coalesced into different interacting sides, has led to the affirmation of specific collaboration dynamics, which have emerged in reference to e.g.: the states' major planning commitments regarding resources distribution capacities; the local governance and associated power devolution functions; and the managerial aspects of capitalistic business plans developed soon after the end of World War II.[10]

Through reconstructive indications, the regional development assessed within spatial knowledge acquisitions has reflected functional strategic plans of post-socialist structural economies that have implemented common leading memberships across, for example, European states. Instead, pluralism determinants and EU Eastern regional nationalism have been translated into rationalist notions adopted as cohesive societal models that appear in accordance to distinctive aggregated elements characterizing comparative national strategies about territorial governance and local planning activities.

One initiative taken up as part of this process is the national integration of regional industrial productions and innovation-related technological programs, which have evolved into progressive undertakings for the contiguous devolution of functions of public regulatory domains. The ordinary elements of national distribution functions have been basically reaffirmed through specific organizational drivers such as: a. prevalence of regional socio-economic innovation determinants, b. implementation of science innovation networks, and c. local technological partnerships associating the structural development changes, which run in parallel to previous planning conditions.[11]

Moreover, the territorial approaches have been supported according to the identification of local governance strategies that can maintain in view the local implementation of industrial production factors. Strategic performance factors, in fact, have indirectly maintained a functional production access, which has been adjusted in combination with the comparative trading networks of regional local suppliers. One regional-level approach is the regulatory facilitation

---

10 See: Keith Gildart, "Coal Strikes on the Home Front: Miners' Militancy and Socialist Politics in the Second World War," *Twentieth Century British History* 20.2/2009, pp. 121–151.
11 See: George M. Guess, "Comparative Decentralization Lessons From Pakistan, Indonesia, and the Philippines," *Public Administration Review* 65.2/2005, pp. 217–230.

including provincial administrations, which can directly operate on modified environmental indications. Similarly, the increasing flow of practical regulatory considerations has also been a concern in terms of the regional local development because of industrial innovation plans reaffirming different states' organizing positions.

Essentially, functional performances of multi-level regulatory entities have been involved at local practice levels for the systematic coordination of related activities. As local responsive adaptations, the evolution of public technocratic agencies has been territorially maintained with a sustainable segmentation of responsibilities encountered at both the national and the local level for the distribution of concerted development plans.

However, the regulatory preparation of sustainable initiatives including normative guidelines has been confronted with the concurrent stages of deliberative negotiations, which have been elaborated through the formalization of ethical approaches adopted according to the characterization of regulatory conditions, which involve the establishment of regional industrial production sites.

Therefore, the spatial organization of technological innovation strategies delivering concerted plans for the monitoring and evaluation M&E schemes has been conducted along the lines of practical implications – with regulative distributions that offer the specific roles and common responsibilities – which have been considered through the material dispositions of the internal capacities, referring to institutionalized bodies and cultural civic-ethic groups, who have been committed in local environmental politics.[12]

Additionally, the actual presence of practical contributions for the adoption of governmental local approaches has been fostered through regional commerce plans based on R&D innovation policies, among other areas, which have particularly remained circumscribed to the physical domain of selective environments. For the simple purpose of planning on multilateral preparation activities, the comparative managerial organizations have also explored common knowledge transfers' processes for the industrial producing areas, which have responded to calculated risks; especially regarding the cases of regional budgets' availability and public financial remittances provided through the local municipalities regulated together with provincial agencies.[13]

---

12   Lagendijk, Boekema, "The Territoriality of Spatial-Economic Governance," pp. 121–140.
13   See: Seung Ho Park, Shaomin Li, David K. Tse, "Market Liberalization and Firm Performance During China's Economic Transition," *Journal of International Business Studies* 37.1/2006, pp. 127–147.

In a retrospective view, some authors like Edwards, 2012; Hollingsworth 1998; and Jessop 2002; have actually examined the combination of participatory policy making and local economic spheres including the public performance of interest groups and local citizens' cohorts for comparative operable transitions. Effectively, the common expression of governmental changing requirements has involved major innovating guidelines on collaborative strategies and territorial political alliances. Whether or not we consider that the resulting evaluation processes can refer *per se* to public and private PPP with implementation directives, a range of domestic logics regarding the reaffirmation of public territorial conducts and managerial arrangements is included. Therefore, it can be justified the notion that legitimate organizational drivers of regional innovation societies have been followed according to mutual exchanging stages of local innovating paths associated with stated declarations and constructive reforming steps, which can implement sectoral economic strategies, also undertaken within the local constituencies.[14]

For instance, the mediation process initiated in order to reach a compromising ground regarding national participatory arrangements for the local provinces involved has been determined by interrelated adapting steps aligned according to the specific coordination aims. For this reason, the intensive area of institutional ruling reforms has been adapted particularly for centralized plans run by the state agencies through validation principles about local transparency and local accountability. At the same time, the economic reforming aims based on science planning activities have been promoted in association with a structural understanding for local provinces and municipalities.[15]

Based on such accounts, one question remains open about the participatory indications drafted for the determination of local contextual measures, concentrating on short-term delivery of governmental programs, publicly affected by internal and external derivative impacts of the public action. The provisional conditions in governance have been related to the faculty of delivering

---

14 See: 1) Jonathan Q. Morgan, "Governance, Policy Innovation, and Local Economic Development in North Carolina," *Policy Studies Journal* 38.4/2010, pp. 679–702; 2) Alan Dignam, "Capturing Corporate Governance: The End of the UK Self-Regulating System," *International Journal of Disclosure and Governance* 4.1/2006, pp. 24–41.
15 Arthur Edwards, "Tensions and New Connections Between Participatory and Representative Democracy in Local Governance," in: *Renewal in European Local Democracies,* eds. L. Schaap, H. Daemen (VS Verlag fur Sozilwissenschaften, Springer Fachmedien Wiesbaden, 2012), Book chapter (3).

promulgation acts affirming the participatory involvement, as well as, the development of structural management policies, which essentially have been directed at the preservation of the common PPPs interests.[16]

In the case of public regulatory trials constituted for the reason of concentrating comparative relations for the inclusion of local participant municipalities, there is the continuous presence of locally involved stakeholders such as civilian associations, the national representative partners, the societal interest groups, and the regional international discussion forums. Through which, the affirmation and maintenance of public dialogue exchanges for a comparative involvement of multilateral societies have remained quite key for the decentralization of representative collective dynamics which emerge at systematic organizational levels.

In fact, in order to draw a more precise picture of the inclusion of respective innovative understanding for the national normative interactions that can support the information diffusion platforms, it is required to introduce technical operable provisions designed by public assigned representatives, through the material presence of local councils' agencies, acting in accordance with accessible development units that pertain to e.g. former and current EU economies such as England with the post-Brexit economic transfer model or the Netherlands with multi-level governance adaptation plans.[17]

Instead, for the reformulation of independent development paths involving different social and political participation environments, the multi-level adaptive regulatory concessions have been modified according to the changing attitudes of public-private interacting members, enabling a systematic configured predisposition. Essentially, the decision-making activity emerging from the acceptance of democratic interactions has been seen in association with the national and local adherence to a normative consensus formed within public reference models. In fact, the mutual involvement of local communities has been addressed in terms of public dialogues and decisional ethical patterns, which have been established in combination with collaborative representatives based as national or regional public-private counterparts.[18]

---

16 See: Soumyadip Chattopadhyay, "Contesting Inclusiveness: Policies, Politics and Processes of Participatory Urban Governance in Indian Cities," *Progress in Development Studies* 15.1/2015, pp. 22–36.
17 Edwards, "Tensions and New Connections Between Participatory," (3).
18 See: 1) Yasumasa Komori, "Evaluating Regional Environmental Governance in Northeast Asia," *Asian Affairs: An American Review* 37.1/2010, pp. 1–25; 2) Sunhyuk Kim, "Collaborative Governance in South Korea: Citizen Participation in Policy Making and Welfare Service Provision," *Asian Perspective* 34.3/2010, pp. 165–190.

Moreover, across the full spectrum of the decision-making process – common engaged participants and national local electorates have formed bonding linkages on deterministic social and cohesive values. Within a common territorial dimension, structural determinant factors and affirmative behavioral correspondences regarding local empowering conditions have remained central key aspects in public regulation and PPP managerial relations. Such progressive relationships have been stimulated through the active presence of political representatives and local civilian counterparts which operate through functional respective agencies.[19]

4.1.2 Territorial Reconstructions

The knowledge-based industrial economies have reflected concurrent PPP models which have recalled networking transitions at systematic level. Across regional economic territories the reconstruction of permeable sustainable models has remained engrained in distinctive simultaneous processes which are not necessarily what public governance grants in terms of environmental appropriation issues.

At the same time, the progressive renewal of programmatic planning options assembled through comparative strategic states' reforms has been evaluated through concurrent regulatory aspects, which involve the governmental participatory arrangements promoting interrelated working dimensions that have been approached through democratic economic relations. For such reasons, the industrial sustainability of regional networked productions in different time periods has been intertwined with global ordered transitions for local commerce distributive systems. On the parallel side, part of political and ideological heritage of regional states' industrial explorations has been undertaken under diffused provisional determinants.

Such transitional knowledge-based economic explorations promoted at state level have been doctrinally categorized according to the distribution of socio-economic plans listing local regulatory characterizations, which have emerged in simultaneous incremental processes delivered in a cooperative manner. This type of transitional call for public participation with a differentiated local involvement has been voiced through the mediation done regarding liberalized market conditions for transnational trading relations, involving an extended degree of complementarity and compromise between public and private institutional actors.[20]

---

19 Edwards, "Tensions and New Connections Between Participatory," (3).
20 See: 1) Dmitry Epstein, "The Making of Institutions of Information Governance: the Case of the Internet Governance Forum," *Journal of Information Technology* 28.2/

A significant study developed in 1998 by Hollingsworth R. J.[21] puts us back to the idea that transitional territorial integration matches better according to defined levels of local governmental understanding, in particular the author stresses out: "social systems of production (have been) constantly changing, they (have not been) converging toward some global system."[22] If we decide to adopt a governance perspective, we have to take into account the fact that transitional consolidation of national socio-economic assets emerging across world regions has been multiplied through local organizational changes for the current modes of transnational production systems with concurrent social adaptations, which have produced in turn both weaknesses and modified influences clearly expressed in comparative state regions.[23]

In the planning environmental process, governmental orientations and multi-level local knowledge adaptations of socio-economic contexts have reflected respective normative inputs also coming from countries mostly identified through compatible delivering assets that can absorb the mutual levels of regional fragmentation, which has been influenced by the flow of actions of transitional competitive production systems remaining embedded in national and local knowledge technology capacities.

In addition, the inclusive methodological approaches – proposed within international liberal exchanges for the strategic planning of diversified supporting activities – have favored the comparative affirmation of (MNC) corporate agencies bringing with them different innovation integrated aspects of the operable competition mechanisms. The standardized implementation mechanisms have also implied a formal and informal organizational restructuring for the targeted local areas of comparable open productions, involving the PPP national operations, which already perform within regional knowledge networks, while eventually absorbed into native producing units.[24]

2013, pp. 137–149 2) David Wilson, "Exploring the Limits of Public Participation in Local Government," *Parliamentary Affairs* 52.2/1999, pp. 246–259.
21 See: Carruthers, B.G., in: *Contemporary Capitalism: The Embeddedness of Institutions*, eds. J. Rogers Hollingsworth and Robert Boyer: *Contemporary Sociology*, 27. 3/1998.
22 Rogers J. Hollingsworth, 1998. "Territoriality in Modern Societies: The Spatial and Institutional Nestedness of National Economies," in: *Territoriality in the Global Society*, eds. S. Immerfall (Springer-Verlag Berlin Heidelberg, 1998), p. 17.
23 See: Jorn Birkmann, Matthias Garschagen, Frauke Kraas, Nguyen Quang, "Adaptive Urban Governance: New Challenges for the Second Generation of Urban Adaptation Strategies to Climate Change." *Sustainability Science* 5.2/2010, pp. 185–206.
24 Hollingsworth, "Territoriality in Modern Societies," p. 17.

Holistic Integration Approaches 95

Nonetheless, this changing environment which exists in governance contexts has been expanded in terms of public political discourses that have been associated to the development of territorial innovative conditions reviewed for the spatial distribution of technology production systems set together with corresponding levels of national public arrangements particularly in the last three decades. Such institutional drivers have been structurally complex and socially challenging because of domestic implications for local enhanced expectations about the generational alignment pursued within industrial innovation mixed models.

In these cases, the local innovation elements and comparative integration programs can be consolidated according to national environment tendencies converging on the application of responsive participatory networks that can create at the same time a common compromising ground. Instead, the process-driven innovation led under industrial transitioning strategies can cause comparative different implications, in particular for the alteration of local programming relations which essentially can affect the respective political directions undertaken on e.g. local environmental policies and regional societal inclusions, by spilling over on international essential productions with R&D dynamics.[25]

Similarly, the author Jessop, 2002[26] has underlined the fact that comparative governmental agencies located across different world regions have shown the material complexity of participatory local adaptive designs. In recent time-periods, the multi-level adaptation process adopted in local governance programs has become particularly functional for the maintenance of serial management guidelines and specialized organizational roles that involve different technical responsibilities, which have been interlinked with effective local governance practices.

In fact, the regional amount of territorial regulatory activities has been combined with the distinctive categorization of sectoral reforms which have been referring to science innovation programs and to operable information networks. The alignment of local material accesses for constructive knowledge-sharing confrontations – including local community's discussions about what can be effectively transferred

---

25 See: OECD report, "National Reports on Uranium Exploration, Resources, Production, Demand, and the Environment," Chapter 3. *Uranium 2011: Resources, Production and Demand*, ISBN 978-92-64-17803-8, OECD 2012.
26 Bob Jessop, "Governance and Meta-Governance in the Face of Complexity: On the Roles of Requisite Variety, Reflexive Observation, and Romantic Irony in Participatory Governance," in: *Participatory Governance in Multi-Level context*, eds. H. Heinelt et al. (Springer Fachmedien Wiesbaden 2002), (2).

in a short-time for the direct translation of normative conditions, has involved the cognitive formulations undertaken at state level. In addition, there are societal cognitive determinants and local mediation research activities applying to the governmental context which have been functional to the simplification of distributed public roles and responsibilities that have been reaffirmed through objective complementary approaches.[27]

At this point, we might note that there are distinctive paradigms regarding the historical formation of institutional and material local architectures separated into public and private organizational spheres. Different historical and regional expressions of power related interactions pursued in technology innovation industries and institutional environments, have indirectly impacted upon the diffusion of distinctive PPP public and private managerial interests, also because of parallel existing conditions that have determined societal tendencies for the creation of interdependent dynamics embedded in public decisional settings, as well as, in a common range of trading exchanging initiatives.

According to author Jessop, 2002[28] the operational daily relations of public agencies establishing direct management measures according to the needs of harmonized regulative frameworks have emerged in terms of comparative unproductive factors, due to the constant presence of regional opposing determinants that are based on public and private conflicting interests. Instead, the promotion of societal integration patterns has led to the inclusion of common programmatic paths for the actual cooperative affirmation of public local exchanges. Similarly, the comparative willingness to conduct public review processes for the actualization of socio-economic incremental stages on local environmental adaptation, has prompted the adoption of constructive approaches, targeting the actual mobilization of societal stakeholders, as well as, their effective participation for the definition of common environmental objectives.[29]

Practically, the enhancement of multi-level governance plans directing regional and local bodies has been assessed toward the achievement of practical aims and

---

27 See: Sebastian Delle Piane, Avella Neda, "Good Governance, Institutions and Economic Development: Beyond the Conventional Wisdom," *British Journal of Political Science* 40.1/2009, pp. 195–224.
28 Jessop, "Governance and Meta-Governance," (2).
29 See: 1) Erik W. Johnson, Jon Agnone, John D. McCarthy, "Movement Organization, Synergistic Tactics and Environmental Public Policy," *Social Forces* 88.5/2010, pp. 2267–2292; 2) Roman Gomez Gonzalez Cosio, "Social Constructivism and Capacity Building for Environmental Governance," *International Planning Studies* 3.3/1998, pp. 367–389.

programmatic goals. To consider also that the level of societal participation coordinated according to specific agencies' dispositions has possibly determined the sense of local cohesion and redistributive justice, through the permanence of mutual relational conditions based on environmental affirmation patterns.

However, there have been cases of major states' practices and transition management issues in governance that point at territorial functional designs of productive capacity systems for a complementary status of local collaboration practices. In such cases, we also need to study the type of combination formulated regarding technology managerial innovations and territorial bureaucratic models, which have been emerging through the national ruling apparatuses – while recalling attention on the changing conditions of PPP trading contexts – which involve the enhancement of comparative governmental processes, for local participating relations.[30] Actually in some conditions, the regional governmental relations and local territorial issues have been connected in reference to the changing delivery of transitional innovation models that have been implying the centralization of multiple knowledge-related activities, which have assimilated an increasing level of sectorialization on concurrent local production relations operated through a systemic differentiation.

Despite the diffusion of critical interpretations about the regional governance tracking process of mutual interdependences incorporating national dynamic factors and local productive identifications, there have been geographical limitation issues that remain to be considered in the comparative analysis. National citizens' participation rights have been fundamentally affected by compatible degrees of national regulatory environments for the distribution of institutional relational powers, which maintain the integration and implementation of common production transfers of local management programs involving, for example, the context of environmental industrial conflicts which have been forming according to distinctive affirmation of states' territorial exchanges.[31]

---

30   See: David J. Brown, John S. Earle, Scott Gehlbach, "Helping Hand or Grabbing Hand? State Bureaucracy and Privatization Effectiveness," *The American Political Science Review* 103.2/2009, pp. 264–283.

31   Santiago Eizaguirre, Marc Pradel, Albert Terrones, Xavier Martinez-Celorrio, and Marisol Garcia, "Multilevel Governance and Social Cohesion: Bringing Back Conflict in Citizenship Practices," *Urban Studies* 49. 9/2012, pp. 1999–2016.

CHAPTER FIVE

5.1 Public Territorial Approaches

For analytic purposes, the identification of common territorial policies has been possible through the characterization of public normative approaches, which have involved common evaluation stages and institutional monitoring conducts implementing the multi-level planning criteria, especially with those concerning the application of governmental district-based measures. In fact, the development of organizational and managerial plans, which entail the elaboration of environmental measures has reflected either direct or indirect local planning spheres rearranged for the applicability of managerial innovation programs that have connected different interdisciplinary domains.

For the case of national energy sustainability, the environmental policies have been related to public and private relational acceptances of PPP for the recurrence of upgraded technology programs which can determine material inputs of organizational expertise committees, also for the use of public monitoring tools and the critical environmental assessments targeting, for instance, the securitization of harmonized energy policies.[1]

In practice, there exists a core set of interrelated territorial specifications which have been primarily mediated through the promotion of managerial strategic goals targeting local environmental issues such as: the implications of public local disposals regarding public hazards and contaminated nuclear waste; or the local nuclear waste siting facilitation process. Public facilitation activities have been commonly conducted in view of addressing the preparation of environmental supporting mechanisms for the nuclear plant NPP siting procedures, related also to the case of nuclear waste disposal programs. In essence, the formation of public nuclear siting policies has been adapted according to the availability of public-private organizational linkages, which have involved main formal stakeholders, as well as, NGOs and local civic groups for the collaboration activities launched in parallel with national and local environmental assessments.[2]

---

[1] See: Xiaofei (Sarah) Li, "Chinese Political Transition Makes Change Possible in Energy Approach," *Oil & Gas Journal* 111.2/2013, pp. 46–52.

[2] See: Harold Kennedy, "A Big Job: Cleaning Up Nation's Nuclear Waste," *National Defense* 83.546/1999, pp. 24–27.

For instance, there are multi-level collaborative regulation issues of the composition of regional managerial and innovation energy programs that have been interrelated with the national delivery capacity of different energy performing systems: such as the case of Canada's energy cooperation and public consultation linkages addressed at multi-level administrative bodies, with technical information publicly discussed at sensitization community forums. Moreover, this highlights the importance of science information exchanges made available across different states, formally arranged in connection with environmental sustainability provisions and development policies as well.

In particular, for the case of national energy's diversification plans, there is a prevailing formal lack of a specific understanding of the changing characteristics of environmental defining provisions that refer to the nuclear waste disposal process and the ways in which this process has been integrated at institutional local level. In addition, local communities have seen the spread of a reaffirming sense of not knowing the systematic identifications of environmental adaptation procedures designed for regional energy innovation plans which include the legal and local requirements. Essentially, for the possible identification of participatory conditions about environmental community's active provisions, it has become feasible to involve the public and private representative groups in order to have informed local community dialogues as an integrative part of the process directed at establishing comparative energy environmental assessments.

Instead, in terms of nuclear safety directives, the nuclear energy development programs have been difficult to reassess and to conduct through the essential participatory public debates set at a community level. That is because of one major critical public concern, already pointed out previously, namely the managerial disposal of nuclear waste for long-term burial locations, which comprise the local operational disposal sites that can be managed by national[3] or transnational corporate groups, and where local governmental organizations can perform in terms of regional territory hubs by allowing comparative implementing measures regarding renewable energy networks.[4]

In this case, the comparable regulatory translation involving multidisciplinary research in science innovation areas can establish technical facilitation mechanisms for the societal inclusion put in relation with the environmental

---

3 See: Brian N. Winchester, "Emerging Global Environmental Governance," *Indiana Journal of Global Legal Studies* 16.1/2009, pp. 7–23.
4 Darrin Durant, "Responsible Action and Nuclear Waste Disposal," *Technology in Society* 31/2009, pp. 150–157.

Public Territorial Approaches 101

decision-making activity. Through common local identifications, the regional countries' development politics have formally been evaluated according to legitimate environmental commitments able to readdress the resolution of long-term industrial disputes about technological energy innovation impacts. In effect, at an environmental government-level, one major aspect remains relevant about the regional coordination path, which can be retraced through collaborative international measures identified across multiple regulatory performing domains, while also including the energy policy plans and local technical aspects of knowledge innovation projects.

Essentially, in systematic development contexts the PPPs with public-private partnering agencies reaffirmed in multilateral governance settings, and put in conjunction with national institutional aggregations, can form a comparative ground for the monitoring and evaluation assessments, which may favor the multi-level cohesive dynamics integrating respective commercial interests, as well as, regulative directives that are already in place. In more specific terms, the definition of dynamic local environmental measures operated in territorial governance has to be aligned according to socio-economic applicable distribution conditions.[5]

At the same time, it is vital to recognize the increase of different types of regional development processes which can affect the concurrent establishment of local organizational industrial programs. Such multi-stage networked innovation processes can lead to further specialized economies that get multi-level comparative knowledge expertise integrated with diffused global technology trading areas, which have inter-crossed multiple operative conditions related to public safety maintenance rules and public regulation measures. Therefore, the multi-level development and local adaptation cycles which have brought significant regional institutional changes, have relied on comparative networked innovation processes, based on the division of managerial technical responsibilities and PPP public-private organizational monitoring reviews that basically can deliver public recognition elements surfacing in common coordination areas that are still built on an uncertain ground.

Fundamentally, we need to consider that the institutional environmental adaptations arranged through functional local organizations polarized on some

---

5   See: Katharina Rietig, "Reinforcement of Multilevel Governance Dynamics: Creating Momentum for Increasing Ambitions in International Climate Negotiations," *International Environmental Agreements: Politics, Law and Economics* 14.4/2014, pp. 371–389.

inclusive aspects of the environmental governance policies, have progressed under common changing dynamics and regulation controlling measures, which have basically targeted the national, federal and local level of common environmental programs.

In fact, according to different levels of national arrangements, the comparative environmental agreements have been stipulated through the involvement of participatory representative members for the local coordination and integration of the agreed sectoral conditions, including comparable local enforcement mechanisms. For instance, in reference to the case of anti-nuclear energy civilian stakeholders and local protesting movements inviting to elaborate on public mediation activities, there is a pressing structural need to put in place local confronting discussions that can focus on common major identified problems, about the pollution investigative issues of nuclear facilities waste disposal operations, and the local regulative maintenance ownerships. Historically, the emerging structural environment conditions have been analyzed in conjunction with comparative national energy designs through increasing supportive energy demands put into place according to different renewable energy scientific fields connected to PPP distribution of decentralized energy assessments.[6]

This material assessment process undertaken in governance has been centered on comparative energy technical requirements and on innovative decision-making tools, which favor public policy cooperation with common procedural linkages for the establishment of local participatory conditions, reaffirmed at regional and provincial level. For the case of Canadian regional provinces, the environmental authorities have produced local impact assessment outcomes while evaluating the regional interplay of local communities involved in comparative climate environmental relations.

5.1.1 Provisional Requirements

In specific territorial locations, the deterministic impact factors regarding local organizational provisions of multi-level nuclear energy industries have concurred to redraw environmental requirements for local sustainability purposes. At critical turning points, the dual formal relationship built up between the public policy regulators and PPP nuclear management organizations has impacted

---

[6] See: 1) Olav Schram Stokke, "Regime Interplay in Artic Shipping Governance; Explaining Regional Niche Selection," *International Environmental Agreements: Politics, Law and Economics* 13.1/2013, pp. 65–85; 2) Holger H. Rogner, "Nuclear Power and Sustainable Development." *Journal of International Affairs* 64.1/2010, pp. 137–163.

on the preparation of decisional selective criteria setting the local internal and external progressive interactions at state level. In order to determine the specific environmental indications that concern the transnational nuclear management plans, the local waste burial activities and regional geological disposals have been commonly targeted on local contested acquisitions.[7]

Moreover, there are territorial, as well as, legal and economic constrains regarding the governmental orientations that enhance the process of public community deliberations, which have been divided on the structural formation of local concerted plans, primarily aiming at the constitution of actual collaborative environmental proposals. The specific involvement of multi-level representative groups in governance has been basically contextualized through the similar constitution of distinctive socio-technical local dimensions.

As an example, for the possible characterization of environmental material issues interrelated with local sustainable plans, the nuclear waste regulatory mandating agencies have been drafting different structural stages for the direct involvement of local cooperating agencies and associated NGO groups. In both political spheres of formalized civic interactions, there have been comparative political networking activities added up to the constructive local environments, which have transitioned according to respective cognitive boundaries including the community's needs on environmental safety standards, and national development aims referred to local energy's production projects. Critically, it has been noted that "under the guise of simply proposing a technical means to deal with a pressing problem, these elites were accused of maintaining a factually dishonest boundary between technical proposals and their public policy implications."[8]

Based on such observations, it can be observed that the material development of either direct or indirect regulatory organizational conditions has influenced public-private polarizing management contexts, because of a divergent intensification of comparative regional interests. Particularly, for the case of nuclear storage programs and the local disposal management activities of nuclear waste, the process *per se* has recalled the attention of public regulatory groups as well as of territorial administrative counterparts and civil deliberative associations, among others, for organizing the regional periodic campaigns on local environmental issues framed on a public confrontation ground.[9]

---

7   Durant, "Responsible Action, and Nuclear Waste Disposal," pp. 150–157.
8   Durant, "Responsible Action, and Nuclear Waste Disposal," p. 153.
9   See 1) Joseph Masco, "Survival Is Your Business: Engineering Ruins and Affect in Nuclear America," *Cultural Anthropology* 23.2/2008, pp. 361–398; 2) Upendra

In a recent historical transition, there have been immediate responses following the nuclear emergency crisis of Japan's' Fukushima Daiichi nuclear energy plant disaster, where the transnational, regional and national institutional counterparts involved with the mutual interested parties have multiplied their efforts to try to identify nuclear energy-related safeguards plans, which can emerge for political discussions confronting different local groups through public orientation debates. At the multi-level public-private regulatory dimension, the distribution of public alternative strategies has been centered on public risk performance evaluation criteria and local organizing relations, also for the actual creation of stable cooperative conducts aligned with public information exchanges, which in turn have been focused on technical procedural grounds, especially regarding the nuclear waste management cycle and the national required licenses' agreements.

However, the regional juxtaposition of comparative socio-economic and political indications that can assess different territorial renewable energy objectives, has been analyzed in terms of resulting programmatic aims and local planning strategies that have been based on the functional determinants of local ecological programs and national energy delivery systems' capacity. These combined environments have been arranged according to the modernization status of local innovative clustering processes, which have been scientifically explored through comparative socio-economic and knowledge information perspectives. Moreover, the formation of complex industrial interacting societies has led to the conjugation of common kn0wledge objectives, which integrate the PPP institutional participation and industrial technology management, with the advancement of local community stakeholders' integration, while effectively facilitating mutual development targets.

Basically, the possible consolidation of PPP with locally associated science and knowledge managerial environments run by multi-level representative sectors has been intertwined with a structural identification of common evaluation procedures, which have been adopted according to the application of constitutive national principles. In this sense, the national and local environmental governance dynamics have potentially favored the interrelation of respective cognitive processes that have been built on practical environmental initiatives,

Choudhury, "The Impact of the Fukushima Daiichi Nuclear Crisis on Anti-Nuclear Movements in India," *Social Alternatives* 31.3/2012, pp. 39–44.

Public Territorial Approaches 105

also characterized by territorial economic strategic performances and regional affirmations of local sustained capacities.[10]

For instance, comparative similarities have been made about the case of East Asian countries with the continuous migrating flow of national societies reaffirming the need of internal transition for labor, which has been addressed through the constructive paradigm of (EM) industrial and ecological modernization process.[11] The regional development of population migration patterns has been affected by managerial and institutional and ecological adaptation plans that have been historically influenced by the application of technological innovation approaches, which can incorporate the common trading areas for specialized sectoral productions, with corresponding levels of intra-regional trading practices.

In addition to this, there have been comparable changes on the sectorialization of aligned production exchanges that involve temporary marketing interactions, which have been essentially led toward the improvement of comparative institutional practices and local aggregation modes, including deliberative associations for the consolidation of material expressions, also related to environmental learning dynamics which can emerge at national but also at local level.[12]

In the case of Hong Kong's semi-independent administrative configuration and a modified regional trading province, the adaptive economic policies and common learning processes regarding EM approaches have been framed at conceptual and practical level. In fact, in order to address the problem of the opening social divide which has been established between the China-related production industries and the territorial local configurations, the roles played by public mediating officers were taken into account, and those include the institutional activity-based bodies for the hierarchical developmental functions.

Essentially, in order to promote sustainable development, the local networking activities of multi-level knowledge learning groups have corresponded to the possible implementation of local governance adaptation plans that have

---

10 See: Luca Salvati, Marco Zitti, "Territorial Disparities, Natural Resource Distribution, and Land Degradation: A Case Study in Southern Europe," *GeoJournal* 70.2–3/2007, pp. 185–194.
11 See: Wai-Hang Yee, Carlos Wing-Hung Lo, Shui-Yan Tang, "Assessing Ecological Modernization in China: Stakeholder Demands and Corporate Environmental Management Practices in Guangdong Province," *The China Quarterly* 213/2013, pp. 101–129.
12 Andrew Gouldson, Peter Hills, Richard Welford, "Ecological Modernisation and Policy Learning in Hong Kong," *Geoforum* 39/2008, pp. 319–330.

been explored in association with significant socio-economic transnational factors. The facilitation of ecological modernization EM elements included in public policy learning exchanges, in turn, has required a mutual correspondence of compatible institutional orientations that can look at the actual initiation of common environmental designs for multilateral local developmental schemes, which in time have been assessed in view of structural environmental changes – that have impacted on the local maintenance processes for local collaborative partnerships, still maintaining a regional industrial technology focus.[13]

The EM approach[14] has been shared in relation to economic innovation processes and environmental sustainability issues that have been analyzed according to the regional technological transitions incorporating local mutual practices in so-called 'modern' industrial societies. Critically, the possible delineation of material sustainability issues, influencing the common transnational innovation markets that can operate through governmental cooperative mechanisms, has been reviewed through the lens of transnational innovation dynamics and local adaptation measures, which can lead to the utilization of national environmental standards based on centralized or decentralized socio-economic proceedings.

In addition, a major emphasis put on the different existing types of public policy coordination systems – performing through deliberative adaptation contexts – has basically led to the need of specifying how to deliver better levels of PPP for common learning identifications, while determining the contingent types of environmental mitigation models, which have been managed by institutional representative entities. Moreover, there are regulatory local bodies which have assumed a specific position regarding the practical construction of societal participatory discussions leading to a common level of acceptance of the environmental promulgation activities that concern local structural reforms. Similar reforming stages in governance have been processed in accordance to the local development of societal learning practices in view of pursuing instrumental organizational dimensions regarding technical ecological transitions and industrial innovation policies.[15]

Broadly speaking, the sectoral production arrangements pursued by multilateral industrial trading partners have supported the transnational terms of

---

13 Gouldson, Hills, Welford, "Ecological Modernisation," pp. 319–330.
14 See: Wiebren J. Boonstra, Florianne W. de Boer, "The Historical Dynamics of Social-Ecological Traps," *Ambio* 43.3/2014, pp. 260–274.
15 See: Benjamin Vail, "Ecological Modernization at Work? Environmental Policy Reform in Sweden at the Turn of the Century," *Scandinavian Studies* 80.1/2008, p. 85-.

multi-level industrial adaptation policies. Such multiple PPP agreements have demanded practical contributions for the search of national policy schemes which can confirm the inclusion of ecological normative tools, associated to collective local terms of systematic optimization practices that can reflect interoperability of different socio-economic dimensions. For instance, regarding the Hong Kong environmental innovation case: in the past there are coordination rules defining the maritime transportation operations which have been put in connection with the water disposal programs, configured according to the respective local managerial plans. This process involved as well the producers' adaptability to implement local activities performed through the instrumental practice tools such as with local divulgation campaigns about international environmental guidelines.

Such progressive coordination guidelines drafted in parallel with environmental politics objectives, have become pivotal elements for matching the local corresponding interests of public-private stakeholders, which have been responsive to the flow of transnational industrial provisions. The harmonization process of managerial innovation conducts lined up for the operationalization as well as for the maintenance of common ecological projects has been identified through selective development targets, such as the reduction of air pollution and water waste disposals.[16] Nonetheless, the regional affirmation of environmental territorial legitimacy for the delivery of public cooperative schemes has remained controversial especially for the involved PPP cases like: Hong Kong development and its China-related international industrial complexes. In such rapid changing scenarios, the local planning activity setting long-term environmental strategies has been associated with the integration of interregional open commerce networks, which have been dependent on characteristic reorganization factors that have been operated at the level of technological requirements identified in comparative co-dependent management projects, and local implementation cycles.[17]

Upon reflection, the distinct national categorization of sectoral economic implications that involve environmental protection assessments has been addressed concurrently in view of providing the local innovation options.

---

16   Gouldson, Hills, Welford, "Ecological Modernisation," pp. 319–330.
17   See: 1) Christine Loh, "Hong Kong-Mainland Innovations in Environmental Protection Since 1980," *Asian Survey* 51.4/2011, pp. 610–632; 2) Lai Tim Wong, Gerald E. Fryxell, "Stakeholder Influences on Environmental Management Practices: A Study of Fleet Operations in Hong Kong (SAR), China," *Transportation Journal* 43.4/2004, pp. 22–35.

Formally, the public emerging framework of functional mediation activities run in governance has been differently perceived at the local level because of corresponding domestic impacts measured through the intra-regional development models.

Similarly, the causal effects of the national regulatory harmonized procedures have been studied for their operable and timely interventions concerning the follow ups on environmental protection regulations managed by established national agencies as the EPAs that operate at field level, (e.g.) with local regulative implications analyzed for the U.S. environmental cases of nuclear waste management programs.[18] For example, the U.S. Department of Energy (DOE) has adopted both national and federal environmental regulative guidelines set in conjunction with local development projects, which have increased the collaboration of multiple PPP stakeholders in order to facilitate the energy development programs targeting the nuclear waste management at local facilities' sites, through the formal involvement of: (e.g.) university research consortia; DOE local units with focused mediation groups; and among others, local community leaderships.[19]

In addition to this, there have been both financial and administrative regulations that have been applied to the case of U.S. (DOE) national standardization rules of respective energy programs, which have involved the annual budgets analyses and the federal resource allocations procedures. Such systematic structural provisions planned at U.S. state or federal levels have been further questioned in relation to the preparation of the annual budgeted projects on energy development and on local planning programs, which have required the necessary identification of technological adaptation measures operable at system level, such as in the case of nuclear waste management facilities. This type of multilateral identification process has been developed according to the common availability of associated managerial innovation conditions which have formed across regional structural networks, supported through the rational compatibility of internalized organizing practices, combined with local administrative activities, while also shifting multi-level knowledge innovation scopes and environmental protection policies.[20]

---

18 Michael Greenberg, David Lewis, Michael Frish, "Local and Interregional Economic Analysis of Large US Department of Energy Waste Management Projects," *Waste Management* 22/2002, pp. 643–655.
19 Greenberg, Lewis, Frish, "Local and Interregional Economic Analysis," pp. 643–655.
20 See: Bryan C. Taylor, Brian Freer, "Containing the Nuclear Past: The Politics of History and Heritage at the Hanford Plutonium Works," *Journal of Organizational Change Management* 15.6/2002, p. 563-

Public Territorial Approaches					109

There have been a number of other institutional contributing assessments of the diffusion of national environmental protection norms and safety and security policies in U.S. federal systems, which have drawn to the concrete realization of the need to adopt deliberative territorial action plans for concerted local environments that can be related to the technical adaptation requirements of local safeguard measures, and to the regional environmental accessibility of collective projects established across time.

Historically, the nuclear waste management facilities have been designed according to the specification of security safety options and environmental structural conditions, which have been dependent on public interrelated elements and critical public risk exposures that refer to the evolution of territorial environmental patterns involving: demographic changing characteristics; internal populations migration waves; local skilled/unskilled manpower supply; direct/indirect upturns of socio-economic impacts; science & technology evolution impacts; and state, federal, or local governance in the case of PPP nuclear waste development programs.[21]

In addition to this, there are into consideration quantitative and qualitative management issues of the nuclear processing facilities[22] which have been dependent on the standardization of mutual operational understanding of the nuclear waste storage security and the project life cycles of nuclear energy facilities that have been coordinated in association with composite technical and financial plans varying at the level of regional, national, federal, and provincial organizations.

In another example, the European Union member states have promoted regional policy studies that discuss the potential similarities of national regulative protection adaptations, which refer to the local levels of environmental nuclear safety, as well as, to financial and business implications.[23] In essence, there have been convergent interrelated issues regarding the environmental safety and security factors regarding national industrial energy plans distributed across transnational innovation sectors, which involve a cohesive understanding of the regional development of EU common guidelines applied in territorial bounded areas, while also recalling attention on multilateral PPP operating groups.

---

21 Greenberg, Lewis, Frish, "Local and Interregional Economic Analysis," pp. 643–655.
22 See: Lawrence Flint, "Shaping Nuclear Waste Policy at the Juncture of Federal and State Law," *Boston College Environmental Affairs Law Review* 28.1/2000, pp. 163–190.
23 See: Diane Ryland, "The Future of Nuclear Energy in Europe: Questions, Problems, and Perceptions," *Managerial Law* 44.4/2002, pp. 91–111.

For the planning activity of distinctive regional regulatory environments have been taken into consideration: 1. the coordination of (R&D) research and development initiatives; 2. the structural infrastructures' maintenance activities; 3. the nuclear spent fuel management with security monitoring accesses; 4. the nuclear energy industry and civil protection goals; 5. the financial local assets and local impact evaluation projects; and 6. the technological innovation consolidation rules; among other domains.[24]

With particular reference to the structural policy distribution factors which have been targeted for the national energy and security implementation of common local activities, the emerging materialization of public environmental concerns has been focused on the territorial management designs for the controlling energy systems and maintenance operations of respective territorial areas; which have included the integration of common knowledge learning practices of regional prospective functions of long-term energy innovation programs, influencing the concerted local actions that have been undertaken in accordance with participant local communities.

At local community level, the contextual cognitive preferences of multi-level stakeholders have been critically elaborated in open political debates pointing at the materialization of local collaborative exchanges based on regional development patterns. On the participatory front, the monitoring and evaluative regulation policies have led to public environmental confrontations for the possible assimilation of local nuclear waste management plans put in connection with financial planned goals, that have further widened national preparatory indications involving science technology programs and local innovation knowledge activities.[25] For comparative purposes, an overview of single countries' public policies on nuclear radioactive waste disposal cases is provided. This has been simplified in the matrix form based on OECD/NEA 2007 study,[26] see Table 4.

---

24 Tom Vander Beken, Nicholas Dorn, Stijin Van Daele, "Security Risks in Nuclear Waste Management: Exceptionalism, Opaqueness and Vulnerability," *Journal of Environmental Management* 91/2010, pp. 940–948.

25 See: Doug Brugge, Jamie L. deLemos, Cat Bui, "The Sequoyah Corporation Fuels Release and the Church Rock Spill: Unplicized Nuclear Releases in American Indian Communities," *American Journal of Public Health* 97.9/2007, pp. 1595–1600.

26 OECD/NEA, "Approaches and Practices in Decommissioning of Facilities and Management of Radioactive Waste from Non-Nuclear Fuel Cycle Related Activities," *Proceedings of the Topical Session of the RWMC 40th Meeting at NEA Offices in France 14th March.* (NEA/RWM 2007/9).

Table 4 Overview Nuclear Waste Management Territorial Contexts

Country	National Plan	Disposal Sites	Decommissioning
France	"France National Agency for Radioactive Waste Management (ANDRA) fulfils a multi-faceted task within the framework of the policy governing radioactive waste management defined by public authorities … Within the whole picture, distinction can be made between the situations of the various radioactive waste producer categories which cover very different conditions in technical and economic terms and the function of waste holders." P. 71	"The waste produced by the nuclear industry (EDF, the CEA and AREVA) accounting for most of the volume and radioactivity of the waste produced is disposed of within the framework of a system of technical specifications and tests in order to produce waste packages controlled by producer-operators." P. 71	"Around 20 sites must be cleaned up due to radioactive contamination. The inventory is limited and very often very old mainly involving radium-contaminated sites, for which contamination normally dates back several tens of years. In most cases, the current owners did not cause the contamination of which they consider themselves the victims, as the owners responsible are either deceased … or are bankrupt." P. 73 "ANDRA's public service mission … sought by the Nuclear Safety Authority as early as 2003, stipulated in article 4.6 of the State-Andra contract, and confirmed by article 14 (1 and 6) of the Act of Parliament of 28 June 2006, focuses on 3 objectives: – Compilation of the inventory of radioactive material and waste – Andra's original mission; – Disposal of family nuclear waste; – Clean-up of sites contaminated by radioactive substances whose owner is bankrupt and disposal of waste produced." P. 74

(*Continued*)

Table 4 Continued

| Belgium | "The foundation of ONDRAF/NIRAS (Belgian Agency for Radioactive Waste and Fissile Materials) on 8 August 1980 is the result of a decision of the Belgian authorities to entrust the management of radioactive waste to one single institution under public control. This was done in order to ensure that the public interest would prevail in all decisions on the subject. … ONDRAF/NIRAS has a centralized waste management policy, by making use of processing and conditioning facilities and interim storage facilities centralized on the Belgoprocess sites in Dessel and Mol. Some waste producers have their own processing and conditioning facilities and transfer their conditioned waste to Belgoprocess for interim storage." P. 43 | "As the agency responsible for radioactive waste management, ONDRAF/NIRAS is in charge of all matters relating to the safety of waste management and the protection of the environment in close cooperation with FANC [Federal Safety Authorities], which is the licensing authority for nuclear installations. A formal agreement organizing all the legal interfaces between the two agencies was signed in 2003 by the Board of Directors of both agencies and is in force since September 2003." P. 44 | "Each owner or operator of a nuclear installation is responsible for the future dismantling of his installations, once they have been definitely decommissioned. ONDRAF/NIRAS verifies that the owner/operator takes timely steps to carry out the dismantling program; the owner/operator must submit his decommissioning program to ONDRAF/NIRAS for approval. The radioactive waste resulting from dismantling is managed by ONDRAF/NIRAS according to the same principles as the waste of other origins. Furthermore, it is part of the missions of ONDRAF/NIRAS to follow up the evolution of the dismantling methodologies and technologies." P. 45 |

Public Territorial Approaches 113

| Netherlands | "The Netherlands has a small nuclear energy programme, with 1 nuclear power plant in operation (Borssele NPP), 1 shut down and put in a safe enclosure for the next 40 years (Dodewaard NPP) and 3 research reactors (HFR and LFR in Petten and the HOR in Delft). … The national policy on radioactive waste management laid down in a position document in 1984 and endorsed by the parliament, envisaged the institution of a single organization responsible for the removal, processing and storage of all radioactive wastes, ranging from NORM waste to vitrified high level waste and spent fuel. As a result COVRA, the Central Organization for Radioactive Waste, founded in 1982, was charged with the implementation of this decision. COVRA is located at a site in the south-west of the country near the Borssele NPP. It was also decided that dedicated surface storage facilities be constructed for each waste category with sufficient capacity to accommodate all radioactive wastes generated in a period of about 100 years." P. 93 | "One of the challenges associated with the long-term storage of radioactive waste is to keep record of the characteristics of individual packages. For practical purposes the waste package identification procedures for management steps of relatively short duration, such as treatment are different from those applicable during the long storage period…". P. 94 "One of the basic principles governing radioactive waste management and also adhered to in the Netherlands is the polluter pays principle. This principle requires that all costs associated with radioactive waste management are borne by the persons or institutes responsible for the generation of this waste. These costs, which include costs for removal, transport, treatment, conditioning, storage and disposal are charged by COVRA to its customers.… COVRA is in practice a monopolist, because it is the only recognized radioactive waste management agency. On the other hand, COVRA has a legal obligation to accept the waste offered for removal by license holders provided that it meets the acceptance criteria, set by COVRA." P. 95 | "The focus of regulatory requirements for decommissioning is clearly on nuclear power plants. … There are no legal requirements for the decommissioning of non fuel cycle-related facilities. However, the license conditions for those facilities stipulate that at closure of a facility the regulatory body should be notified, radioactive materials be disposed of or transferred to a third party with a valid license and any residual radioactive contamination be removed. After the licensee has demonstrated compliance with the license conditions, clearance of the facility is obtained from the regulatory body." P. 97 |

(*Continued*)

Table 4 Continued

Canada	"The Atomic Energy of Canada Limited (AECL), which is a Government of Canada Crown Corporation, has conducted nuclear R&D on behalf of the federal government over the past 50 years, resulting in significant legacy waste and decommissioning liabilities at its two research laboratories: Chalk River Laboratories (CRL) in Chalk River, Ontario and Whiteshell Laboratories (WL) in Pinawa, Manitoba. These nuclear legacy liabilities include accumulated waste (buried and stored) and contaminated buildings and land. CRL is operational, whereas WL is shut down and undergoing decommissioning." P. 55	"In July 1996, the Government of Canada announced its Radioactive Waste Policy Framework. The Framework sets the stage for the further development of institutional and financial arrangements to implement long-term management of radioactive waste in a safe, environmentally-sound, comprehensive, cost-effective and integrated manner. The federal government has the responsibility to develop policy, to regulate, and to oversee radioactive waste producers and owners in order that they meet their operational and funding responsibilities in accordance with approved long-term waste management plans. It is recognized that there will be variations in the general approach for the different waste types i.e., nuclear fuel waste, low-level radioactive waste and uranium mine and mill tailings.... In Canada irradiated fuel taken out of nuclear reactors at the end of its useful life is considered as waste. There are no plans to reprocess and recycle this fuel, so current plans are based on direct long-term management of the nuclear fuel waste.... In total, about 1.5 million bundles of nuclear fuel waste are currently in safe storage at the reactor sites, where it can be kept for decades, in pools or in dry concrete canisters. Canada's entire nuclear power program produces about 60000 bundles annually." P. 61	"Regarding WL, the site-wide decommissioning plan was subjected to an environmental assessment under the Canadian Environmental Assessment Act (CEAA). The environmental assessment addressed the partially decommissioned WR-1 research reactor and all buildings, infrastructure, waste management areas and affected lands. It was completed in March 2002, and following public hearings in September and November 2002, the CNSC issued a decommissioning license in December 2002 for a 6-year term.... AT CRL approximately 20 buildings are shutdown and in various states of decommissioning... Some shutdown buildings have been decommissioned and dismantled, whereas other buildings have been decontaminated and made available for other uses." P. 55

Japan	"The Atomic Energy Commission formulated the Framework for Nuclear Energy Policy on October 11, 2005. Based on the framework, the national policy on nuclear energy was decided by the cabinet of the government. … Radioactive waste has to be appropriately treated and disposed of for each classification under the principles of 1) the liability of waste producers, 2) minimization of the amount of radioactive waste, 3) rational treatment and disposal, and 4) implementation based on public understanding and acceptance." P. 89	"Currently, RI [radioisotopes] waste is collected, partly treated, and stored by Japan Radioisotope Association (JRIA); however a disposal facility has not been constructed. Most of the research waste is stored by the producers. JAEA which is a dominant contributor to generate research waste in Japan is storing the waste, a part of which has been treated for volume reduction. The total amount of stored RI and research waste in Japan is about 446 thousands of 200-liter drums as of March 2005. … Regarding RI waste, JRIA continues to collect, store, and treat it. JAEA also continues to store, and treat its own research waste." P. 90	"As for the decommissioning, it is important to undertake the activities relevant to decommissioning of nuclear facilities, such as commercial power reactors, test and research reactors and nuclear fuel cycle facilities, with the first priority on the safety assurance, under the responsibility of the installer, according to the amended Nuclear Reactor Regulation Law and the safety regulations of the Government, while promoting the understanding and cooperation of the local community." P. 89
Australia	"Spent fuel management is the responsibility of the Australian Government [Commonwealth Government] Commonwealth facilities are regulated by ARPANSA (Australian Radiation Protection & Nuclear Safety Agency). The Commonwealth is the only jurisdiction that has a requirement to manage spent fuel. Commonwealth management policy is to send spent fuel overseas for processing" P. 29–30	"Waste management facility to be developed for Commonwealth use only – to be constructed on existing Commonwealth land – known as the Commonwealth Radioactive Waste Management Facility, or CRWMF – States and Territories expected to implement appropriate arrangements for waste within their own jurisdictions. … Territory laws purporting to ban or regulate the Facility have been disallowed. All Commonwealth regulatory processes must be followed. The legislation enables a community to nominate a volunteer site in its own local area" P. 33–34	"Decommissioning: Current best practice … – Stage 1. Fuel is removed… – Stage 2. Care and maintenance stage, where a regime of monitoring and maintenance remains in place until arrangements and a decommissioning licence is issued – Stage 3. Decommissioning itself, covering the process of removal of all radioactive wastes – Stage 4. the final stage, remediation leading to the site being permitted to return to use for other purposes" P. 39

Source: OECD/NEA, Radioactive Waste Management Committee, 2007. "Approaches and Practices in Decommissioning of Facilities and Management of Radioactive Waste from Non-Nuclear Fuel Cycle Related Activities." Proceedings of the Topical Session of the RWMC 40th Meeting at NEA Offices in France 14th March 2007. (NEA/RWM 2007/9).

The actual verification conducted for the inclusion of civil nuclear energy facilities into the structural components of national production systems has been implemented according to regional technology energy assets and local economic configurations, which have been monitored in terms of transitional organizational stages. Essentially, an overall transition of national renewable energy compounds has taken place, which led to the prioritization of industrial innovation sectoral goals, facilitating the actualization of comparative national management arrangements and local organizational approaches.

Fundamentally, the strengthening of industrial transforming orientations has emerged in combination with distinctive cultural development patterns and territorial strategic policies that have offered a confronting dimension, which can incorporate public reflections, in particular based on western or eastern ideological contexts, with liberal trading dynamics that have been explored through intra-regional policy relationships.[27] For the case that we have considered about environmental governance regulations addressing environmental safety activities of nuclear reprocessing facilities, the socio-economic domains have reflected multilateral inclusive approaches due to the PPP specifications about collaborative functional roles of industrial partnering bodies and civilian activism networks.

5.1.2 Community Attention

Independent comparative regional studies focused on best local practices and local adaptive designs in governance have highlighted the local dynamics of community risk perceptions about the nuclear waste management issues for the implementation of alert monitoring systems designed locally. This attention has been crucial for providing mutual cognitive exchanges developed on a contested environmental regulatory ground.

In fact, the constitution of identified representative roles in governance has been acquired according to the dual characteristics of regulatory practice models, historically linked to regional energy transition relations, which have been reaffirmed through the availability of oligopolistic and monopolistic conditions of resources-based production systems. In turn, the regional differentiation activity of complex energy innovation systems has led to systemic regulatory

---

27 See: Ivanov Kalin, "Legitimate Conditionality? The European Union and Nuclear Power Safety in Central and Eastern Europe," *International Politics* 45.2/2008, pp. 146–167.

patterns for the technical knowledge provisions, which have been specifically undertaken by the western and eastern parties of world regions.[28]

Under changing institutional terms, both domestic and regional spheres that follow on the regulation process for local knowledge management operations, have been practically arranged for the energy managerial facilitation of domestic nuclear power plants, related also to the maturation of local community acceptance issues. In fact, the consolidation of multi-layered participatory relationships has essentially been aligned with the culture of nuclear power regulation based on the local industrial development that has been set in accordance with regional PPP factors combined with public agencies normative frameworks, supporting the technical understanding in local industrial contexts.[29]

But at the same time, the presence of environmental advocacy groups based within intra-regional newly developed organizations has been increasing, thus building a cohesive base for civilian associations that rely on common ethical and environmental considerations, while supporting the public aggregated coalitions and public social movements, which can express the critical terms of practical adaptability models and local distributive dimensions. Moreover, in order to achieve deeper levels of critical understanding of which modern tools are better suited to enhance regional nuclear power strategies and common environmental designs, the future willingness to do so remains quite key for the territorial changing dimension. It ought to be considered that domestic policy regulations, PPP knowledge innovation acquisitions, and local integration transfers, have performed according to the legal and civilian environments, which in turn have been confronted with formal conditional boundaries based on traditional perspectives about the state and local empowering implications.

The discussion made about the rationalization of managerial industrial assets involving the nuclear power plants and the energy reprocessing facilities has been addressed through the analysis of regional institutional partnerships PPP, which have been formed through public-private energy companies for the

---

28 See: Claire A. Watkins, "Nuclear Power Rate Regulation after Eastern Enterprises: Are Ratepayers Being Taken for a Ride?," *Boston College Environmental Affairs Law Review* 28.1/2000, pp. 191–228.

29 See: 1) Matthews R. Bruce, "Nuclear Safety: Expect the Unexpected," *Professional Safety* 50.12/2005, pp. 20–27; 2) Daniel Nohrstedt, "The Politics of Crisis Policymaking: Chernobyl and Swedish Nuclear Energy Policy," *Policy Studies Journal* 36.2/2008, pp. 257–278.

affirmation of comparative renewable energy sustainable activities that reflect prospective common directions in science technology development.[30]

Fundamentally, the possible evolution of transnational energy sustainability programs modified in accordance with concurrent legal interpretative implications related to the national public and private interactions, has acquired a comprehensive public normative spectrum by including local regulation policies. In fact, the industrial configuration of nuclear power reprocessing facilities with waste management disposal services has been established through long- or short-term cooperative development objectives identified for public renewable energy policy schemes, produced under normative periodic terms.[31]

According to local territorial contexts of regional economic development areas, the respective statutory agencies and the implementing decisional groups have given assistance in order to initiate collaborative programs for the case of public nuclear waste disposal and public protection safety schemes. As a matter of importance, national energy performances and innovation technology acquisitions have tentatively led to the intervention of flexible governance paradigms, particularly about the objective reasons of enhancing the nuclear energy security and the local risk reduction targets.[32]

Moreover, in EU countries or in the United States, historical environmental concerns related to energy sustained provisions have followed in view of the democratic identification of progressive orientations, characterizing the materialization of environmental regulation responses in the case of multi-level regional innovation distribution schemes. The corresponding institutional bodies that favor public local interactions have been alternated with grassroots initiatives and civic confrontation platforms, which involve the formation of anti-nuclear campaigning groups that among other civilian counterparts, have aimed at the solid reduction of nuclear power energy facilities and future development plans.

This competitive process has remained engrained in public risk perceptions about the local mitigation and implementation of governance procedures with interrelated transfer mechanisms that have been affected according to different territorial scales. For instance, the national and international nuclear policy programs and regulatory control frameworks have been classified according to

---

30 Per Hogselius, "Spent Nuclear Fuel Policies in Historical Perspective: An International Comparison," *Energy Policy* 37/2009, pp. 254–263.
31 See: Nina Tannenwald, "The Nuclear Taboo: The United States and the Normative Basis of Nuclear Non-Use," *International Organization* 53.3/1999, pp. 433–468.
32 See: Harold A. Feiveson, "Faux Renaissance: Global Warming, Radioactive Waste Disposal, and the Nuclear Future," *Arms Control Today* 37.4/2007, pp. 13–17.

the level of management construction costs, related to the energy capacity of nuclear power plants NPPs, which in turn have also been reviewed under public risk reduction plans and local prevention programs.[33]

Nowadays, the actual industrial increase of regional technological concentrations for new generations of nuclear power plants NPPs emerging within national responsive models, for future renewable energy capacities' growth – based on international interdependencies meeting the global energy demands – has indirectly created divergent aspects between the effectiveness of territorial local trajectories and private industrial energy's landscapes in OECD and non-OECD countries.[34]

In terms of territorial governance, the divergent path about local implementation activities on nuclear energy programs can increase the level of public environmental concerns about NPP safety issues. Divergent regulatory issues can indirectly cause the reduction of technology and innovation management diffusion on nuclear energy activities and local maintenance systems, which have included the technical as well as operational comprehensive performances of different managerial operating systems located in specific administrative areas.[35]

As comparative overview, there have been stringent debates about nuclear power plants NPPs in the science policy literature, (mostly focused on technical engineering capacities and regional maintenance activities for operational security changes in nuclear facilities, developing waste disposal programs) which have highlighted the problematic interactions regarding the civilian environmental risk acceptance orientations toward nuclear waste disposal regulations, and the local processing operations of industrial waste components at nuclear facilities management sites.[36]

For instance, in the United States the growing environmental political debates and local identified obstacles, about the nuclear repository case of the

---

33   A. Adamantiades, I. Kessides, "Nuclear Power for Sustainable Development: Current Status and Future Prospects," *Energy Policy* 37/2009, pp. 5149–5166.
34   See: 1) Teresa Hansen, "Positive Image Fuels Nuclear Energy's Resurgence," *Power Engineering* 110.9/2006, pp. 18–30; 2) Karl Grandin, Peter Jagers, Sven Kullander, "Nuclear Energy," *Ambio* 39/2010, suppl. Special Report: Energy 2050, pp. 26–30.
35   Adamantiades, Kessides, "Nuclear Power for Sustainable Development," pp. 5149–5166.
36   See: 1) Richard G. Kuhn, "Social and Political Issues in Siting a Nuclear-Fuel Waste Disposal Facility in Ontario, Canada," *The Canadian Geographer* 42.1/1998, pp. 14–28; 2) Daniel P. Aldrich, "A Normal Accident or a Sea-Change? Nuclear Host Communities Respond to the 3/11 Disaster," *Japanese Journal of Political Science* 14.2/2013, pp. 261–276.

Yucca Mountain have mounted over the years.[37] In such case, public scholars have retraced commonalities of ideas and characteristic orientations about the (NIMBY) *not in my back yard* syndrome which has emerged for the lack of consensus and public trust of local national communities.

The critical resurgence of nuclear innovation performance systems has been retraced also according to the affirmation of public cognitive patterns, which have mediated on the disposal management activity of stored radioactive waste for the short-term. At the same time, the refusal to recognize a potential nuclear environment crisis has caused an eventual separation among local communities that have been attached to respective territorial boundaries retracing the expansion of nuclear waste disposal sites.[38]

As proximate meaning, technical industrial innovation plans have basically characterized national or regional industrial orientations for energy innovation policies that involve g technical reprocessing cycles of nuclear energy technologies. For which case, we may point out the frame of a systematic coordination process able to cope with innovating managerial approaches, which have also been adopted specifically for the local integration of nuclear waste disposal programs.[39]

Additionally about the nuclear power industry, the prospective ruling dispositions have been related to the making of nuclear energy policies, which have been validated through permanent participatory processes. For this reason, permanent environmental issues have essentially configured PPP public-private organizational aims for the industrial delivery process of nuclear energy facilities and nuclear waste reprocessing plants.[40]

5.1.3 Industrial Concentration

Through an actual industrial concentration of renewable energy facilities, the major commitments related to public safety and security plans about nuclear disposal sites have been debated worldwide for the upgrade of national ageing

---

37 See: Jeff Tollefson, "Battle of Yucca Mountain Rages On," *Nature* 473.7347/2011, pp. 266–267.
38 Marvin Backer Schaffer, "Toward a Viable Nuclear Waste Disposal Program," *Energy Policy* 39/2011, pp. 1382–1388.
39 See: Dmitri A. Titoff, Caitlin A. Buckley, Dmitry Novak, Richard Weitz, "Assessing Kazakhstan's Proposal to Host a Nuclear Fuel Bank," *UNISCI Discussion Papers* 28/2012, pp. 99–125.
40 Backer Schaffer, "Toward a Viable Nuclear Waste," pp. 1382–1388.

infrastructures with internal knowledge mechanisms applied in comparative assessment areas.

Moreover, there are regional environmental risk analyses, country-based, and technology innovation research reports[41] which have been at the core of institutional reforming activities that propose the renewal of structural energy innovating plans. The planning organizational objectives have been advanced in actual terms with the involvement of regulatory proceedings applying to the nuclear power management systems at a global level, with interrelated safety and security risks, regionally updated for nuclear waste disposal modalities.

In essence, the practical integration of international regulatory measures has evolved in time through comparative managerial identifications of energy security programs. Public and private industrial energy goals of PPP have been paralleled to the nuclear energy technology programs through internal knowledge integration and local innovation processes, which have changed in terms of mutual responsive adaptations for the national energy agreements. For instance, public environmental debates combined with common modified orientations have reflected the institutional prioritization given to renewable energy policies with national interdependent strategies, which have led to local cooperative management blueprints. Especially, in consideration of future technology incremental prospects associated with regional industrial programs concurrently emerging in countries such as the U.K., France, and Finland,[42] during the last three decades.[43]

However, the affirmation of regional alternative markets about the production of renewable nuclear energy sources has created public critical trends increasing an open grounded opposition, when fundamental questions such as the national implementation dynamics of nuclear power plants NPP have emerged in connection with the local management of operational maintenance systems. In theory, alternative nuclear energy developing programs that include NPP

---

41 See: 1) Craig Summers, Donald W. Hine, "Nuclear Waste Goes On the Road: Risk Perceptions and Compensatory Tradeoffs in Single-Industry Communities," *Canadian Journal of Behavioural Science* 29.3/1997, p. 211; 2) Gary Taubes, "Whose Nuclear Waste?," *Technology Review* 105.1/2002, pp. 60–67.

42 See: 1) Brian Heap, "Man and the Future Environment," *European Review* 12.3/2004, pp. 273–292; 2) Rogner H-Holger, "Nuclear Power and Sustainable Development," *Journal of International Affairs* 64.1/2010, pp. 137–163.

43 Tuula Teravainen, Markku Lehtonen, Mari Martiskainen, "Climate Change, Energy Security, and Risk – Debating Nuclear New Build in Finland, France and the UK," *Energy Policy* 39/2011, pp. 3434–3442.

activities have been supported according to different types of institutional local capacities involved, while fostering a participative community reciprocation in terms of collective understanding, which has been reaffirmed through cultural democratic debates that also include progressive public directions.

The confidence and trust status granted for associated public reforms established through mutual governmental strategies specifying the actual interplay of policy reforming directives, which regulate the national energy delivery process, has been coordinated through national diversification activities concerning the energy resource distribution capacities. In particular, about the cases of the U.K., France, and Finland, the renewable energy sectors have represented important material sources of regional growth for their respective economies.

In addition, more information is now available on the annual production costs of electricity supplied through nuclear energy plants, and in accordance to the development of national nuclear processing facilities and correlated industrial innovation projects, which have been resized case-by-case according to the nuclear managerial capacities facilitating the local aggregated expansions.[44]

As mentioned, there have been academic thematic issues that have been discussed about technology changes and multi-level innovation management elements, which have been related to nuclear industrial risks at the local energy facilities. The surfacing of critical public risk-perceptions and public local acceptances which converge into societal environment movements, has ignited nuclear policy debates becoming important components for the environmental decision-making process that can respond to the organizational changes of alternative energy diversification plans, including public cooperative requirements. At the same time, the diffusion of nuclear risk perception studies has been focused on the syndrome of [NIMBY] because of the direct implications emerging from local risk exposures' perceptions, which determine local security concerns about national radioactivity accidents and land-changing pollution aspects.[45]

We may stress that the combination of different explanatory factors regarding for instance: global climate change (GCC), global energy *securization*, and local

---

44 Teravainen, Lehtonen, Martiskainen, "Climate Change, Energy Security, and Risk," pp. 3434–3442.
45 See: 1) Andrew J. Evans, Richard Kingston, Steve Carver, "Democratic Input Into the Nuclear Waste Disposal Problem: The Influence of Geographical Data on Decision Making Examined Through a Web-Based GIS," *Journal of Geographical Systems* 6.2/2004 pp. 117–132; 2) Marci R. Culley, Holly Angelique, "Participation Power, and the Role of Community Psychology in Environmental Disputes: A Tale of Two Nuclear Cities," *American Journal of Community Psychology* 47. 3–4/2011, pp. 410–426.

community risk's perception, has been interrelated to public levels of attentive decisional deliberations promoted by regional states for the facilitation of local adaptive regulatory capacities. Similarly, public participation debates mainly focused on the nuclear power development, nuclear waste disposal process, and the energy production availability for commercial uses, have been explored according to the dimension of public participation that fosters a regional collective understanding that has been identified through public consensus.

For instance, the European Union member's countries such as Finland, and France have modified respective national regulative norms and derogatory policy processes, set in line with public exploratory adaptations undertaken at governmental level in parallel with public mediation activities. Basically, in order to promote a common participative ground for technical engaged discussions based on local environmental determinants, the respective inclusion of socio-economic priorities, importing local community's inputs provided by both social opposing sides in order to grant democratic equal accesses was recognized. Public confrontational orientations for institutional discussions sustained in the longer term have been relevant for the identification of common future growth prospects which define strategic local sustainable guidelines, focused also on the civilian energy security programs.[46]

---

46 Teravainen, Lehtonen, Martiskainen, "Climate Change, Energy Security, and Risk," pp. 3434-3442.

CHAPTER SIX

6.1 Socio-Technical Transitions

The institutional practices matching with political party systems and public environmental bodies have been elaborated for national and transnational sustainability purposes combined with structural ownerships of technological energy systems. The industrial managerial systems have been reaffirmed in terms of technical innovation components, that can determine the institutional formation of local regulatory measures. The launch of environmental cooperative arrangements for the compatible renewable energy transitions, has been structurally pursued according to the identification of long-term sustainable objectives reflecting the socio-economic national conditions.

For instance, the material occurrences of national local politics with corresponding national and local parties' power shifts have led to the sectoral implementation of provincial and local industrial production targets associated with performing energy systems. In Canada's case the industrial renewable energy implementation has been based on progressive local adaptation with territorial cooperative changes, including the promotion of community's dialogues and public environmental forums that can secure a mutual level of local correspondence about the formalization of regional managerial transitions operated in the case of industrial energy changing domains.[1]

Essentially, the public intercalation of managerial industry innovation operations has been characterized by respective levels of distributive environmental policies defined in connection with local embedded practices and national planning dynamics comprising socio-economic and technical development spheres, also coordinated for the actual maintenance of nuclear energy reprocessing facilities.[2] In a critical way, political and economic regional studies tackling the organizational process of PPP with national electoral party systems and PPP with

---

1   See: Severin Borenstein, "The Private and Public Economics of Renewable Electricity Generation," *Journal of Economic Perspectives* 26.1/2012, pp. 67–92.
2   Margot Hulbert, Kathleen McNutt, Jeremy Rayner, "Pathways to Power: Policy Transitions and the Reappearance of the Nuclear Power Option in Saskatchewan," *Energy Policy* 39/2011, pp. 3182–3190.

national business groups[3] have specifically indicated some of the reasons about the leading territorial choices made through the formation of structural local arrangements undertaken in view of renewable energy innovation strategies.

In addition, the regional consistence of development transformation cycles indirectly affecting the institutional collaborative spheres has also emerged across different local governing agencies and multilateral energy production consortiums, which have transitioned by adopting divergent territorial management practices, linked to transnational dynamics of multilateral PPP energy accesses which can offer a different combination on local innovation programs.[4]

For instance, the Canadian federal states[5] reviewed from national and local policy perspectives have respectively encountered territorial socio-technical transitions, which have been dependent on the environmentally sustainable plans and local knowledge innovation policies undertaking structural regulatory paths. Related to this, the programmatic action in governance has been historically determined through the sectoral availability of regional industrial businesses and local entrepreneurial circuits operating within national distribution frameworks that follow common territorial policy targets.

Nonetheless, the possible implementation of national regulatory conditions has affected the local changing environments intertwined with public-private partnerships PPP through the presence of e.g. nuclear power regulative agencies; industrial management organizations and technical investment bodies, operating according to the political interplay of different environmental political parties – whose cooperative local initiatives have been proposed at federal and provincial governmental level – due to organizational components of multilateral economic transition strategies.[6]

For the Canadian case, the environmental cooperation parties have particularly emphasized the municipal and provincial roles of local administrative counterparts that perform the public regulatory duties through the involvement of distinct knowledge information networks, which offer a dynamic interplay of common states' relations about e.g. local administrative performances

---

3 See: Alexander Hertel-Fernandez, "Who Passes Business's 'Model Bills?' Policy Capacity and Corporate Influence in U.S. State Politics," *Perspectives on Politics* 12.3/2014, pp. 582–602.

4 See: Lea Gynther, Irmeli Mikkonen, Antoinet Smits, "Evaluation of European Energy Behavioural Change Programmes," *Energy Efficiency* 5.1/2012, pp. 67–82.

5 See: Miranda A. Schreurs, "Federalism and the Climate: Canada and the European Union," *International Journal* 66.1/2011, pp. 91–108.

6 Hulbert, McNutt, Rayner, "Pathways to Power," pp. 3182–3190.

at Canada's uranium deposits, and in nuclear energy production sites, where local technical transitions for the nuclear energy domain have differently applied because of divergent agencies' results; notably, "environmentalists advocate for an expansion of renewable energy options, pushing for greater commitment to wind, solar, and...hydro. Public opinion in the province is generally unclear on a preferred option but there is a strong support for...corporation to continue playing the lead role.."[7]

These observations can allow us to further scrutinize the societal constitution of divergent necessities put in relation with local community transitions and multilateral energy technological systems, involving local administrative requirements. Also, historically the political and economic energy patterns have been formed through a critical decision-making process largely motivated on socio-economic grounds; as consequence the public environmental concerns have locally evolved in conjunction with territorial changing expectations, by also presuming the reconfiguration of national industrial technology platforms, with respective operable management capacities.

6.1.1 Local Practice Projects

Socio-economic reconfiguration patterns structured in terms of regional industrial production interests have been based on the possible inclusion of comparative regulatory networks operating through alternative decentralized bodies, which have been essential for the development of national integrated dynamics that emerge from the regional and transnational economies.[8]

What can be perceived as an energy regulatory pattern of technological production systems, has been related to the fragmentation of networked duties that become functional in terms of multi-level governance, which emerges in different economic capacities, including the formation of national participation organizations and public distribution platforms, favoring comparative civilian

---

7 Hulbert, McNutt, Rayner, "Pathways to Power," pp. 3182–3190. p. 3184.
8 See: 1) Andrew Kakabadse, Nada K. Kakabadse, Alexander Kouzmin, "Reinventing the Democratic Governance Project Through Information Technology? A Growing Agenda for Debate," *Public Administration Review* 63.1/2003, pp. 44–60; 2) Adrian Smith, Al Rainnie, Mick Dunford, Jane Hardy, Ray Hudson, David Sadler, "Networks of Value, Commodities and Regions: Reworking Divisions of Labour in Macro-Regional Economies," *Progress in Human Geography* 26.1/2002, pp. 41–63.

interactions.⁹ Social change reaffirmed through progressive national assessments with a critical understanding of the nuclear energy science policy, has been intertwined with embedded societal dynamic components characterizing the industrial information societies. The information societies have undertaken a progressive exploration process about the application of spatial methodologies related to widespread technology fields and local innovation accesses within dimensional schemes.

Accordingly, the different adoption of societal development approaches delivered across national democratic or socio-technocratic regimes has emerged according to the aggregation of specific institutional settings,¹⁰ with the decision-making activities taking place at federal, national, and provincial level, while implementing regulatory technical environments.¹¹

It can also be considered that the process of sectoral energy policy development that incorporates the technical industrial assessments for the evaluation procedures has been deepened through the progressive facilitation of nuclear energy power partnerships PPP which imply growing interrelated aspects, identified across the national cooperative spheres of nuclear energy supply associations. Multilateral regional industrial partners have cooperated to the preparation of nuclear power policies due to regional industrial partnerships that have been dependent on institutional mutual positions led within the governmental regulative assets, which perform through local distributive agencies that are directed according to specified public organization targets.¹²

For instance, the U.K. government has adopted independent formal regulative perspectives for the renewal of energy planning objectives which can also refer to the industrial energy risks of nuclear processing facilities. To consider that NPPs industrial adaptation operated across the last decades has gone through a sectoral ageing process, which has required a continuous support put in place through structural and operational programs incorporating PPP functional roles with public local commitments. Such types of public local commitments have

---

9   See: Josh Whitford, Cuz Potter, "The State of the Art. Regional Economies, Open Networks and the Spatial Fragmentation of Production," *Socio-Economic Review* 5/2007, pp. 497–526.
10 See: Yasumasa Komori, "Evaluating Regional Environmental Governance in Northeast Asia," *Asian Affairs: An American Review* 37.1/2010, pp. 1–25.
11 Hulbert, McNutt, Rayner, "Pathways to Power," pp. 3182–3190.
12 See: Etel Solingen, "Macropolitical Consensus and Lateral Autonomy in Industrial Policy: The Nuclear Sector in Brazil and Argentina," *International Organization* 47.2/1993, p. 263.

been pursued through local services provided by multinational nuclear energy providers, while meeting the domestic users' energy demands, with the involvement of environmental local agencies. The elaboration of progressive participatory stages has been subscribed through the actualization of local participative associations, as well as, climate mitigation programs combined with local action plans, and local environmental risks reduction activities.[13]

From quantitative energy studies that highlight the U.K. socio-economic strategies about the renewable energy diversification effects interlinked with comprehensive regional exporting programs of technological innovation capacities – mentioned by the authors Pidgeon, Lorenzoni, and Poortinga 2008 – has emerged that comparative public interrelated engagements on nuclear energy power designs have still been perceived as a problematic choice thus reducing public acceptance levels on the use of nuclear energy power for climate risk reduction policies.

Essentially, the past constitution of dramatic historical events about radioactive nuclear accidents experienced in the U.K., but also in the U.S., Japan, and in Ukraine with the Chernobyl disaster,[14] has significantly marked public opinions' orientations, according to comparative environmental studies. Such types of public cognitive perceptions that reflect the nuclear policy orientations have moved towards the integration of local protection mechanisms for the environment and the communal life. Through which, the establishment of local public safety and local anti-risk protection programs has been advanced with national and transnational campaigns focused on environmental preparedness and disaster prevention plans, commonly arranged for the possibility of nuclear industrial hazards and nuclear risks' radiological exposures.[15]

For instance, there is now a need to effectively identify national geological locations that have been proposed for various nuclear waste disposal sites, e.g.

---

13 Nick F. Pidgeon, Irene Lorenzoni, Wouter Poortinga, "Climate Change or Nuclear Power – No Thanks! A Quantitative Study of Public Perceptions and Risk Framing in Britain," *Global Environmental Change* 18/2008, pp. 69–85.
14 See: 1) Sung Ju Cho, (2009). "From Proliferation to Renunciation: Why Some States Give Up Nuclear Ambitions While Others Do Not," PhD Dissertation submitted at the Graduate Faculty of the University of Virginia, The Woodrow Wilson Department of Poltics, University of Virginia, United States.; 2) Akira Omoto, "Where Was the Weakness in Application of Defence-in-Depth Concept and Why?," in: *Reflection on the Fukushima Daiichi Nuclear Accident*, eds. J. Ahn et al. (chp. 8, 2015) DOI 10.1007/ 978-3-319-12090-4_8.
15 Pidgeon, Lorenzoni, Poortinga, "Climate Change or Nuclear Power," pp. 69–85.

U.S. Nevada state, in order to build new processing nuclear facilities[16] which have been planned for the long-term period. Regarding the public environmental acceptances on long-term management of nuclear disposal sites, the public awareness created through leading civilian opposition groups has brought instead to the constitution of local communities' assemblies which have catalyzed attention of national policy representatives and local environmental groups, in order to express common perceived concerns about the new organizational plans for the nuclear waste disposal constructions.

For public evaluation purposes, the progressive regulatory changes brought within the managerial operationalization of nuclear energy's power sites have established local associative conducts interlinked with territorial contextual conditions, which have included the study of catastrophic weather events. Thus, the preparation of industrial organizing activities sustained under uncertain environment conditions has been drafted in conjunction with the preparation of structural implementation goals targeting the use of alternative renewable energy sources.[17]

For such reasons, we can consider that the tendency to foster a managerial adaptation process about the use of renewable energy, is preferred in order to favor an incremental availability of alternative structural energy choices. Nonetheless, there has been critical environment acceptance at a comprehensive level about the formation of science innovation policies and nuclear energy innovation strategies. Through public cognitive interactions, a divisive regional element about the open sharing of collective territorial responsibilities still persists. Especially for those concerning the regional conditional scopes of local energy identified priorities, which have had major implications for the development process and for the technological innovation diffusion of systematic practical assessments.[18]

Effectively, on the same table there are national local governance indications about the practical informal steps and local monitoring tools advised for the case of renewable energy implementation. With corresponding support strategies that enable both technological and industrial science affirmation[19] within local distributive schemes, have been included sectoral initiatives for local risk

---

16 See: Siobhan Sutherland, (2002). "First Nations and Nuclear Fuel Waste Management: An Analysis of Stakeholder Position," Master of Arts Thesis submitted at the Faculty of Graduate Studies of the University of Guelph, Canada.
17 Pidgeon, Lorenzoni, Poortinga, "Climate Change or Nuclear Power," pp. 69–85.
18 Pidgeon, Lorenzoni, Poortinga, "Climate Change or Nuclear Power," pp. 69–85.
19 See: Joseph B. Boland, (2002). "The Cold War Legacy of Regulatory Risk Analysis: The Atomic Energy Commission and Radiation Safety," PhD Dissertation submitted at the

planning activities and local managerial distribution of national or transnational renewable energy programs. In essence, such types of multilateral cooperative plans in governance PPP have either directly or indirectly affected the changing terms of respective institutional collaborative policies.

Particularly, what has come into cause has been the relationship created between the nuclear power transnational industrial authorities and the public environmental regulation bodies, which have redefined national structural directives about energy policies in different ways. Common regulative measures have been designed through a complex mix of provisional understanding formed about the nuclear fuel cycle programs, but also about the organizational maintenance of local innovation processes. Through which, the involvement of public environmental representatives has led to the formulation of sustainable development provisions, which refer to multi-level renewable energy projects assessed according to corresponding governmental adaptation models – reflecting common states' industrial productions reforming stages.

As example, in modern China the industrial case about nuclear energy operational facilities has suggested a critical rethinking about multilateral coordination guidelines of regional energy programmatic activities, which have been established for nuclear power delivery with long-term objectives. Regarding the civilian uses of nuclear energy reprocessing cycles, "China's decision to reprocess its spent fuel could be made with an absence of transparency and a lack of public and outside expert input."[20] The actual possibility of dynamic policy interactions created between technical energy expertise bodies and national management planners has provided ground for additional discussions about the clarification of local operational conditions. Considering regional respective levels of mutual understanding and long-term commitments to the diffusion of alternative renewable energy sources, in China's case this process has also involved the construction, maintenance, and operationalization of nuclear processing facilities, which have been located within internal territories of the country.[21]

Department of Political Science and the Graduate School of the University of Oregon, United States.
20 Yun Zhou, "China's Spent Nuclear Fuel Management: Current Practices and Future Strategies," *Energy Policy* 39/2011, pp. 4360–4369. P. 4361.
21 See: Erich William Schienke, (2006). "Greening the Dragon: Environmental Imaginaries in the Science, Technology, and Governance of Contemporary China," PhD Dissertation submitted at the Graduate Faculty of Rensselaer Polytechnic Institute, New York, United State.

When we shift the attention on comparable nuclear energy countries, that follow the regional nuclear policy scenarios, there were a number of confrontational demonstrations about the influence played by the local changing dynamics for deterministic strategies pursued through respective regulatory balancing bodies. Practically, the environmental orientation schemes have also been translated into local integrative processes, because they have facilitated a functional understanding of the material implications of industrial reprocessing stages of nuclear fuel facilities, as well as, on the final destination of nuclear waste storage sites.[22]

In fact, public energy policy and PPPs linked with official supportive orientations have been central to the functional path toward security and public acceptance models established in the case of long-term nuclear management strategies and technological innovation activities. For instance, in the case of Canadian federal states, the historical progressive undertakings about nuclear energy policy have reflected the diffusion of participatory knowledge transfers for local integrative mechanisms, which concern local nuclear waste issues, including spent nuclear fuel programs, which have produced a direct involvement of local communities, as well as, of provincial district administrations.[23]

The Canadian Nuclear Waste Management Organization (NWMO) has progressively acknowledged the need for the adoption of public evaluation approaches able to respond to changing innovation requirements of Ontario and Canadian local governments, especially for the process of management of nuclear waste disposal activities.[24] Technically, the nuclear fuel cycle has integrally entailed local knowledge safety and local maintenance operations of clustered disposal innovation systems. The coordination process for local knowledge safety and upgraded managerial systems[25] has been debated, at governmental

---

22 See: Bazan J. Romero, (2000). "Sovereignty or Environmental Inequity: The Complexities Surrounding the Proposed Controversial Nuclear Waste Facility at Mescalero," PhD Dissertation submitted at the Graduate College of Bowling Green State University, Ohio, United States.
23 M.V. Ramana, "Shifting Strategies and Precarious Progress: Nuclear Waste Management in Canada," *Energy Policy* 61/2013, pp. 196–206.
24 See: Charles Hostovsky, (2002). "Integrating Planning Theory and Waste Management: A Critical Analysis of Current EIA Practice in Ontario," PhD Dissertation submitted at the University of Waterloo, Planning Degree, Ontario, Canada.
25 See: John F. P. de Grosbois, (2011). "The Impact of Knowledge Management Practices on Nuclear Power Plant Organization Performance," PhD Dissertation submitted

level, in order to provide a mutual ground on environmental risk assessments, which have involved the supervision of federal administration capacities, despite the fact that local resistance factors to the nuclear waste disposal process have remained at a critical point.[26] As ulterior example, Canadian historical perspectives about public-private energy partnerships PPP involving local cooperative changes, related to the nuclear waste disposal programs, have emerged according to the increasing complexities of national information societies focused on environmental protection norms and local safety needs, which have required public regulatory consultations.

What has come into cause this time has been the local industrial capacity of managerial storage systems incorporating technical knowledge and local expertise which can support nuclear power plants' local teams and national management practitioners in view of alternative organizational plans. Parallel to this, there have been concerted efforts to identify public legitimacy paths put in place through public policy interactions, which have directly or indirectly involved the affirmation of local assimilation processes, interlinked with nuclear energy distribution patterns, which rely on technical impact assessments for PPP reviews. In fact, "substantial sections of Canadian society continue to be ambivalent about an expansion of nuclear power, their support for geological disposal of nuclear waste might be available only when such disposal is part of a commitment not to construct any new reactors."[27]

Moreover, there are comparative science literature studies about local environmental governance and regional policy development that integrate public cognitive analyses about public nuclear risk perceptions, referred to the nuclear science policy, mentioned by: Stoutenborough, Sturgess, and Vedlitz, 2013;[28] Rosa and Rice, in 2004.[29] Some highlights have been pointed at the development of nuclear power facilities in terms of progressive consolidated stages, which have brought unforeseen elements in reference to societal cohesive motivations that have justified public opposition grounds.

---

at the Faculty of Graduate and Postdoctoral Affairs, Carleton University, Ottawa, Ontario, Canada.
26 Ramana, "Shifting Strategies and Precarious Progress," pp. 196–206.
27 Ramana, "Shifting Strategies and Precarious Progress," pp. 196–206. p. 204.
28 James W. Stoutenborough, Shelbi G. Sturgess, Arnold Vedlitz, "Knowledge, Risk, and Policy Support: Public Perceptions of Nuclear Power," *Energy Policy* 62/2013, pp. 176–184.
29 Eugene A. Rosa, James Rice, "Public Reaction to Nuclear Power Siting and Disposal," *Encyclopedia of Energy* 5/2004. pp. 181–194.

To consider also the fact that nuclear civil technologies and industrial commercial facilities for civil power energy plants generating nation-wide electricity have experienced potential distribution variations according to each country's territorial levels of public identifications. On such aspects, the actual management knowledge transfer process transmitted from the agrarian to the industrial societies has determined a certain degree of national objective explorations about the possible implementation of risk management models, which can incorporate local environmental analyses with the provision of innovation knowledge platforms, developed for the identification of environmentally sustainable conditions.

At the same time, the democratic and communitarian societies have reshaped the institutional lines of local strategic development connected with the energy evolution path, through representative legitimacy models, that have been locally intertwined with complex vulnerability issues. Such programming issues have included the existence of multi-level nuclear waste disposal policies, associated to the affirmation of public environmental concerns involving the case of e.g. uranium mining areas with radioactive contamination.[30]

For instance, the European Union countries and the Post-Soviet republics have been specifically confronted with radioactive contamination releases e.g. in land mining areas, where governmental regulatory disposal policies have framed the actual imposition of control and surveillance mechanisms involving national and local party pressures put on the nuclear industry.[31] In addition, in the United States, the implementation of local risk assessments and local performance evaluation reviews has pressured the technical scientific experts to re-establish public programmatic directions, which favor improved local level interactions about nuclear waste disposal programs, also by including national and international cooperative meetings, public local deliberations, national environmental representations, national interest groups, and local advocacy associations.[32]

---

30 See: Jim Harding, "Living behind the uranium curtain," *Briar Patch* 37.8/2008, p. 13 -15.
31 M.D. Siegel, and C.R. Bryan, "Environmental Geochemistry of Radioactive Contamination." in: H.D. Holland & K.K. Turekian *"Environmental Geochemistry,"* eds. Barbara Sherwood Lollar (2005). (Elsevier Ltd. Sandia National Laboratories, Albuquerque, NM, USA, 2005, chp. 9.06).
32 See: PR Report, "Groups: Nuclear energy institute is pushing NRC to shortchange environmental review of long-term radioactive waste storage," *PR Newswire Association LLC*, New York, U.S, 2013.

6.1.2 Nuclear Waste Disposal

According to case-based nuclear waste disposal policies, nuclear power energy programs have already shown the ageing signs of scientific industrial configurations which have required continuous national assistance and regulatory understanding of the technical incorporated approaches and the distributed territorial plans. Through similar trends, the local and international civilian environmental groups have in time lost their confidence about public regulative propositions on local waste disposal programs, which target specific geological locations for the identification of terminal nuclear repositories. The civic local groups have fundamentally continued to criticize the public representatives' interactions on nuclear siting policies, also related to local potential arrangements about nuclear waste disposal locations, such as the Yucca mountain's case of HLW in U.S.[33] As described in the study of Siegel and Bryan "there are several classes of nuclear waste; each type is regulated by specific environmental regulations and each has a preferred disposal option."[34] Table 5 provides a classification about nuclear waste elements identified in scientific research and testing grounds.

---

33 Siegel, and Bryan, "Environmental Geochemistry of Radioactive Contamination," chp. 9.06.
34 Siegel, and Bryan, "Environmental Geochemistry of Radioactive Contamination," p. 211.

Table 5: Definition of nuclear waste disposal

Classification Summary
Spent Fuel **(SF)** "consists of irradiated fuel elements removed from commercial reactors or special fuels from test reactors. It is highly radioactive and generates a lot of heat; therefore remote handling and heavy shielding are required. It is considered a form of HLW because of the uranium, fission products and transuranics that it contains. [...]"
High level wastes **(HLW)** "includes highly radioactive liquid, calcined or vitrified wastes generated by reprocessing of SF [...]."
Transuranic waste **(TRU)** "is defined as waste contaminated [...] TRU is primarily a product of the reprocessing of SF and the use of plutonium in the fabrication of nuclear weapons. In the US, the disposal of TRU at the Waste Isolation Pilot Plant in southeastern New Mexico is regulated by 40 CFR Part 194 (US EPA, 1996) [...]."
Uranium mill tailings "are large volumes of radioactive residues that result from the processing of uranium ore. In the US, the DOE has the responsibility for remediating mill tailing surface sites and associated groundwater under the Uranium Mill Tailings Radiation Control Act (UMTRCA) of 1978 and its modification in 1988."
Low level wastes **(LLWs)** "are radioactive wastes not classified as HLW, TRU, SF, or uranium mill tailings. They are generated by institutions and facilities using radioactive materials and may include lab waste, towels, and lab coats contaminated during normal operations."

Source: M.D. Siegel, and C.R. Bryan, 2005. "Environmental Geochemistry of Radioactive Contamination." Sandia National Laboratories, Albuquerque, NM, USA. Chapter 9.06, Book publication: H.D. Holland & K.K. Turekian "Environmental Geochemistry" edited by Barbara Sherwood Lollar (2005). Elsevier Ltd. Amsterdam, The Netherlands. (pp. 211–212).

CHAPTER SEVEN

7.1 Preferential Ecosystems

Fundamentally, the ecological preferential domains have been re-built through local monitoring activities on land-use change and on land-soils geological composition. Therefore, the practice of introducing environmental economic policies that implement technological production targets has been applied according to the national or regional dispositions for public ecosystems' management.[1] In case of comparison about the constitution of multiple ecological dispositions, the combined territorial local mechanisms can be based on information systems coordination, data diffusion analysis, and preferential legal statements, also interrelated with other issues such as individual property and land rights; environmental controls validity; or pollution and monitoring tools; it can be consequentially questioned the similarity of local emerging processes about comparative ecosystems' identifications classified by specific development cooperating agencies setting common organizational objectives.

Environmental regulatory assigned agencies have also established the structural integration programs for diversified national energy systems with functional distributive levels of collaborative networking capacities. Moreover, according to defined energy policy initiatives undertaken by single regions, identifying transnational managerial accountability, which has been classified through the programmatic activities of responsible organizing institutions, establishing either direct or indirect technical restrictions, according to the distribution of environmental local control risks and local design agreements, is now a need that needs to be addressed. As result, the cohesive formation of land-changing policies has also caused potential long-term implications from local ecological planning schemes.[2]

In essence, for the case of long-term public monitoring agreements that have been renewed for the regulation of industrial nuclear wastes stored at local disposal facilities, the functional local ecosystems have responded with the

---

1 P. J. Loveland, P.H. Bellamy, "Environmental Monitoring," *Reference Module in Earth Systems and Environmental Sciences* 2005, pp. 441–448, from Elsevier Encyclopedia of Soils in the Environment.
2 See: Virginia H. Dale, Rebecca A. Efroymson, Keith L. Kline, "The Land Use-Climate Change-Energy Nexus," *Landscape Ecology* 26.6/2011, pp. 755–773.

introduction of diversified levels of combined energy uses, which have been developed under uncertain conditions; it has been pointed out that "the distinctive nature and potential hazard make nuclear wastes not only the most dangerous waste ever created by humanity, but also one of the most controversial and regulated with respect to disposal."[3]

In terms of more pressing issues, according to an over-imposition of systematic local environmental requirements, there are two major determinants that need to be considered: 1. the material process of local waste storage facilities and 2. the nuclear power plants' operations, which have included the PPP with multilevel integration of systematic conditions about the environmental radioactivity pollution impacts. National protection and safety protocols in turn have relied on both the licensing agreements and categorization of processing functions, which have been structured according to different types of renewable energy production systems. Where the nuclear energy waste has been defined according to the following applicable categories:

1. *High-level waste (HLW) and industrial reprocessing of spent nuclear fuel;*
2. *Transuranic waste (TRU) and nuclear weapons;*
3. *Low-level waste (LLW) and lower radioactivity;*
4. *Uranium Mill Tailings and groundwater;*
5. *Mixed waste with hazardous chemicals and radioactive waste.*

Source: Gee, Meyer, and Ward, 2005. "Nuclear Waste Disposal," pp. 56–63. p. 57.(See Table 5 for NW definition categories)

On an actual basis, these categories about nuclear waste classification criteria have been set for the progressive implementation of regional states' policy measures combined with regulative schemes operating for the national and international technical energy cooperation, which has involved responsive local planning adaptation including environmental co-management directives.

---

3   G. W. Gee, P. D. Meyer, and A.L. Ward, "Nuclear Waste Disposal," *Reference Module in Earth Systems and Environmental Sciences* 2005, pp. 56–63. P. 56, from Encyclopedia of Soils in the Environment, Elsevier.

In fact, the case of multi-level local adaptation process has basically involved the presence of inter-governmental technical interfacing agencies interacting in relation with standardized mutual compliance terms and effective local approving criteria.[4]

In a similar way, the gradual advancement of environmental policy and regional energy management systems that refer to the national distribution of supportive renewable innovation resources, has been re-arranged in association with the local safety proceedings of e.g. the nuclear power plants NPPs.[5] Across time, an incremental need to build up national programmatic accounts that surround the consistent organizational levels of peoples' managerial interactions has developed to enable the promotion of multiple regulative innovation processes that should bind together national, but also international counterparts for the concretization of public commitments about the diffusion of compatible legitimacy roles and local administration measures.

The overall purpose has been to bring in view the mutual organizational changes interconnected with local regulatory spheres up to a level of national networking conditions, for the facilitation of energy innovation industrial programmes. Concerted planning has led to the affirmation of structural management linkages functional to the local coordinating groups, which have been in charge of interrelated socio-economic innovation domains. The development of comparative managerial assessments and managerial knowledge areas has integrated the effective renewable energy compliance issues, including local collaborative dynamics, which can facilitate the sharing ability of responsible territorial agencies, also through the material convergence of common environmental operating activities.[6]

For instance, in the case of north-west areas of the Russian Federation, as Honneland's and Jorgensen's study reports,[7] that there are some emerging

---

4 See: The Federal Register Report, "Licenses, Certifications, and Approvals for Nuclear Power Plants," Find72, 166, Office of New Reactors, U.S. Nuclear Regulatory Commission, Washington D.C., U.S, 2007.
5 Geir Honneland, Anne-Kristin Jorgensen, "Implementing International Agreements in Russia: Lessons from Fisheries Management, Nuclear Safety and Air Pollution Control," Global Environmental Politics 3.1/2003 by the MIT (Massachusetts Institute of Technology). pp. 72–98.
6 Honneland, Jorgensen, "Implementing International Agreements in Russia," pp. 72–98.
7 Honneland, Jorgensen, "Implementing International Agreements in Russia," pp. 72–98.

practical hindrances about the institutional combination of different industrial production networks that operate in relation to the nuclear energy safety protocols, associated with environmental trading distributions. In this case, the creation of international collaborative agreements, for the national renewable energy sector has been followed through the institutional mediation activity oriented at the broader implementation of multi-level transnational, national, and local administrative agreements, which have been negotiated through the possible adoption of multi-year knowledge innovation programs, based on policy cooperation factors and local developmental approaches.[8]

This process has been translated into the over-imposition of different levels of public legitimacy norms and corresponding managerial regulations that have led to the systematic public-private configurations of PPP with dependent common interrelations. Cooperative institutional relations have either directly or indirectly been reaffirmed due to the cognitive specification of regional environmental programs, which have introduced procedural implementing statements, for example, regarding the Russian Federation about the: "declaration of Artic Military Environmental Co-operation (AMEC)[9] and Cooperative Threat Reduction (CTR)[10] programs [put] in connection with discarded nuclear submarines and storage of spent nuclear fuel and other nuclear wastes."[11]

For instance, a common coordination profile was initiated, with a major focus on local environmental safety and public protection activities; which have essentially been put in place through the combination of joint managerial characteristics functional for the local implementation programs regarding the alternative uses of national energy platforms relating with comparative cases of transnational and local regulatory assessments. In essence, the national operating performances and regional commercial transitions of multi-layered energy control systems have catalyzed attention because of the increasing regulatory nature of innovative modified systems, managed in accordance with programmatic

---

8 Honneland, Jorgensen, "Implementing International Agreements in Russia," pp. 72–98.
9 Defence International Environmental Program archive (AMEC), available at: https://www.denix.osd.mil/international/archives/amec/
10 (CTR) Directorate Agency, available at: https://www.dtra.mil/Mission/Mission-Directorates/Cooperative-Threat-Reduction/
11 Honneland, Jorgensen, "Implementing International Agreements in Russia," pp. 72–98. p. 80.

changing shifts, which have been supported across national and local territorial boundaries.[12]

In fact, common critical notions about the effectiveness levels of local responsive environments and interdependent administration networks, have been defined according to mutual public-private spheres, including the local community needs. For the public energy domains, we may emphasize the complex interdependent relationship established between the public-private implementing organizations, which have brought into place concerted protection mechanisms that need to be set in order to support the renewable energy planning projects. Similar PPP designs have been implemented through local assessment dynamics with comparative mutual orientations concerning local environmental disputes and practical contributions that have been provided for local conflicts' resolutions.

Essentially, the public regulatory environmental schemes based on concerted renewable energy transitions have involved local, as well as, national, and transnational organizations deploying representative groups, which have relied on deterministic common elements about the provision of a shared ground for public level cooperation and local environmental acceptance issues. As a result, cognitive reforming elements have been included, as they emerged due to the recurrent patterns concerning local interest groups, which have voiced their integration to national or transnational environmental bodies, already shaped within local administrative frameworks. As result, public local exchanges and mutual environmental dialogues have also created complex levels of fragmented regulatory interactions within respective territorial dimensions.[13]

Analogously then, we can underline that the activity of policy-making stretched between the defining institutional cognitive boundaries has been marked by distinctive cooperative adaptation guidelines, which have been run across the governmental regulatory bodies through their activities, and the national local or individual groups, already engaged for the communal expression of local interests, which tend to polarize on the effective management of natural environmental resources.[14] Regarding the common spatial policy

---

12 See: Alexis Jonathan Conrad, (1999). "Assessing the Adequacy of Intergovernmental Collaboration as an Organizing Principle for Environmental Protection: A Case Study of the Canada-Wide Accord on Environmental Harmonization's Environmental Standards Sub-Agreement," Master of Arts, Thesis submitted at the Department of Political Studies, Queen's University, Kingston, Ontario, Canada.
13 S. Hayden Lesbirel, "Project Siting and the Concept of Community," *Environmental Politics* 20.6/2011, pp. 826–842.
14 Lesbirel, "Project Siting," pp. 826–842.

dimension, cooperative development projects focused on nuclear waste disposal programs involving civilian-military repositories have been embedded with the resulting socio-economic innovation patterns established across democratic or hierarchical transitional systems. The local institutional interactions, therefore, can be interpreted in relationship with the promotion of local communitarian approaches, which have expanded beyond the domain of local environmental interests, depending on respective states' reformed integration measures that are identified for regional development agreements.

7.1.1 Local Adoption Agreements

The comparative formal knowledge acquired from the characterization of local energy transition systems has determined a wider diffusion about the organizational processing stages, which have evolved in relation to communitarian dynamics and public preferential choices. Essentially, parallel theoretical investigations have specifically emerged about the local dynamic process that can reflect both operable trading policies, and sectoral adaptation of technological production systems. Another significant issue appearing in recent discussion is the normative identification of collective regulation patterns, which can better specify the changing understanding that has been involved for practical evaluations of local environment conditions. In addition, the actual identification of societal interest groups which operate within states' organizational trading framework, for the inclusion of common knowledge platforms, while facilitating public local exchanges, has been aimed at the diversification of sectoral integrated prospects in regard to institutional developed requirements.

In terms of public managerial policies that involve the public actors' knowledge exchanges associated with the regulative as well as delivery capacity spheres, the particular programmatic implementation related to regional ecological ecosystems have been commonly shared or inductively reduced, which presents local adaptation issues of the territorial planning measures.[15] The often referred to as 'ecological divide' between local citizens having collective responsibilities and centralized knowledge-based actors that operate in view of regional programmatic duties, has been considered in reference to pre-regulatory cyclical

---

15 See: 1) Robert Deyle E., "Conflict, Uncertainty, and the Role of Planning and Analysis in Public Policy Innovation" *Policy Studies* Journal 22.3/1994, p. 457; 2) Ismar Borges de Lima, Leszek Buszynski, "Local Environmental Governance, Public Policies and Deforestation in Amazonia," *Management of Environmental Quality: An International Journal* 22.3/2011, pp. 292–316. Emerald Group Publishing Limited.

conditions of risk-taking innovation societies. Industrial collectivities have been in constant transformation through the implementation of public-private consumption patterns, leading to the transnational production of proximity innovation and adaptation models.

Consequentially, we can assume that the possible identification made about public-private consumption behaviors associated to local motivational elements has affected as well technological semi-integrated supportive environments. So that, the actual possibility of maintaining cultural collective memberships, that also refer to science and technology spheres, has therefore fostered the domestic prevalence of public-private networked production domains. Through which, at government-level, a progressive normative expression of coherent common identifications was included together with the involvement of regulatory coordinating entities, which have taken charge of technology distribution models developed on a national territorial dimension.[16]

In fact, through the formal aggregation of public networked functions of performing agencies, it is clear that collective distribution choices set at the micro-economic and macro-economic level while influencing the participatory promotion accesses need to be formed. This composite promotion has involved the cultural commonalities and the material intentions for the convergence of practical organizational steps, which have been undertaken in view of participative regulation processes, oriented in favor of the local implementation of administrative innovation areas that have been politically engaged through multi-year regional investment programs.

Under such conditions, we can recognize the prevalence of inherent levels of societal assimilation through structural mobilized capacities seen in reference to the natural environment resources, and especially to the local environment protection and local conservation aims, which have been pursued through cumulative multi-stages of national/regional collaborative projects.[17] However, the political discourse developed locally about the materialization of public environmental engagements has either directly or indirectly led to communitarian planning logics, while also showing a propensity for the inclusion of local grassroots members with different ideological backgrounds. The affirmation of local grassroots initiatives has introduced additional cognitive determinants about the

---

16 Emily Huddart Kennedy, "Rethinking Ecological Citizenship: the Role of Neighbourhood Networks in Cultural Change," Environmental Politics 20.6/2011, pp. 843–860.
17 Huddart Kennedy, "Rethinking Ecological Citizenship," pp. 843–860.

fragmentation of multilateral environment distributive tendencies, that also correspond to social internal dilemmas.

In other words, the comprehensive inclusion of specified societal determinants about the rational local concerns that evolve according to the availability of natural environment resources, which have been associated with aggregative distribution schemes, has become interlinked with the actual prospective interpretations based on comparative regulatory technological domains. In sociological terms, the systematic provision of natural material resources has invited reflections toward a progressive environmental dynamism for the facilitation of local political decisions and environmental local activism. Through which, it can be supported a comparative societal and ecological mutual consensus, which may provide the interaction of institutional transnational organizations, involving public participatory capacities and domestic transferring tools.[18]

7.1.2 Formal Ethic Aggregations

Essentially, calling for formal aggregated solutions of transnational industrial societies has been analyzed in relation to the regional political transfer of legitimacy components that have been calibrated in terms of local accessibility and local recognition. As a matter of fact, the ecological integration route defined in view of modernizing approaches has been central to modified institutional attempts for the development of aggregative production patterns, focused on science and technology programs, among other investment areas.

From the social community perspective, an ethical investment involving the actual diffusion of common ethical relation activities and responsible local conducts has entailed the local sharing of legitimate societal norms with the integration of cultural identities.[19] Nonetheless, the conventional understanding that has emerged about the inclusion of public environmental exchanges promoted through local participation in governance with associated transparent identified measures, has also been the subject of criticism because of the perception of local vulnerability influencing national cooperative responses.[20]

---

18 Huddart Kennedy, "Rethinking Ecological Citizenship," pp. 843–860.
19 See: Benjamin J. Richardson, "Keeping Ethical Investment Ethical: Regulatory Issues for Investing for Sustainability," *Journal of Business Ethics* 87/2009, pp. 555–572.
20 See: 1) Marci R. Culley, Joeph Hughey, "Power and Public Participation in a Hazardous Waste Dispute: A Community Case Study," *American Journal of Community Psychology* 41.1-2/2008, pp. 99–114; 2) Fitzmaurice Malgosia, "Public Participation in the North American Agreement on Environmental Cooperation," *The International and Comparative Law Quarterly* 52.2/2003, pp. 333–368.

Comparative reduction should of public local accesses with less environmental viability options should, similarly, also be taken into account, particularly in regard to rural and local communities, as it has often been associated with a concrete coping inability connected to an increasing stratification of public environment duties intertwined with local participation levels. In fact, in the case of comparative industrial innovation systems with intra-governmental operating bodies, the legitimate collaborative directives have not been immediately able to align functional adaptation planning measures related to the local environmental changes and climatic modifying events of comparative neighboring regions, despite the adoption of public mitigation plans and local resilience strategies.[21]

For the case study about Australia, the national water distribution schemes have been a challenging regulatory ground because of the increasing levels of climatic local areas affected with drought and co-related environment issues, where regional local communities, regional economic actors, and federal technical agencies, have been present in terms of organizational consumption patterns and local farming policies, also by providing a corresponding support on PPP public-private irrigation schemes and federal territorial distribution plans, despite controversial PPP local impacts.[22-23]

Under similar modifying conditions, the search for a public-private compromise in reference to the national modelling environments due to drastic weather conditions as in Australia's[24] case, has involved the functional development of local legitimacy mechanisms, technically operated across national socio-economic assets and local distribution systems. As example, the regulatory concentration of national irrigation water policies combined with local distribution plans that favor water license programs, as well as, decentralized

---

21 Margaret Alston, Kerri Whittenbury, "Climate Change and Water Policy in Australia's Irrigation Areas: A Lost Opportunity for a Partnership Model of Governance," *Environmental Politics* 20.6/2011, pp. 899–917.
22 Gabrielle Chan, "We Have Lost Control:' NSW Farmers Battle Private Irrigation Companies for Water," 2019. The Guardian Australia Edition, available at: https://www.theguardian.com/australia-news/2019/apr/30/we-have-lost-control-nsw-farmers-battle-private-irrigation-companies-for-water
23 Alston, Whittenbury, "Climate Change and Water Policy," pp. 899–917.
24 See: 1) Thomas Jenkin, "Exploitation or 'Wise Use' of the Coongie Lakes, South Australia: Issues Arising from a Petroleum Exploration Proposal," *Australian Geographer*, 30.3/1999, pp. 355–371; 2) Anna Lukasiewicz, Geoffrey J. Syme, Kathleen H. Bowmer, Penny Davidson, "Is the Environment Getting Its Fair Share? An Analysis of the Australian Water Reform Process Using a Social Justice Framework," *Social Justice Research* 26.3/2013, pp. 231–252.

managerial accesses, has been critically put into question by sectoral cooperating partners, government members, national union farmers, and private industrial enterprises. In fact, because of local uncertain conditions about sustainable converging measures promoted for water distribution regulative goals, there has been an additional lack of cohesion leading to a decisional fragmentation among main territorial stakeholders, who have been engaged in co-management environmental projects.[25]

As a matter of fact, respective institutional steps and environmental impact policies have been implemented through national engagements for the targeted local communities. However, national cooperation efforts have also determined a multi-layered framework of decision-making initiatives, which have tended to marginalize progressive environmental debates for comparative understanding, related to limited levels of local interactions and social integration, put in relation with environmental public choices,[26] that nonetheless have been identified for sustainable local plans and long-term recovery economic partnerships.[27]

Essentially, the critical roles adopted by institutional responsive bodies associated with local management programs and local governance policies has become evident, though it remains comparatively underestimated in terms of producing a regional impact through local environmental regulations. Such practical regulative approaches have basically been affected by comparative trends of social communication patterns and local participative aggregations, which tend to form civilian protests' groups and social environmental movements, which have acted for instance in eastern and western European countries.

Moreover, depending on the actual level of local institutional responsiveness to rapid environmental and developmental changes, concerted civil parties and public interests' groups, set in western and eastern European states[28] have put in place the so-called local green parties adhering to transnational

---

25 Alston, Whittenbury, "Climate Change and Water Policy," pp. 899–917.
26 See: Ade Peace, "Anatomy of a Blockade. Towards an Ethnography of Environmental Dispute (Part 2), Rural New South Wales 1996," *The Australian Journal of Anthropology* 10.2/1999, pp. 144–162.
27 Alston, Whittenbury, "Climate Change and Water Policy," pp. 899–917.
28 See: 1) Ana Iglesias, Sonia Quiroga, Marta Moneo, Luis Garrote, "From Climate Change Impacts to the Development of Adaptation Strategies: Challenges for Agriculture in Europe," *Climatic Change* 112.1/2012, pp. 143–168; 2) Juan Casado-Asensio, Reinhard Steurer, "Integrated Strategies on Sustainable Development, Climate Change Mitigation and Adaptation in Western Europe: Communication Rather Than Coordination," *Journal of Public Policy* 34.3/2014, pp. 437–473.

changing societies. The multilateral industrial transformation of newly developed regions has also given the possibility of establishing public communication networks combined with local information activities. Through which, the protection of national and transnational socio-economic industrial interests has been confronted with public environmental practices and local participatory dialogues that can provide common transparent explanations, mainly concerning the right to assess local information transmission and local verification processes – which may address socio-ecological changes and public resolutions reducing socio-environmental concerns.[29]

The prevailing regulatory status in governance which can define some core aspects of the social environmental activism and public stakeholders' engagement has been intertwined in comparative structural dimensions that refer to local development strategies undertaken in regional joint processes, with legitimacy characteristics put already in place. Basically, the pursuance of regulatory collaborative programs that might directly impact on institutional economic policies, has been attained according to the presence of multi-level sectoral counterparts which can facilitate the actual flow of innovative industrial productions.

Such type of differentiation based on local action-led industrial management processes has been adapted according to national local parties and policy collaborative formations, which have been confronted with the managerial acceptance models exercised on theoretical and practical dimensions.[30] The practical evolution of socio-ecological distribution choices, including the local environmental provisions, has fundamentally been the result of competitive development organizations, which have maintained into focus the comparative socio-economic advantages.

In particular, there is the technological transnational availability of PPP in multi-level production systems that integrate the diffusion of local material opportunities, which can be translated into regional trading routes which can support international functional directives for the liberal and communitarian

---

29   Jon Burchell, Joanne Cook, "Banging on Open Doors? Stakeholder Dialogue and the Challenge of Business Engagement for UK NGOs," *Environmental Politics* 20.6/2011, pp. 918–937. p. 921.
30   See: 1) Christopher Rootes, "Climate Change, Environmental Activism, and Community Action in Britain," *Social Alternatives* 31.1/2012, pp. 24–28; 2) Pablo del Rio, Xavier Labandeira, "Barriers to the Introduction of Market-Based Instruments in Climate Policies: An Integrated Theoretical Framework," *Environmental Economics and Policy Studies* 10/2009, pp. 41–68.

mixed economies.³¹ In the case of European Union regional trading agreements, the practical affirmation of social innovation perspectives about the industrial environmental performances has been discussed in association with the increase of industrial economic exchanges, which have brought to the use of additional evaluation and monitoring E&M tools for the political transition, favoring a particular local cohesion for the constitution of green parties' regional agendas. These green parties' globalist agendas have already been formed during the 1980s and 1990s, through the formal and informal establishment of environmental policy discussion forums having also spread on a transnational basis.³²

From a socio-ecological perspective, transnational political strategies have reflected, on the one hand, the logic incremental direction of industrial production systems. While on the other, social protection measures and local environmental practices were constructively included and have substantially led to public local engagements of regional business organizations, civilian environmental parties, and institutional regulatory actors involved; in fact, it has been pointed out that "for both business and NGOs there is undoubtedly greater pressure to go beyond the traditional war of words and actually work towards practical change."³³

In effective ways, when thinking about the direct integration of social environmental spheres with regional business partners and societal stakeholders for local environmental corresponding relations, the convergent relational dynamics have been usually constructed through respective functional cognitive stages. Public environmental frameworks have essentially been fragmented between local practical aims and national valid norms, of democratic or hierarchical modifying regimes which have granted a corresponding legislative dimension. In this case, the social appropriation of multi-sectoral environmental programs has been associated with comparative local knowledge exchanges interlinked with systematic regional innovation markets proposing technological commercial prospects, which can determine local evolution dynamics.³⁴

---

31 Burchell, Cook, "Banging on Open Doors?" p. 918–937.
32 See: Anton Ming-Zhi Gao, "Development of a Legal Framework for Climate Change in Taiwan: Lessons from Europe and Germany," *Carbon & Climate Law Review: CCLR* 7.1/2013, pp. 54–70.
33 Burchell, Cook, "Banging on Open Doors?" pp. 918–937. p. 932.
34 See: Shobita Parthasarathy, "Whose Knowledge? What Values? The Comparative Politics of Patenting Life Forms in the United States and Europe," *Policy Sciences* 44.3/2011, pp. 267–288.

CHAPTER EIGHT

8.1 Sectoral Participation

In connection to the nuclear industry, development scholars have observed that the participatory convergence has been displayed by the national, as well as, international governmental advocates, who have been engaged with the managerial development of nuclear power facilities and national energy supporting programs as in France collaborative case.[1] However, any other particular type of state's industrial distributed configuration about the national nuclear energy policies, has been confronted with socio-economic adaptation approaches taking into consideration: 1. the continuous local organizational changes, and 2. the different combination of multi-sectoral nuclear science projects and technology innovation developments.[2]

As consequence, the local governmental availability for the structural monitoring process of national energy industrial assets has required the progressive inclusion of standardized regulatory measures which can support determinant participatory procedures about the local safety and security practices adopted at NPPs in regional sites.[3] The practical assessment and local monitoring process, in addition, have been regulated according to the emergence of critical environmental factors, concerning local normative areas and local organizational domains of national and international l political landscapes.

In terms of industrial environmental regulation, the major representative players involved into the sectoral affirmation of local energy production strategies have defined common policy and planning actions across different time periods, in order to configure regional ecological schemes, set in line with local environmental trajectories. At the same time, the industrial inter-operable and

---

1   See: Sarah Elise Wiliarty, "Nuclear Power in Germany and France," *Polity* 45.2/2013, pp. 281–296.
2   Joseph Szarka, "From Exception to Norm – and Back Again? France the Nuclear Revival, and the Post-Fukushima Landscape," *Environmental Politics* 22.4/2013, pp. 646–663.
3   See: 1) Fabienne Gralla, David J. Abson, Anders P. Moller, Daniel J. Lang, Ulli Vilsmaier, Benjamin K. Sovacool, Hnrik von Wehrden, "Nuclear Accidents Call for Transdisciplinary Nuclear Energy Research," *Sustainability Science* 10.1/2015, pp. 179–183; 2) Atsuyuki Suzuki, "Toward a Robust Nuclear Management System," *Daedalus* 139.1/2010, pp. 82–92.

inter-exchangeable technological programs have been concretized through the formation of transnational nuclear energy networks with national/local clusters, transitioning into comparative operable settings managed for international business transfers, while safeguarding the protection of common adaptive ecosystems.[4]

Nonetheless, it is important to stress the fact that the diversity aspect of industrial energy cooperation patterns reaffirmed regionally has influenced respective managerial performances of interchanging production models, which have had particular policy implications both nationally and locally, but at the same time the affirmation of ecological and industrial implementation programs has been based on provisional local energy sectors' activities.[5] At the end of the day, we can refer to the combination of complex regulative initiatives which have been interrelated to the industrial cross-cutting energy areas and local innovation approaches, experienced in common local economies.

In fact, the governmental energy policies have been dependent on distinctive affirmative dynamics, which have emerged from local practical contributions of corresponding regulatory institutions, as well as, regional labor environmental representations, national science innovation agencies, and public information providers, which tend to converge on interdependent regional collaboration activities for a material cohesiveness of the local organizational practices.[6] In fact, public planning policies have reflected a closer affiliation to multiple intergovernmental interests identified within decentralized networking spheres of territorial stakeholders, who have supported environmental development initiatives pursued in association with 1 mutual cognitive designs and progressive decisional criteria, that are also established in view of promoting ecological regulative norms.

Regarding the public policy cooperation activities related to local environmental planning, and industrial management programs, regional policy studies provided by Davoudi, 2009,[7] Davies, 2009,[8] and Wolsink and Devilee,

---

4 See: Daniel Nohrstedt, "The Politics of Crisis Policymaking: Chernobyl and Swedish Nuclear Energy Policy," *Policy Studies Journal* 36.2/2008, pp. 257–278.
5 Pauline Deutz, and David Gibbs, "Industrial Ecology and Regional Development: Eco-Industrial Development as Cluster Policy," *Regional Studies* 42.10/2008, pp. 1313–1328.
6 Deutz, and Gibbs, "Industrial Ecology," pp. 1313–1328.
7 Simin Davoudi, "Governing Waste: Introduction to the Special Issue," *Journal of Environmental Planning and Management* 52.2/2009, pp. 131–136.
8 Anna R. Davies, "Clean and Green? A Governance Analysis of Waste Management in New Zealand," *Journal of Environmental Planning and Management* 52.2/2009, pp. 157–176.

Sectoral Participation 151

2009,[9] have interestingly reported on cumulative issues that have been explored in reference to the economic and industrial *contingent green transitions* performing under common levels of local accountability, for the public-private stakeholders involved, also in the case of nuclear waste management with innovating industrial practices, which have been interlinked to a sectoral differentiation of regional production technologies.

Importantly, at the national and sub-national level, the introduction of governmental legal reforms as in the New Zealand's case[10] has been modified according to the environmental resources' distribution plans and local management policies, which have determined organizational changing implications. The provision of local adaptive plans has been drafted both in content and scopes according to territorial regulatory policies, which have been arranged in line with the legal operational background of the country involved; due to concurrent managerial changes, district-level environmental waste disposal programs have also been implemented locally, as in the case of New Zealand.

However, despite legislative and organizational adaptations, the development of public policies targeting municipal waste programs, for instance, has not been fully planned in a comprehensive manner, because of national planning restrictions and local reduction of management activities, which have been programmed according to multilateral regulatory patterns and local organizing settings.[11] Moreover, the levels of access to the land planning policies that are diversified in terms of public cognitive approaches have also created fragmented structural programs associated to local managerial preferences for the distribution of municipal waste policies.

In return, the local communities have been eventually aggregated into civilian opposition fronts in order to choose in particular about the maintenance of public arranged programs for the urban waste treatment sites, with the inclusion of collective environmental dispositions, which favor the local application of social justice principles. Public local conducts have also been based on the recognition

---

9 Maarten Wolsink, and Jeroen Devilee, "The Motives for Accepting or Rejecting Waste Infrastructure Facilities. Shifting the Focus from the Planners' Perspective to Fairness and Community Commitment," *Journal of Environmental Planning and Management* 52.2/2009, pp. 217–236.
10 See: Marcel Eusterfeldhaus, Barry Barton, "Energy Efficiency: A Comparative Analysis of the New Zealand Legal Framework," *Journal of Energy and Natural Resources Law* 29.4/2011, pp. 431–470.
11 Davies, "Clean and Green?" pp. 157–176.

of common emerging policies that can support the participatory interactions of communities involved for the protection of collective environments.[12]

8.1.1 Public Stratification

The regional public proceedings drafted in accordance to collective environmental initiatives, and local stratified programs, have led to the operability of local knowledge transfers, which do not necessarily capture the distinctive development needs of local communities engaged within regulatory constitutional schemes.

Through constructive reviews, the social justice norms and environmental relations associated with the issues of nuclear waste disposal facilities have been observed through case-based analytic interpretations[13] which have been affirmed according to the renewal of dichotomous processes that have incorporated inductive elements of industrial technological organizations.[14] However, there are structural difficulties about industrial innovative management outcomes that have been processed according to regulatory adaptation models, which develop in analogy with democratic governance paths.

For instance, in North-eastern EU countries such as in Sweden and Czech Republic, political environmental relationships have developed between institutional cooperating members and public networks of local citizens' groups, resulting in common collaborative mechanisms interrelated to respective decision-making systems, which have created a confidence ground, and social trust for the combined relations.[15] Supposedly, the public participatory policy introduced in national decision-making cycles has been reaffirmed in view of realizing societal organizational aims by forming social trusting conditions. At the same time, the disseminated information activities in turn have been

---

12   Wolsink, and Devilee, 2009. "The Motives for Accepting or Rejecting Waste," pp. 217–236.
13   See: 1) Pius Krutli, Michael Stauffacher, Dario Pedolin, Corinne Moser, Roland W. Scholz, "The Process Matters: Fairness in Repository Siting for Nuclear Waste," *Social Justice Research* 25.1/2012, pp. 79–101; 2) Luther J. Carter, Thomas H. Pigford, "Confronting the Paradox in Plutonium Policies," *Issues in Science and Technology* 16.2/2000, pp. 29–36.
14   See: Grace Mcglynn, Gregg Butler, Alan Pearman, "Stakeholder Preference Mapping—Seeking a Way Forward for the Processing of Spent Nuclear Fuel," *The Journal of the Operational Research Society* 66.2/2015, pp. 219–230.
15   See: Martin Dusinberre, Daniel P. Aldrich, "Hatoko Comes Home: Civil Society and Nuclear Power in Japan." *The Journal of Asian Studies* 70.3/2011, pp. 683–705.

independently aggregated to the political formation processes established according to respective levels of local distribution powers.[16]

Fundamentally, there are a number of different theoretical and empirical findings about the societal empowerment increased through social trusting conditions. Multiple social environmental conditions have favored the institutional objective orientation of local knowledge diffusion, that comprises the national and local representation assemblies, which have been in charge of regulating the variability of common social environmental settings, still interrelated with distinctive local expanding relationships.

Regarding the cases of Czech Republic's post-socialist background and Sweden's democratic institutional consolidation, we can take into account a series of historical determinants, which have offered the distinctive approaches undertaken about national substitution paths of ascribed societies, that may assume similar characters of neighboring countries. In fact, there have been historical dependency paths of semi-democratic models, which have been established according to pre-existent political conditions of national territorial frameworks. Structural frameworks at country level have been actively marked by the geographical specificities of transnational power diffusion – especially when compared to neighboring areas shared with satellite post-socialist republics, such as the former Soviet Union alliances.[17]

For these reasons, the determining distribution factors regarding social acceptance models have reflected the democratic escalation of local participatory responses formulated in common organizational terms, involving at the same time the national disposal of nuclear waste activities as in Sweden's case.[18] This material waste disposal in NPP has depended on the reliability of pre-existing local planning policies run by correspondent national institutions. So that, the contribution of reliable environmental policies has followed through the local procedural understanding of the decision-making process, which has been retraced within the common democratic context, fostering among other participatory elements: the common trust, mutual transparency and accountability norms, evolving in both directions: toward the local participation, and toward the integration of people acting together for the institutional delivery of

---

16 Jane I. Dawson, and Robert G. Darst, "Meeting the Challenge of Permanent Nuclear Waste Disposal in an Expanding Europe: Transparency, Trust and Democracy," *Environmental Politics* 15.4/2006, pp. 610–627.
17 Dawson, Darst, "Meeting the Challenge," pp. 610–627.
18 See: Lennart Sjoberg, "Antagonism, Trust and Perceived Risk," *Risk Management* 10.1/2008, pp. 32–55. Palgrave Macmillan Ltd.

collaborative governance activities, also related with the renewable energy maintenance programs, which have been designed and promoted at country level.[19]

This double direction taken on common public policy relations, whether or not reciprocated in mutual democratic terms set in other countries, has also been established among the national representative counterparts, who have developed relational diplomatic conditions emerging, for instance, between the Western and Eastern European states due to territorial dividing lines of regional trading zones.[20] [The 2022 Ukraine's invasion perpetrated by the government of the Russian Regions has added uncertainty and public alliances' shifts to regional economic exchanges influenced by open conflicts and realpolitik logics]. The public societal conducts associated with relational territorial recognitions in regional politics have essentially been marked by serious challenging dilemmas about, for example, the long-term security management issues on permanent industrial disposal of NPP nuclear hazardous waste.

Concerning the relational constitution of larger regional communities' interests, we can stress the fact that public environmental organizations, as well as, local civilian advocacy groups[21] have been embarked on cultural ideological transitions impacting on the national policy orientations, which have comparatively emerged in multilateral regulatory areas, such as in the United States. The democratic collective representations for environmental protection programs that embody the national public interest have influenced the course of normative legal actions, in particular, about local renewable energy development programs and structural containment policies.[22]

In essence, the public unionization of programmatic local environmental activities has remained grounded on disputed cultural and ideological connections. So far, the possible distribution effects for representative local communities, have retrospectively increased local contrasting views and local practical orientations about the normative adaptation on multi-level environmental governance that involves local integrative affirmations, while determining the

---

19   Dawson, Darst, "Meeting the Challenge," pp. 610–627.
20   See: David Turnock, "Environmental Problems and Policies in East Central Europe: A Changing Agenda," *GeoJournal* 54/2001, pp. 485–505.
21   See: David Bosold, Wilfried von Bredow, "Human Security," *International Journal* 61.4/2006, pp. 829–855.
22   Matt Grossmann, "Environmental Advocacy in Washington: A Comparison With Other Interest Groups," *Environmental Politics* 15.4/2006, pp. 628–638.

course of organizational advocacy programs.[23] The promotion of national environmental programs at community level can be redefined through the degree of effectiveness of socio-economic dynamics tracing a possible diversification of national industrial policies and environmental anti-nuclear campaigning targets.

In some ways, it can be specified that current participatory implications emerging within democratic governance systems have been systematically linked to progressive local knowledge dissemination factors, which are also related with well-defined national ideological orientations. Essentially, the recomposition of multi-level regulative governing assets[24] can be brought to the point of mutual public affirmations and local independent discussions intertwined with common knowledge aggregative elements. In fact, the policy environmental recognitions may be introduced in view of maintaining an open diversity and social mediation, in order to facilitate the multilateral direction of national citizens environmentally committed to local territorial programs.

At the same time, the local societal aggregations have been aligned according to respective socio-economic determinants interrelated with national constitutional rights and institutional local responsibilities, consolidating across different collaborative spheres, which have also been connected with the local environmental dimension. Practically, the political established notions about local citizens providing a direct involvement through civilian environmental participation, netted across different time-periods, have implied the convergence of local institutional assets that are responsive to the reconstruction of collaborative democratic models.

In fact, the promotion of local institutional dialogues has been able either to favor or to reduce fundamental gaps about the characterization of participatory adaptive environments. The attempted resolutions formulated in order to enhance the local civilian dialogues have tried to envision comparative participatory mechanisms, in order to be able to recreate a level of mutual regulatory interconnectedness, through the exploration of territorial innovation policies

---

23 See: 1) Brian A. Ellison, "Intergovernmental Relations and the Advocacy Coalition Framework: The Operation of Federalism in Denver Water Politics." *Publius* 28.4/1998, pp. 35–54; 2) Bryan C. Taylor, Brian Freer, "Containing the Nuclear Past. The Politics of History and Heritage at the Hanford Plutonium Works," *Journal of Organizational Change Management* 15.6/2002, pp. 563–588.
24 See: Robert C. Lowry, "All Hazardous Waste Politics Is Local: Grass-roots Advocacy and Public Participation in Siting and Cleanup Decisions," *Policy Studies Journal* 26.4/1998, pp. 748–759.

addressing, for instance, the environmental management of land and of natural resources available for the collectivities involved.

However, there are dominant cultural models carrying out specific identity patterns and social progressive definitions that have been developed in association with the evolution of public exclusionary factors and societal diversity issues; which essentially have been framed within the dimension of public local organizations and national interconnected bureaucracies,[25] maintaining common relational functions. On these aspects, the mainstream local regulatory decisions, at government-level, have been conducted under the problematic contextual duality of close and open societies addressing the local knowledge exchanges.

In essence, the social work coordinated in single nations associated with the search of common environmental planned solutions, has been projected across time through the reflective lenses of social, economic, and environmental conditions,[26] which have led to a collection of different sustainable policies and technological innovation transitions. But the local sustainability issues associated with collective responsible aims at organizational level, have also produced an oversimplification of multiple territorial achievements, enhanced across the local communities involved. In fact, the systematic overlapping of comparative behavioral modes referring to local participatory governance has been targeted within the subsidiary complexity of democratic networks.[27]

Fundamentally, the emerging comparative capabilities developing within democratic conditions through environmental protection systems,[28] have been directly implemented according to respective degrees of effectiveness and efficiency criteria, which have involved spatial instrumental elements, particularly adapted to local decision-making processes, concurrently arranged. However, we need to consider that analytical recurrent modes about local civilian conflicts

---

25 See: Richard Price, "Transnational Civil Society and Advocacy in World Politics," *World Politics* 55.4/2003, pp. 579–606.
26 See: Angela C. Halfacre, Albert R. Matheny, Walter A. Rosenbaum, "Regulating Contested Local Hazards: Is Constructive Dialogue Possible Among Participants in Community Risk Management," *Policy Studies Journal* 28.3/2000, pp. 648–667.
27 Janet McIntyre, "Part 1: Working and Re-Working the Conceptual and Geographic Boundaries of Governance and International Relations," *Systemic Practice and Action Research* 18.2/2005, p. 173–220.
28 See: Robert F. Durant, "Sharpening a Knife Cleverly: Organizational Change, Policy Paradox, and the 'Weaponizing' of Administrative Reforms," *Public Administration Review* 68.2/2008, pp. 282–294.

formed around the direction of political protection strategies, have recalled attention on the sustainability of mutual coordination paths. Both the legislative process and transnational managerial assessments that have been functional in terms of regulatory environmental measures, have become integral components of local administrative mechanisms, establishing a fragmented sectoral process, which has been explored according to key determinants such as: social adaptability, contextual visibility, as well as, organic reliability.

At this stage of analysis, we can highlight that managerial structural identification in governance has been run in parallel with public networked distribution systems and environmental implementation processes, which all together have been influenced by public participation policies intertwined with actor-based organizational knowledge duties. In addition, civilian environmental organizations have expressed the constant need for changing public policy relations by favoring local environmental recognitions and local institutional engagements which involve the communities' legitimacy status.[29]

Nonetheless, the actualization of effective local civilian interactions at public policy level has been approached in terms of geographical boundaries and regional trading sites. Therefore, the affirmation of different socio-economic transitions has been characterized by newly-formed national political regimes incorporating temporary regulatory approaches, at the local level.[30]

Consequentially, comparative different degrees of public involvement for civilian participatory organizations, have been integrated through the actual forms of procedural administrative understanding.[31] Structural insights on procedural undertakings have also pointed at the potential for environmental conflict that has been identified in association with public cyclical decisions and local information strategies, which have reflected comparative local accesses to distributive business resources, associated to local management choices that are promoted in line with formal planning orientations.[32]

---

29   See: Herve Corvellec, Johan Hultman, "From 'Less Landfilling' to 'Wasting Less.' Societal Narratives, Socio-Materiality, and Organizations," *Journal of Organizational Change Management* 25.2/2012, pp. 297–314.
30   Alexandra Sauer, "Conflict Pattern Analysis: Preparing the Ground for Participation in Policy Implementation," *Systemic Practice and Action Research (SPAR)* 21/2008, pp. 497–515.
31   Patrick d'Aquino, "Empowerment and Participation: How Could the Wide Range of Social Effects of Participatory Approaches Be Better Elicited and Compared," *CIRAD (Centre de coopération internationale en recherche agronomique pour le développement)*, 2007, France.
32   Sauer, "Conflict Pattern Analysis," pp. 497–515.

In other words, the combination of technological business transitions, local transformation guidelines, and national development plans, involving societal stakeholders, has been operated in accordance with multilateral strategic trading interests, that also correspond to science advisory committees' orientations, due to the complexity of environmental governance issues, particularly about comparative local distribution logics, including public local representations that perform under uncertain conditions.[33] In essence, the growing practical expectations that have formed for ensuring common ethical paths on local adaptive legitimacy conditions, and local institutional provisions, have been put together in combination with public environmental networking programs.[34]

---

33 Sauer, "Conflict Pattern Analysis," pp. 497–515.
34 See: Shui-Yan Tang, Carlos Wing-Hung Lo, "The Political Economy of Service Organization Reform in China: An Institutional Choice Analysis," *Journal of Public Administration Research and Theory* 19.4/2009, pp. 731–767.

CHAPTER NINE

9.1 STS States Innovation Patterns

From a territorial perspective, the transmissible formation of public local policies has been extended according to the application and regulation of technological innovation areas' requirements, which have been integrated into common states' sectoral planning strategies that target science and society development programs. At the same time, local territorial regulations and practical proceedings expressed at national level have been operated in conjunction with local societal interests and regional programmatic aims, put into relation with the essential constitution of public decentralized interfaces and community-based receiving organizations able to maintain open the debating process about respective collaborative orientations.[1]

In which case, societal public attitudes about science and technology regulations that can modify territorial adaptive planning have presented controversial positions about the national acceptance models with local embedded relations. Because, the public innovation and technology drivers have indicated stratified connections among local states' distribution agencies, regulating across different political landscapes and functional domains. In order to be able to redefine the technology local programs in parallel with public ICT appropriation models, is a series of consensual regulatory directives that coordinate, for instance, the private-public enterprises and territorial communities, which have been engaged in the multi-level adaptation process, already shaped according to comparative regulative agendas of transnational state economies.[2]

As example, the technology innovation ground-based PPP rules for Eastern Asian regions, have been linked to the actual constitution of public regulative interests, which have indirectly required a type of analytical approach, functionally responsive to socio-economic evolution patterns. In addition, the local changing patterns about STS technology innovation domains have been introduced through the facilitation of networked governmental initiatives, which

---

1 See: Eran Vigoda, "From Responsiveness to Collaboration: Governance, Citizens, and the Next Generation of Public Administration," *Public Administration Review* 62.5/2002, pp. 527–540.
2 See: Wendy L. Currie, Matthew W. Guah, "Conflicting Institutional Logics: A National Programme for IT in the Organizational Field of Healthcare," *Journal of Information Technology* 22.3/2007, pp. 235–247.

offer stability for material occurrences within public and private productive and innovative dimensions.

In fact, through the formal association of statutory regulatory systems with legal measures and collective norms, the regional expansion of networked trading relations, put in line with science and technology transfer policies is sustained. Respective qualitative and quantitative degrees of transnational trading knowledge areas have also reflected a comparative mutuality and local communication projects, for the maintenance of monitoring and evaluation processes (M&E) established at the local government-level.[3]

Moreover, political and legal theoretical orientations have been reaffirmed according to regional states' regulatory dimensions, which have included the alternation of assigned operational roles with related exploratory functions, based on the reformation of STS contextual adherences, as well as on public acceptance choices that have been performed within local IT interests' areas, through systemic formation processes. The open characterization about IT alternative public choices has been taken into account according to respective societies' environmental and collaborative components, which emerge through the affirmation of cultural local recognitions and incremental organizational designs.

Overall, the material affirmation of multi-referential innovative associations has determined a progressive inclusion of specific normative expressions, either favoring or disfavoring public cohesive policies arranged in view of technological and scientific advancements, which have been performing in conjunction with local institutional and legal settings.[4]

The intertwined political relationships that have emerged within local clustered innovation models and public acceptance groups' approaches, have formalized the regional process of technological innovation activities, while developing incremental adaptation plans that have also caused local environmental concern. There have been into consideration the societal trusting issues put in relation with common regulatory patterns, which across the last four decades have influenced technological consumption models applied to the multi-level sectoral

---

3   Ching Ching Leong, Darryl Jarvis, Michael Howlett, Andrea Migone, "Controversial Science-Based Techonology Public Attitude Formation and Regulation in Comparative Perspective: The State Construction of Policy Alternatives in Asia," *Technology in Society* 33/2011, pp. 128–136.
4   Leong, Jarvis, Howlett, Migone, "Controversial Science-Based Techonology," pp. 128–136.

industries such as: renewable energy development projects, and the nanotechnology implementation programs, verified through multilateral diversification systems.[5]

At the same time, there are civilian mobilization campaigns focused on local environmental issues and public health concerns, emphasizing the case of safety risk perceptions exposed at regional and local level. For such reasons, the science regulatory path for national leading conducts and local organizational drivers of socio-economic programs, which have been planned, for instance, in the case of specific health innovation areas, involving technological implementation activities has remained unsettled.[6]

From the normative perspective, the transnational territorial linkages constructed through respective political communitarian leaderships engaged in environmental governance, have had at their base supportive societal information networks, reflecting the common cognitive orientations. But, there are concerning aspects of the evolution of local regulatory explorations, which have primarily conducted to the constitution of managerial innovation systems focused on: 1. science and technology involvement; 2. local environmental sustainability; 3. public direct participation networks; and 4. technical regulative incremental initiatives. In essence, public local awareness in science and technology evolution has fostered the affirmation of supporting normative rules, reinforcing indirectly public accepting attitudes, for long-term science knowledge affirmative plans and local cognitive transfers defined according to specific time-frames.

Essentially, the governmental coordination of structural STS determinants related to territorial industrial linkages has been operated through the involvement of normative transferring formulations. But in view of contributing to international collective expressions, there are some particular highlights about comparative regional knowledge systems, that have been implemented through societal reforming steps, associated with knowledge adaptation frameworks, which can lead toward the construction of technologically innovative components – while basically favoring public levels of common accessibility and local open opportunities.

---

5   See: Sylvester Douglas J., Abbott Kenneth Wayne, Marchant Gary E., "Not Again! Public Perception, Regulation, and Nanotechnology," *SSRN Working Paper Series* 09/ 2009. Social Science Research Network.
6   Leong, Jarvis, Howlett, Migone, "Controversial Science-Based Techonology," pp. 128–136.

However, the development of industrial innovation and technology opportunities has been interrelated with regional interdependent factors, which have been connected to the regional affirmation of managerial innovation plans and local implementation mechanisms, combined with national political representations, as well as, integrated industrial knowledge activities.[7] For the specific case of nuclear power energy, the public-private industrial implementation programs PPP, reflecting the availability of national monitoring frameworks, have been involved with the construction of nuclear waste disposal systems, and repository facilities' sites targeted for comparative local settings.

9.1.1 Public Analytic Settings

The comparative governance analysis elaborated in the case of modified technology innovation models, interrelated with comparative industrial energy areas, has been designed according to the need of improving regional industrial open settings, which have shown a propensity for local self-sustenance plans, and environmental aggregated actions for the countries involved.

Still at government-level, the local multi-hazards and anti-risks environmental preservation programs have been characterized by the national constituencies with political normative implications. Basically, it can be considered that the role of domestic regulatory institutions coordinating in cooperation with multi-layered influential bodies has been functional to distributed levels of local power assessments for public participatory dynamics. Essentially, environmental local expressions have been affirmed according to collective collaboration networks aligned with the succession of environmental socio-economic interests for the development of national energy policies and local safety planning.

The type of national regulatory interdependency about energy policy configurations has suggested an inherent need to establish specific responsibilities and course of actions, put in conjunction with local national capacities. The diversification of public local initiatives has been related to the involvement of collaborative national platforms, that have been managed in connection with the regional work conducted by the consultation regulatory bodies and environmental management proponents, which operate according to different sets of public participation tools[8] such as through the Canadian governmental

---

[7] See: Rostam J. Neuwirth, "'Novel Food' for Thought' on Law and Policymaking in the Global Creative Economy," European Journal of Law and Economics 37.1/2014, pp. 13–50.

[8] Catherine Alexander, Joshua O. Reno, "From Biopower to Energopolitics in England's Modern Waste Technology," Anthropological Quarterly 87.2/2014, pp. 335–358.

adaptive approaches; the UK's consultative public forums; or Sweden's progressive involvement of local communities.[9]

As a matter of fact, public choices made about environmental concerted efforts and affirmative regulatory relations among technical administrative agencies have been established in connection with the federal, national, and local governments,[10] which have been organized within multi-level environmental programmatic paths, when possible. In such terms, the evolving relationship for the nuclear waste management agencies and industrial corporation groups focusing on the maintenance process of nuclear power energy systems, has come, either directly or indirectly, to terms with national sectoral orientations, that integrate public responsiveness frameworks.

In fact, there are political environmental guidelines expressed for the application of regulatory nuclear energy plans that have commonly addressed the local NPP case of deep waste burial management of nuclear deposits, as these managerial innovating plans have required constant geological assessments that include the local monitoring activities of multilevel ecology projects, reaffirming local participatory conditions that demand public communication reviews.[11]

However, there have also been some major controversies which have materialized about the location preference of local geological deposits and the provincial assessments that have suggested collective emerging needs for collaborative interactions, on nuclear safety risk principles, becoming more relevant, for the possible determination of specific course of actions, elaborated in relation to local environmental choices. In a multi-level governing context, the industrial technology aggregations, technical innovation institutes, and local regulatory agencies have included sectoral managerial bodies, which have maintained comparative mutual distances about respective political and commercial involvement, particularly when related to the regional nuclear industry, disciplined in terms of public management risks and collective regulative objectives.[12]

---

9   Darrin Durant, "Responsible Action and Nuclear Waste Disposal," *Technology in Society* 31/2009, pp. 150–157.
10 See: Anne Jerneck, Lennart Olsson, Barry Ness, Stefan Anderberg, Matthias Baier, Eric Clark, Thomas Hickler, Alf Hornborg, Annica Kronsell, Eva Lovbrand, Johannes Persson, "Structuring Sustainability Science," *Sustainability Science* 6.1/2011, pp. 69–82.
11 Durant, "Responsible Action," pp. 150–157.
12 See: Kristian Kallenberg, "Operational Risk Management in Swedish Industry: Emergence of a New Risk Paradigm," *Risk Management* 11.2/2009, pp. 90–110.

Fundamentally, the correspondent material aggregations concretized in association with governance practical approaches combining defining roles and compatible responsibilities, have been functional for establishing public local adaptation and local compliance rules, that adhere to states' organizational layers, which have included local distributive regulation conditions, defined within territorial boundaries and socio-economic changing transitions. Common socio-economic transition plans have concurrently led to the development of transnational science and technology programs, also identified in relation with distinct sets of environmental guidance provisions.

This type of local adaptive association in governance has been able to redefine collective operable interests corresponding to the cultural education objectives of multi-level coordination programs, integrating societal knowledge transfers, which have been functional to collective environmental policies intertwined with participatory directives, which involve targeted populations in countries such as: China,[13] India,[14] and the United States. Through the actual affirmation of respective regional and transnational innovation trading areas, the multilateral provision of common management services and local distribution networks has been assessed according to global production dimensions, and through national facilitation measures drafted for the implementation of comparative science and technology supportive systems.[15]

At country level, there have been specific objectives carried out through the national development programs and international innovation initiatives,

---

13 Note: For instance, in terms of higher education, China in 1911 after the collapse of the Qing dynasty established in Nanking a University through democratic institutions because of the Guomindang party under the leadership of Sun Yat Sen which affirmed democratic rule on the southern parts of the country. (Ratchford, Blanpied: p. 217). In 1949 under the People's Republic of China (PRC) it was also established the Chinese Academy of Sciences (CAS) while most universities controlled by the Ministry of Education remained confined to undergraduate instruction level. (Ratchford, Blanpied: p. 217).

14 Note: Instead, in India from 1950 through 1975 the country after the Nehru's leadership assumed the Science and Technology practices, and developed technology infrastructures also during Indira Gandhi's Premiership role through for instance a National Committee for Science and Technology in line to the establishment of economic guidance and pilot atomic energy programs during the 1970s. (Ratchford, Blanpied: p. 221).

15 J. Thomas Ratchford, William A. Blanpied, "Paths to the Future for Science and Technology in China, India and the United States," *Technology in Society* 30/2008, pp. 211–233.

interrelated to industrial communication technologies favoring knowledge-based networked transfers. The actual resulting outcomes have reflected comparative levels of national and local integration, which have been assessed in parallel with science and innovation modifying systems, affecting comparative large countries such as China, India, and the United States. However, the formation of distinct socio-economic blueprints has included regional and international managerial decentralized plans for science development.

Overall, respective cultural backgrounds and societal participative trends have been characterized by nationally-based progressive preservation policies, where the scientific and technological development activities have been pursued according to local institutional collaboration processes. In addition, there are mixed managerial public-private models for collaboration and coordination, that also favor the implementation of regional innovation standards, which have been calibrated in connection with regional industrial clustered sites, for major reasons related to local collective interests and cooperative actions that, in any case, have been differently arranged.[16]

Moreover, regional development initiatives have been explored in view of public institutional commitments shifting toward the prioritization of science and technologies programs, which have often implied the inclusion of balanced environmental knowledge conditions, established between public local protection agencies and national economic institutions. This type of planned directions combined with local environmental adaptation pursued in governance, has evolved through the renewal of regional industrial alliances, as well as, the availability of multilateral management innovation partners. Transnational industrial partners, in fact, have moved beyond the national borders in order to gain an international visibility set on different production contexts, including local structural options and common cooperation targets.

9.1.2 Processing Country Targets

Moving forward, the institutional presence of functional management organizations interrelated with the process of local governance adaptation, has been interposed according to concurrent policy-making activities, including local environmental campaigns, which have been able to establish respective interconnecting roots linking different sectoral associations and societal mutual

---

16 Rodney W. Nichols, "Innovation, Change, and Order: Reflections on Science and Technology in India, China, and the United States," Technology in Society 30/2008, pp. 437–450.

counterparts, also according to the size of larger economies such as: China, India, and the United States.[17]

Moreover, a specific technological characterization is intertwined with the national regulatory mechanisms that rely on public local administrations, which have been connected to the regional changing process adapted in view of transitional science and technology upgraded activities. The formation of educational knowledge programs supported by the regional innovation research departments has included a specific type of systematic guidance plan for the local managerial organizations and the multilevel independent advisory agencies, which have been directly mobilized for the implementation of top-down independent innovation approaches, targeting selective local policies and competitive socioeconomic cyclical reviews.[18]

For comparative organizational case studies, the contextual research areas of science information and innovation with decentralized participatory objectives, incorporating the contemporary cultural dimension of countries like China, India, and the U.S., have been independently influenced across more recent decades: as pointed out, we can consider that "the pattern is crystal clear. The Chinese economy sustains its boom. The Indian economy continues to accelerate. The American economy advances despite teetering toward a recession… Relentless competition drives innovation and trade. The welcome economic growth with increased productivity is difficult to manage. It disturbs the order, jobs change or disappear, markets tighten, and technological change disrupts"[19]

On such accounts, the need to explain the cultural reasons for the association of competitive international transnational aggregating coalitions, which have largely emerged from progressive networking interactions, based on day-by-day correspondence of industrial local capacities with concentrated technical operations, has been analyzed under mutual intra-regional conditions. Analogously, when we focus on the nuclear science policy issues, there are respective local organizational specifications, for the case of technical nuclear waste management, which have been subjected to a number of comparable significant factors, corresponding for instance to: a.) the formal constitution of civilian nuclear energy programs; b.) the health management sectors evolving in relation with local maintenance activities; c.) and the local operational infrastructures compatible with transnational highly skilled managerial availability; in substance,

---

17 Nichols, "Innovation, Change, and Order," pp. 437–450.
18 Nichols, "Innovation, Change, and Order," pp. 437–450.
19 Nichols, "Innovation, Change, and Order," pp. 437–450, p. 441.

these effective elements when combined together have inherently implied local knowledge-based costing activities, maintaining the separation dimension of short, very short, or long-term, nuclear waste disposal management processes.[20]

Instead, unspecified regional aspects of the constant involvement of civilian nuclear industries have emerged in connection with local managerial operating restrictions, associated with energy development programs. The development adaptation process has meant local restructuring through governmental cooperative scenarios, interrelated with industrial managerial innovation factors. In particular, we can underline that single states have been present as institutional control players, put in charge of laying out the conditional local implications, which refer to the direct employment of international technical specifications, concurrently identified by specialized nuclear energy agencies such as: the IAEA or EURATOM.[21] For instance, in terms of national precautionary surveillance and local security measures of diversified projects that correspond to the nuclear production, reprocessing, and disposal of nuclear energy waste products,[22] the regional single states have either directly or indirectly subscribed normative controlling conditions, also in view of getting a future industrial assets' correspondence.

In particular, comparative nuclear management processes have been favored in order to reach objective industrial goals of management efficiency in energy sectors with local servicing provisions. By redrawing mutual boundaries about states' internal efforts for local adaptability and cooperation partnerships established in governance, there have also been composite organizational elements either internalized or externalized in both economic and political spheres, intertwined with the material application of mutual legal agreements, which can deliver concurrent arranged implications.[23]

---

20   Tom Vander Beken, Nicholas Dorn, Stijn Van Daele, "Security Risks in Nuclear Waste Management: Exceptionalism, Opaqueness and Vulnerability," *Journal of Environmental Management* 91/2010, pp. 940–948.
21   See: 1) Justin Alger, Trevor Findlay, "Strengthening Global Nuclear Governance," *Issues in Science and Technology* 27.1/2010, pp. 73–79; 2) Arunas Molis, Justina Gliebute, "Prospects for the Development of Nuclear Energy in the Baltic Region," *Lithuanian Annual Strategic Review* 10.1/2012, pp. 121–150; 3) Regina S. Axelrod, "The European Commission and Member States: Conflict Over Nuclear Safety," *Perspectives* 14.1/2006, pp. 5–22.
22   Beken, Dorn, Daele, "Security Risks in Nuclear Waste Management," pp. 940–948.
23   See: 1) Sylvain Brouard, Isabelle Guinaudeau, "Policy Beyond Politics? Public Opinion, Party Politics and the French Pro-Nuclear Energy Policy," *Journal of Public Policy* 35.1/

Regarding the technical reprocessing of nuclear fuel cycle adapted within dynamic hazardous risk conditions, a critical vulnerability was exposed due to local entrepreneurial arrangements characterized by long-term weaknesses. In this case, there are to consider past and present issues of civilian and military 'nuclear' legacies, which have been inherited together with dynamic institutional relations and national organizational changes. Parallel to this, there are evolving aspects of local technical requirements, which have been specified for comparative nuclear states that have been entitled to reform international evaluation processes, through constant monitoring and regulatory work conducted by institutional nuclear bodies, such as the U.K. (NDA) Nuclear Decommissioning Authority.[24]

Through this convergent ecosystem, we can note that in concrete terms the management of local innovation provisions including the nuclear power industry and the nuclear waste disposal programs, has essentially contributed to the increase of local information exchanges, that may also determine potential knowledge gaps for the decision-making changing bureau as well as interlinked processing units, that have been expanded in size due to the safety and security regulation practices requiring separate sub-committees advisory support. On the other hand, at local community level, the public environmental risks and common knowledge perceptions have led to associative ethical interactions, that address again the nuclear regulatory information mechanisms.[25] For instance, the enhancement of interrelated technical management capacities, for the commercial expansion of national electricity production projects, has been characterized by both socio-economic and political issues, which have affected nuclear states' management directions such as in the United States. Therefore, a respective monitoring propensity to regulate nuclear energy programs has emerged, while at the same time, putting closer attention at the local operational knowledge process, that has been developed for renewable energy plans, which target territorial planning expansions operated nationally.[26]

2015, pp. 137–170; 2) Michael G. Faure, Karine Fiore, "The Coverage of Nuclear Risk in Europe: Which Alternative?" *The Geneva Papers* 33.2/2008, pp. 288–322.

24   Beken, Dorn, Daele, "Security Risks in Nuclear Waste Management," pp. 940–948.
25   See: 1) Gregory Rose, Ben Milligan, "Law for the Management of Antarctic Marine Living Resources: From Normative Conflicts Towards Integrated Governance?" *Yearbook of International Environmental Law* 20.1/2010, pp. 41–87; 2) Nicolas de Sadeleer, "Environmental Governance and the Legal Bases Conundrum," *Yearbook of European Law* 31.1/2012, pp. 373–401.
26   See: Ove Eikeland, "EU Energy Policy Integration – Stakeholders, Institutions and Issue-Linkages," *FNI Report* at Fridtjof Nansen Institute, 2012, pages 141.

9.1.3 Supporting Technical Views

The multiple interoperability of territorial integrated institutions has required a complex type of understanding of specific direct decisional options that refer e.g. to the disposal of nuclear waste programs based on multilateral technical reviews managed globally. In particular, there have been public environmental concerns that have been raised about the specific definitions provided for radiation exposure levels and radioactivity value factors of nuclear energy programs. The eventual adoption of public environmental approaches has been followed through the inclusion of operable institutional options, which have been foreseen for the establishment of local environmental protection mechanisms, that can implement the necessary protection mandates for both local residents and local urban citizens e.g. located in the U.S., with federal cases assessed through the provision of widespread knowledge-based linkages and public information platforms targeting the public local schools and cultural education centers.[27]

Similarly, the potential affirmation of more integrative experiences reaffirming a collaborative pursuance of local environmental commitments, has remained focused on the safety management of nuclear industrial facilities, and on the commercial aims of renewable energy plans, such as in U.S. NPP cases. Internationally, there are progressive institutional steps that maintain the direction of mutual local cooperation for intra-governmental agreements, undertaken through multiple representative stakeholders, which have been responsible for the regional participatory inclusion concerning multilateral technical domains, as well as, public environmental anti-risks programs, which sustain local cooperative assessment analyses.

Comparatively, this cooperative process has led to the consolidation of governmental innovation objectives about the nuclear energy production systems, having been merged according to the availability of managerial risk operators, legal public agencies, and regional funding entities,[28] converging on further construction plans of nuclear power plants NPP, while also technically supporting alternative sustenance programs for short-term renewable energy transitions, which stimulate sectoral management deployments.[29] Nonetheless, the fact

---

27 Charles W. Pennington, "Comparative Population Dose Risks from Nuclear Fuel Cycle Closure and Renewal of the Commercial Nuclear Energy Alternative in the U.S.," *Progress in Nuclear Energy* 51/2009, pp. 290–296.
28 See: Alan W. Wolff, "China's Drive Toward Innovation," *Issues in Science and Technology* 23.3/2007, pp. 54–62.
29 Pennington, "Comparative Population Dose Risks," pp. 290–296.

remains that the local organization of nuclear waste facilities, designed under regional monitoring processes of the nuclear operational fuel cycles, has regularly divided the public opinion with different opposing orientations as well as classified open confrontations.

One of the reasons of such public local divide has been connected to the material local storage activity of nuclear waste cycles. This activity has been conjugated with local operational functionalities, which have been collectively reconsidered in relationship with local environmental radiation events and potential radioactive local risks' exposures. Reconsidering the embedded comparative methodological issues of technical protection plans about the nuclear energy health risks' cases, it remains a primary concern for social policy experts[30] to highlight the need of reducing distance created between local collectivities and technical local operators, that manage scientific knowledge-sets of common practice tools applied through cognitive modifying processes, which have been serving the public involved.[31]

For instance, another question concerns about the issue of nuclear waste management in the case of nuclear fuel cycle closures,[32] involving local regulatory restoration plans, and regional environment politics, for assessing the public affairs related to socio-economic distributive configurations. The regional states' regulative indications on potential nuclear facilities' decommissioning activities have essentially differed across distinctive territories due to the formation of

---

30 See: 1) Lars Hogberg, "Root Causes and Impacts of Severe Accidents at Large Nuclear Power Plants," *Ambio* 42.3/2013, pp. 267–284; 2) Yuri Rojavin, Mark J. Seamon, Ravi S. Tripathi, Thomas J. Papadimos, Sagar Galwankar, et al., "Civilian Nuclear Incidents: An Overview of Historical, Medical, and Scientific Aspects," *Journal of Emergencies, Trauma, and Shock* 4.2/2011, pp. 260–272.

31 Pennington, "Comparative Population Dose Risks," pp. 290–296. p. 295.

32 Note: It is specified that "in closed fuel cycles, the remaining uranium in spent fuel, along with different isotopes of plutonium will be removed so that it can be reused as fresh fuel, but of course the extracted plutonium also carries proliferation threats. Proliferation is by far the most important concern when reprocessing. Security is one of the main reasons why the US, which has about one fourth of all world's nuclear reactors, does not reprocess. The major stockpiles of plutonium in the US derive from the nuclear warheads that were dismantled after the Cold War era. The idea of producing more plutonium is generally considered to be highly undesirable in the US." Behnam Taebi, "Additional security concerns in conjunction with plutonium," chp. 15.3.2. in the Series: Radioactivity in the Environment, in *Social and Ethical Aspects of Radiation Risk Management*. (Published by Elsevier Science, 2013). eBook ISBN: 9780080914299.

comparative country-to-country analyses, which have been developed in countries such as Japan, South Korea, or Canada.[33] About such countries, public participatory policies undertaken through the promotion of local adaptive collective choices, have brought the national controlling operators to select specialized local environmental indications, also for the development and maintenance of nuclear power programs, similarly connecting multi-level processing agencies. Comparable distribution agencies have been able to re-establish concomitant essential factors regarding this regulatory process, set also according to an overlapping of distinctive evaluative dimensions, referring to societal, technical, and regulative activities.

Moreover, in view of theoretical approaches about nuclear policy impacting aspects, some authors such as Valentine, and Sovacool 2010; and Hulbert, McNutt, and Rayner 2011[34] have provided corresponding analytical explorations tackling the prevalence of similar but divisive organizational tendencies, that refer to comparative states' multi-level large energy capacities, assessed in consideration of respective ideological orientations, eventually put into practice. At local context level, the natural permeation of social development norms and relational interconnecting dynamics has influenced the progressive regulatory changes for national energy projects focused on science technology with possible decentralized alternative paths. Through the regulatory agreements for science development programs, the multilateral innovation systems, including regional clusters of health management networks, have basically supported local implementation models, which connect the structural local management initiatives, also undertaken for the case of nuclear power energy programs.

At the same time, the effectiveness of nuclear energy programs modelled in line with diversified management systems has been collectively perceived as an integral strategic component of policy safety protection services and socio-economic planned schemes, including regulatory aggregated cases based in e.g. Japan or the South Korean Peninsula.[35] In addition, at historiographical level, a significant distinctive line can be traced between the national and transnational industrial innovation organizations, and regional technological capacities complemented with corresponding local management for contemporary energy

---

33 See: Chen Kane, Stephanie Lieggi, Miles A. Pomper, "Time for Leadership: South Korea and Nuclear Nonproliferation," *Arms Control Today* 41.2/2011, pp. 22–28.
34 Scott Victor Valentine, Benjamin K. Sovacool, "The Socio-Political Economy of Nuclear Power Development in Japan and South Korea," *Energy Policy* 38/2010, pp. 7971–7979.
35 Valentine, Sovacool, "The Socio-Political Economy," pp. 7971–7979.

assets, which have been expanding after the period of World War II[36] with its raging conflicts.

Particularly, for regional states operating within institutionalized planning economies, searching for socio-cultural correspondences, new strategies have been put in place in order to facilitate the diffusion of science technology initiatives promoting comprehensive renewable energy projects, which have produced mutual public assertive positions, validating environmental distribution programs that have formed through common societal cognitive frameworks. In terms of mutual collective empowerment norms and local emancipation doctrines, the institutional functional powers and internal distribution organizing networks have been historically coordinated according to selected planning objectives, and national regulatory directives adapted in view of recurrent socio-economic issues connected, for instance, to the fact that "nuclear energy was ideologically linked with visions of military autonomy and strength as well as economic competitiveness."[37]

As a result, the emerging types of structural historical consolidations for strategic competitiveness have been shaped in capitalist and semi-capitalist regime models, having fundamentally favored the centralization of institutional assertive cores for policy-making processes. Such composite assertive systems have in time allocated the development and production of nuclear power industries according to the alignment of managerial construction designs, which have been classified through the national spectrum of technological advanced plans, which involve the incidence of local collective groups and social opposition movements.[38]

For example, about the case of the Canadian province Saskatchewan, some authors as Hulbert, McNutt, and Rayner, 2011[39] have been able to specify the magnitude of historical changes across the regional provinces, which have favored political and relational participatory interactions for nuclear managerial plans and converging socio-economic transitions. The associated recognition manifested about contextual nuclear policy issues, and environmental regulative elements pertaining to Canada's nuclear policy, has included comparative

---

36 See: T. V. Paul, "The US-India Nuclear Accord: Implications for the Nonproliferation Regime," *International Journal* 62.4/2007, pp. 845–861.
37 Valentine, Sovacool, "The Socio-Political Economy," pp. 7971–7979. p. 7975.
38 Valentine, Sovacool, "The Socio-Political Economy," pp. 7971–7979.
39 Margot Hulbert, Kathleen McNutt, Jeremy Rayner, "Pathways to Power: Policy Transitions and the Reappearance of the Nuclear Power Option in Saskatchewan," *Energy Policy* 39/2011, pp. 3182–3190.

STS States Innovation Patterns 173

institutional analyses about the adoption of local converging measures that have been structured within multi-level energy technological plans, regularly opting for the uses of alternative renewable energy sources, which include the nuclear power development related to local renewable energy projects. For a summarized historical overview about Canada's nuclear power plants, with reporting status see: Table 6 titled 'Overview IAEA Nuclear Power Plants' 2004.

Table 6 Overview IAEA Nuclear Power Plants 2004

Country	Location	Owner	Capacity (MW (e))	Date of Operation	Date of Shutdown	Status
Canada	Kincardine, Ont.	Bruce Power	769	1977/1	1997/10	Permanently Shut Down
Canada	Kincardine, Ont.	Bruce Power	769	1976/9	1995/10	Permanently Shut Down
Canada	Kincardine, Ont.	Bruce Power	790	1977712	2017	Operating
Canada	Kincardine, Ont.	Bruce Power	790	1978/12	2018	Operating
Canada	Kincardine, Ont.	Bruce Power	790	1984/12	2024	Operating
Canada	Kincardine, Ont.	Bruce Power	790	1984/6	2024	Operating
Canada	Kincardine, Ont.	Bruce Power	790	1986/2	2026	Operating
Canada	Kincardine, Ont.	Bruce Power	790	1987/3	2027	Operating
Canada	Clarington, Ont.	Ontario Power Generation	881	1990/12	2030	Operating
Canada	Clarington, Ont.	Ontario Power Generation	881	1990/1	2030	Operating
Canada	Clarington, Ont.	Ontario Power Generation	881	1992/12	2032	Operating
Canada	Clarington, Ont.	Ontario Power Generation	881	1993/4	2033	Operating
Canada	Tiverton, Ont.	AECL	206	1967/1	1984/5	Permanently Shut Down

(*Continued*)

Table 6 Continued

Country	Location	Owner	Capacity (MW (e))	Date of Operation	Date of Shutdown	Status
Canada	Quebec	AECL	250	1972/4	1977/6	Permanently Shut Down
Canada	Becancour, Quebec	Hydro-Quebec	635	1982/12	2022	Operating
Canada	Pickering, Ont.	Ontario Power Generation	515	1971/4	1997/12	Permanently Shut Down
Canada	Pickering, Ont.	Ontario Power Generation	515	1971/9	1997/10	Permanently Shut Down
Canada	Pickering, Ont.	Ontario Power Generation	515	1972/4	1997/12	Permanently Shut Down
Canada	Pickering, Ont.	Ontario Power Generation	515	1973/5	2013	Operating
Canada	Pickering, Ont.	Ontario Power Generation	516	1982/12	2022	Operating
Canada	Pickering, Ont.	Ontario Power Generation	516	1983/11	2023	Operating
Canada	Pickering, Ont.	Ontario Power Generation	516	1984/11	2024	Operating
Canada	Pickering, Ont.	Ontario Power Generation	516	1986/1	2026	Operating
Canada	Point Lepreau, NB	New Brunswick Power	635	1982/9	2022	Operating
Canada	Rolphton, Ont.	Ontario Power Generation/ AECL	28	1962/4	1987/8	Permanently Shut down

Source: IAEA, International Atomic Energy Agency, 2004. "Status of the Decommissioning of Nuclear Facilities around the world." IAEA, Vienna, Austria. STI/PUB/1201. Annex I "Nuclear Power Plants," pp. 3–4.

9.1.4 Public Sustenance Applications

In connection to this, it's crucial to bring attention to the co-determination of national socio-economic goals that have established an amplified tendency about the affirmation of science and technology politics that entail constructive programmatic aims, which have been identified through multi-level governance feedbacks, with publicly based local acceptance programs, which have been assessed according to normative conditions conducting to a substantial transformation.

Therefore, the constructive regulatory boundaries between states supporting science innovation development fitting with regional policy objectives, have been approached across time according to parallel levels of public local engagements and spatial territorial transitions, which can incorporate specialized commercial division areas within trading networked exchanges, operating through industrial energy sectors.[40] Historically, the recognition of comparative states' normative ethical framework has been progressively adopted by democratic as well as communitarian societies, which have converged toward the conservative lead for the promotion of national industrial policies moving in favor of high-technology production clusters. In turn, collective socio-economic adaptation conditions have changed because of respective country-based technology integration patterns that have been implemented through the rearrangement of international and transnational reforming processes, that reflect science and technology development for contingent industrial knowledge-based innovation projects, reviewed under diversified national information provisions.[41]

From a technological standpoint, national and local industrial innovation business activities seen in transition, have been organized in parallel with the specialization of collaborative governmental agencies – seeking to collect larger commercial benefits and reduce public costing implications, for instance, about renewable energy systems' controls and local maintenance issues, which have been evaluated often beyond the national capacities' offerings, in particular for the use of renewable nuclear energy options. On such aspects, the combination of public regulatory commitments, together with indigenous cultural beliefs, and peoples' environmental rights, have formed multiple societal tendencies leading toward territorial environmental commitments, which have contributed

---

40   See: Lincoln L. Davies, "Beyond Fukushima: Disasters, Nuclear Energy, and Energy Law," *Brigham Young University Law Review* 6/2011, pp. 1937–1989.
41   Michael A. Dennis, "Scientific and Technical Knowledge and the Making of Political Order," *History and Technology: An International Journal* 28.4/2013, pp. 415–421.

to the redirection of a regulative focus targeting the freedom and legitimacy of political engaging actions.

Through the collective adaptation of multilateral governance models, the decision-makers in charge have contributed to rearrange the composition of national environmental knowledge issues, that refer to local environmental warning systems, local environmental engagement programs, and local environmental participatory dynamics, regionally negotiated under public confrontation platforms and statutory diffused declarations. In such context, there have also been concurrent normative implications put together with local cultural interpretations about collective normative aims that have affected the constitution of mandatory regulatory paths, specifically pointing at industrial nuclear energy projects, operable under PPP management policies.

In essential terms, when the ruling pragmatism is at hand, basically it can touch upon different types of civilian organized transitions, engaging for example participatory environmental groups, which have either directly or indirectly expressed national value systems, that include the civilian nuclear safety ethical matters, associated with local cooperative designs and translated into public democratic undertakings, when the case.

Nonetheless, global market regionalism trends and national pragmatic orientations of comparative transnational societies, have determined commercial nuclear policy umbrella programs, transfused in terms of national collective acceptances, when possible, in order to increase built-in civil-military capacities of nuclear technology stockpiles and nuclear power energy reactors. However, this does not imply that in retrospect the national critical positions reflecting public environmental concerns can be set aside in terms of national credibility, especially when critical collective needs have still remained in evidence despite strategic countries' safety protection plans and reassuring security visions.

Rhetorically, the civilian nuclear energy sector has been associated with concrete organizational devolution processes carried out through science and technology innovative formulations, that presuppose the public affirmation of common societal protection values. However, the formation of methodological processes based on the expansion of renewable energy trading configurations at the regional level, has resulted into progressive interdependent exchanges with socio-economic interactions – tentatively matching the prevalence of local development schemes – while also incorporating the multi-level regulative arrangements established according to characteristic aspects of public policy agendas.

Even so, the comparative entrepreneurial states have partially failed to deliver the correct measure on science and innovation policy incentives, because of associated structural changes affecting the economic GDP growth rates and operational transition policies. The resulting structural economic policies, in the end, have reduced material access to public collective local capacities, for the broader PPP with commercial gains and affirmation of respective PPP value-based advanced opportunities.[42]

For instance, the mutual functional roles supported in multilateral governance systems for the centralization or decentralization of regional governmental agencies, have been essentially combined with affirmative technical skills and local requirements, including prospective limitations for what concerns the formulation of science policy directives and internal development mechanisms that can address, for example, the direct establishment of comparative transnational energy programs.

Essentially, the possible identification of regional collective interests going in favor of structural prioritization areas that promote multi-level technological and renewable energy choices, has been followed up through the science development regulations. At the same time, complex decision-making activities run at the local level have been a matter of domestic intra-state relations extended according to governance collaborative conditions. Diverse public collaborative networks emerging in the form of public facilitation programs – still evolving on research science and technology applications – have similarly implied precise and sustained multi-level relational adaptations, also focused on the annual budgetary exercises.

On such modifying issues, the institutional dependency based on mutual networked relations has been rearranged according to the material inputs of national adaptive bodies which maintain comparative institutional characterizations, particularly about the public delivery of ICT regulatory paths, put together according to public distinct needs in order to rationalize the status of contemporary ICT local adaptation systems, reaching a common normative consensus.[43]

In the case of formalized societal recognitions, we can take into account the concept of a public normative consensus characterizing the regional, transnational and national, propensity of libertarian societies, when put into relationship with science technology trading dynamics and with local development planning. However, it has become more difficult to question the type of interpretation that

---

42   Barry Bozeman, and Daniel Sarewitz, "Valuing S&T Activities. Public Values and Public Failure in US Science Policy," *Science and Public Policy* 32.2/2005, pp. 119–136.
43   Bozeman, Sarewitz, "Valuing S&T Activities," pp. 119–136.

should be given in reference to democracy groups' representations and applied technological schemes for modern distributive adaptations, due to the lack of a full awareness about the structural organizational changes that have been basically occurring while juxtaposed to the people's self-determination rights for civilian legitimacy exercised e.g. in the U.S. federal states.[44]

In the United States' case, there have been structural changes that have occurred at the national and international level about scientific transition policies, which have created a multilateral adaptation ground for reaffirming comparative levels of diffused local power relations. The bureaucratic regulation of distributed national power relations at the federal level has in turn facilitated an instrumental local transfer about the comprehensive knowledge-based scientific innovation agreements. But the ambivalent societal aggregative elements remain uncertain, and these elements have pointed at critical issues of common environmental adaptation practices, eventually leading to nation-based social confrontation approaches, which favor better public understanding and open debating stages.[45]

Therefore, a multiple understanding focused on decisional supporting systems which adhere to a transnational or national democratic or/and communitarian evolution of systemic regulatory transfers through local knowledge-based adaptive mechanisms, that also target different technological innovation domains, has been associated with governmental analyses conducted about decentralization processes, which incorporate the local action-planning schemes, that are shaped in view of local policy arrangements.

In retrospect, then it can be questioned the type of relational evolution acquired in connection to the status of governmental knowledge processes interrelated to the national productive capacities, which have been regrouped for creating responsive multilateral adoptions of alternative technological learning systems. The knowledge transfer management models have been selectively decentralized through multi-level development stages of production on S&T (science & technology)[46] activities; where science innovation policies interrelated with local distribution domains have included, for instance, the preparation of

---

44 Clark A. Miller, "Science and Democracy in a Globalizing World: Challenges for American Foreign Policy," *Science and Public Policy* 32.3/2005, pp. 174–186.
45 See: Janet Atkinson-Grosjean, "Canadian Science at the Public/Private Divide: The NCE Experiment," *Journal of Canadian Studies* 37.3/2002, pp. 71–91.
46 See: 1) Hongli Wang, Zhenlong Peng, Feng Gu, "The Emerging Knowledge Governance Approach Within Open Innovation: Its Antecedent Factors and Interior Mechanism," *International Journal of Business and Management*, 6.8/2011, pp. 94–104; 2) Nicolai J. Foss, 2011. "Knowledge Governance: Meaning, Nature, Origins, and Implications,"

comparative renewable energy monitoring and implementing initiatives, which have emerged across different regions such as in the Russian Federation, or the European Union region as analyzed in: Soubbotina, Weiss 2009;[47] Nielsen, Knudsen 2010;[48] Parthsarathy, 2010.[49]

9.1.5 Public Diversification Plans

The governmental advocacy projects undertaken in view of modern applications of S&T through science policies[50] supported beyond the central aspect of human sustainability, have identified functional development areas involving both public and private technical expertise of regional industrial partners. At regulatory level, incremental available suggestions have facilitated common participatory approaches, especially regarding the designing and planning levels of local compatible development blueprints, in order to get formal or informal supportive societal accesses based on e.g. public environmental activism. However, there have been comparative domestic industrial sectors, through which the increasing degree of organizational adaptive strategies has not necessarily reflected public distributive aims due to the changing governmental provisions related to e.g. local environmental adaptation domains; structural regional innovation programs; or commercial technological learning tools.

Therefore, the considerable amount of inter-sectoral regulatory requirements which have been introduced for the appropriation of common technical knowledge transfers, for instance, on nuclear civilian technologies, has also meant legitimation conditions and public open recognitions established in accordance with country-level transferring initiatives. As such, the comparative knowledge transfer initiatives have been undertaken in parallel with progressive multilateral geopolitical interests, which have been combined as well with the rapid

---

in: Anna Grandori, eds., *Handbook of Economic Organization*. Edward Elgar, 2012. SSRN Working Paper Series. (Social Science Research Network).
47 Tatyana Soubbotina, and Charles Weiss, "A New Model of Technological Learning for Russia," *Science and Public Policy* 36.4/2009, pp. 271–286.
48 Henry Nielsen, and Henrik Knudsen, "The Troublesome Life of a Peaceful Atoms in Denmark," *History and Technology* 26.2/2010, pp. 91–118.
49 Shobita Parthasarathy, "Breaking the Expertise Barrier: Understanding Activist Strategies in Science and Technology Policy Domains," *Science and Public Policy* 37.5/2010, pp. 355–367.
50 See: Jorge M. Soberon, Jose K. Sarukhan, "A New Mechanism for Science-Policy Transfer and Biodiversity Governance?" *Environmental Conservation* 36.4/2009, pp. 265–267.

distribution of technological innovation capacities particularly diffused in welfare states, for instance, in Finland, Norway, and the Netherlands, often associated with complex transnational implications.

On that note, it is now evident that the consolidation of multi-level organizational approaches shapes not only the perception of 'modernity' in social and liberal democracies, but also the recognition of *adaptability* as a concept for some core innovating production regions, with the aggregation of local industrial patterns, transferring at the same time, cross-cultural environmental education programs and interrelated science development policies.[51]

Put in a comparative perspective, regional, transnational or national, renewable energy innovation strategies have been addressed at open learning societies, which have adhered to traditional socio-economic patterns, while either increasing or decreasing regional corporatism waves and political multi-party systems, possibly leading to the material inclusion of participatory collaborative domains that in any case have been difficult to directly involve in more comprehensive ways.

For instance, in the case of public demonstrative orientations about nuclear environmental regulations, the civilian nuclear cooperation networks have been cyclically formalized, even though associated with full changing dynamics of multilateral trading states. Multilateral governmental partnerships assessed through PPP have also maintained regional hegemonic conditions, catalyzing attention on correspondent national developments of nuclear power energy programs. Therefore, incremental PPP cooperative exchanges able to favor the direct flow of mutual formal relationships, undertaken between international and national decision-making representatives, have remained critical for what concerns the promotion of nuclear power energy pursued for civilian peaceful aims.

In essence, technological assistance programs and international security politics have been promoted in terms of nuclear energy international memberships, such as for the legal agreements of Nuclear Non-Proliferation Treaty (NPT) formulated against the acquisition of nuclear weapons systems. Such security awareness programs have been authorized in view of global ideological fundamentalism based on post-conceptual realism[52] issues elaborated about the essential characteristics of security categorization, which has been exploited in

---

51 Nielsen, Knudsen, "The Troublesome Life," pp. 91–118.
52 See: 1) Marvin Miller, Lawrence Scheinman, "Israel, India, and Pakistan: Engaging the Non-NPT States in the Nonproliferation Regime," *Arms Control Today* 33.10/2003, pp. 15–20; 2) Campbell Craig, Jan Ruzicka, "The Nonproliferation Complex," *Ethics & International Affairs* 27.3/2013, pp. 329–348.

terms of international cooperative relations materializing during the past five decades.[53]

Regarding the international security effects and national mutuality conditions, the national enforcement of civilian nuclear energy cooperation programs is adapted for the identification of non-proliferation security assistance models, which have been specifically adopted according to structural development environments, evolving in view of transnational nuclear energy programs.

Nonetheless, ambivalent existing factors regarding the local construction processes of nuclear power energy plants (NPPs) have been in time encountered again and again, because of increasing resurgent safety issues and dynamic local managerial changes, which have converged toward an attentive procedural understanding promoted through public environmental engagements. Local environmental civic exchanges have been diffused through regional networking societal platforms that can address the delivery of compatible operational tools which generate local risk reduction initiatives, when especially put into relation with recent climatic changes, as well as, with nuclear power plants disaster crises, such as: during the 2011 Japanese earthquake, and the regional emergency radioactivity crisis occurring at the Fukushima's NPP plant with nuclear reactors shutting down. In this case, "while the debate is likely in the first instance at least, to be dominated by technical and safety considerations, it seems timely to re-examine existing social science research on nuclear energy and think through some of the key socio-political dimensions in light of this accident."[54]

The extensive environmental political dimension based on local participatory conditions for public acceptability about the diffusion of renewable energy measures envisaged at the level of regional communities, has been directly addressed for the case of local nuclear safety and integrated monitoring & evaluation risk activities. Such types of environmental PPP management programs have been related to the nuclear technical compliance issues, which have been identified in line with the level of decision-making, that may allow a consolidation of local processing rules, specifically about national nuclear energy projects[55] – which have entailed associated safety risk conditions running in parallel with the

---

53   Christoph Bluth, "Correspondence. Civilian Nuclear Cooperation and the Proliferation of Nuclear Weapons," *International Security* 35.1/2010, pp. 184–200.
54   Catherine Butler, Karen A. Parkhill, and Nicholas F. Pidgeon, "Nuclear Power after Japan: the Social Dimensions," *Environment Magazine* 53.6/2011, pp. 3–14. p. 5.
55   See: 1) Sulfikar Amir, "Nuclear Revival in Post-Suharto Indonesia," *Asian Survey* 50.2/2010, pp. 265–286; 2) J. Mohan Malik, "China and the Nuclear Non-Proliferation Regime," *Contemporary Southeast Asia* 22.3/2000, pp. 445–478.

development of science technology and comparative knowledge transfers, that reflect respective states' techno-economic configuration assets.

9.1.6 Local Demonstrative Cases

In essence, the regional configuration of societal trust models re-framing environmental cooperative maintenance programs in the case of nuclear power energy plans has been confronted with comparative multilateral regulatory independent paths,[56] as for the case of Japan and the Fukushima's post-crisis organization for the recovery situation. In this type of public adaptation path have emerged dichotomous aspects related to the transnational affirmation of states' commercial interdependent relations.

In addition, environmental concerning issues that emerge about the local immediate responses that are provided for field recovery actions, have entailed transparent regulation based on the quite practical nature of local coordination mechanisms, which have been designed for supporting the systematic monitoring control cycles, referring to nuclear power energy technology and industrial maintenance systems.[57]

In terms of fundamental accountability reasons, there have been other incidental factors in relation to technical standardization designs fulfilling the local safety requirements of renewable energy programs. Alternative renewable energy sources have been set in association with environmental preventive awareness plans, which include the local governance transfer mechanisms for public distribution measures. Moreover, the nuclear re-conversion and decommissioning processes of nuclear power energy plants NPP have been assessed in view of developing civilian protection identifications, because of common provisioning implications interrelated to the need of promoting peaceful solutions for nuclear energy environmental safety and local remediation activities.

For instance, depending on case-by-case structural management conditions verified within national radioactive industrial areas (e.g.) U.S. Rocky Flat, a legacy[58] has been left related to nuclear waste policy management, which has

---

56 See: 1) Trevor Green, "Meeting the Rising Energy Demands of a Greener Future," *Cost Engineering* 50.12/2008, pp. 15–16; 2) Koji Omi, "Alternative Energy for Transportation," *Issues in Science and Technology* 25.4/2009, pp. 31–34.
57 Frank Uekoetter, "Fukushima, Europe, and the Authoritarian Nature of Nuclear Technology," *Environmental History* 17/2012, pp. 277–284.
58 US Fed News Service, "U.S. Fish and Wildlife Service Establishes Rocky Flats National Wildlife Refuge," *US FED News Service*, 2007, including US State News (Washington, D.C.), July.

commonly led to the identification of specific environmental programmatic targets; especially for the public-private management of contaminated fields on local construction sites, which have been handled together with local land-uses and local water reservoirs that have been found in proximity of nuclear power plants NPP.[59]

The actual adoption of institutional ecology approaches[60] formulated in favor of the elimination of radioactive hazardous waste has essentially become intertwined with the territorial transformation of nuclear power processing sites, which can also be targeted for the recovery of waste disposal practices.[61]

In comparative political reviews on the environmental nuclear clean-ups issues, Krupar (2012) points toward the example of the U.S. Department of Energy (DOE), which in 2007 transferred the management of the Rocky Flats to the Department of Interior's U.S. and Wildlife Service (FWS), with the aim of recovering the nuclear construction sites within surrounding areas for local environmental clean-ups purposes.[62] To add to this, there have been some interrelated aspects of the effective level of applicability concerning public environment conservation and local preservation policies, through which the national regulatory energy schemes have been drafted in combination with environmental management activities, which incorporate incremental local knowledge transfers.

In essence, public-private partnerships PPP based on technology innovation transfers have redrawn (when possible) respective organizational boundaries, also according to the elaboration of more ethically-centered societal frameworks, which have been able to constructively bring back to memory the community's land struggles that have been inherited from a critical past. Regional ecological risks with direct effects and local temporary exposures have therefore been analyzed more in terms of national recovering operations, through for example: the public waste treatments that encourage collective structured actions, which can consequentially lead to social change with maintenance of common environmental aims, also based on local conservation policies.[63]

---

59 Shiloh R. Krupar, "Transnatural Ethics: Revisiting the Nuclear Cleanup of Rocky Flats, CO, Through the Queer Ecology of Nuclea Waste," *Cultural Geographies* 19.3/2012, pp. 303–327.
60 See: Seungho Choi, Hoonseok Jung and Dongheup Lee, "The Basic Planning for the Environmental Relationship of Improved Nuclear Power Plant," GENES4/ANP2003, September 15–19, 2003, Kyoto, Japan. Paper 1097. Webpage available at: https://www.ipen.br/biblioteca/cd/genes4/2003/papers/1097-final.pdf
61 Krupar, "Transnatural Ethics," pp. 303–327.
62 Krupar, "Transnatural Ethics," pp. 303–327. p. 308.
63 Krupar, "Transnatural Ethics," pp. 303–327.

In a contextual way, the nuclear energy technologies promotion identified through regulative inferential schemes for specified local territories, have been integrated by states for the actualization of essential local coordination practices. The institutional coordination activity conducted within a multilateral regulatory environment, involving either centralized or decentralized administrative bodies, has been basically enhanced for the development of national environmental relations. In fact, associated environmental protection policies have been proposed in order to mediate the regional trading process of nuclear power states, which favor the security and protection proposals followed up through scientific documentation plans of international safeguards systems. Essentially, the nuclear power energy facilities NPP presenting local operationalization issues have required a constant maintenance of increased standardized technical activities, which have been performed under the skilled supervision of transnational and international managerial entities.[64]

Consequently, the public-private provision of instrumental knowledge innovation practices and local management operations has been conformed according to the socio-economic background of regional communities, which have consolidated deep traditional roots within respective regulatory practice environments. For the comparative managerial transfers of technical assessment practices adapted within formalized institutional mechanisms, it has been almost necessary to push the boundaries between theoretical comparative approaches and science and technology programs, for the validity of regulatory identifications based at the transnational national and local level.[65] In more critical terms, the international public-private partners and regional environmental nuclear activists have additionally confronted the classic intersection established between knowledge innovation diffusion, decentralized management practices, and competitive firms' innovation dynamics, which have been optimized in terms of centralized practical applications that have included local environmental determinants for complementary issues of national regulative operations.

The managerial adaptation process that has been activated according to the changing regulative dimension of regional industrial communities, has fundamentally corresponded to national political alignments with mutual empowering mechanisms, which have been coordinated according to different distribution

---

64 N. Al-Rodhan, "The Politics of Emerging Strategic Technologies. Implications for Geopolitics, Human Enhancement and Human Destiny," Palgrave Macmillan. 2011.
65 See: Lisa Rumiel, "Getting to the Heart of Science: Rosalie Bertell's Eco-Feminist Approach to Science and Anti-Nuclear Activism," *Journal of Women's History* 26.2/ 2014, pp. 135–159.

conditions. On the one side, the environmental governance adaptation framework set in line with local knowledge innovation areas has been shaped through the availability of participatory accesses, including common environmental learning platforms reaffirmed at societal environmental level. From the other side, the regional reconstruction of progressive management sectors that promote knowledge sharing activities for science technology and local innovation, has been supported through the diffusion of transparent mechanisms, which have been fostered through the national innovation processes based on local accountability, science rationalization strategies, and technology operability of interconnected sharing practices.[66]

In essence, the visible assessment process undertaken by common regulative entities engaged into regional PPP partnerships has been depicted through cooperating orientations, regarding the local planning and programming activities, which have resulted into specific social environmental strategies, which have been reviewed in terms of practical mobilization issues. For these purposes, the corresponding national implementation of environmental programmatic actions formed in governance has been the result of field-based studies, which identify the comparative operable areas for communities' participation regarding environmental distribution objectives.

In fact, comparative structural identifications have been followed through public systematic attempts that have been pursued in order to build up the local integrated accesses to knowledge-based innovation schemes and compatible organization transfers' stages, and therefore by incorporating local community's dynamics into common adaptive practices for the safety risk management organizations.[67] See the Table 7 extracts on institutional nuclear management orientations highlighting NPP regulatory ecosystem.

---

66 Norman Makoto Su, Hiroko N. Wilensky, and David F. Redmiles, "Doing Business With Theory: Communities of Practice in Knowledge Management," *Computer Supported Cooperative Work* 21/2012, pp. 111–162.
67 Su, Wilensky, Redmiles, "Doing Business With Theory," pp. 111–162.

Table 7 Nuclear Waste Management related to Institutional Orientations and Managerial Plans

Integrated Operational Systems	Crisis Communications	Management NPPs
Nuclear Regulatory Organizations	Proactive approach/Planned actions to crisis in different stages	Continual improvement approaches for optimization of the process
Regulatory Implementation	Implementation of a graded approach for cases of radiological exposures	Common timing and increase of common designs with states' vicinity and common challenges
Programs	Documentation of the decision-making process, transparency of the process, facilitation of conducts' reviews	Best Available Techniques (BAT) for the adoption of best solution for discharge management to protect the environment and adopt mitigation plans
Industry Representatives	Dialogue and stakeholder involvement to the waste management process	Consolidation of nuclear power plant designers and vendors across time
Public Transfer Mechanisms	Review of knowledge and technical safety issues for assessment and consensus/ Learning and adaptation to societal demands	Implementation of public requirements in regional and local areas concerning both siting and geological repositories

Theoretical Model Adapted. Sources: (a) Report OECD/NEA, 2011. "Road Map for Crisis Communication of Nuclear Regulatory Organisations – National Aspects." (NEA/CNRA/R(2011)11). (b) Report OECD/NEA, 2012. "Good Practice in Effluent Management for Nuclear Power Plant New Build. A report from the CRPPH Expert Group on BAT (EGBAT)." NEA/CRPPH/R(2012)3. (c) Report OECD/NEA, 2014. "CFD for Nuclear Reactor Safety Applications (CFD4NRS-4) Workshop Proceedings." NEA/CSNI/R(2014)4. (d) Report OECD/NEA, 2012. "Geological Disposal of Radioactive Wastes: National Commitment, Local and Regional Involvement." NEA/RWM(2011)16.

9.1.7 Environmental Diffusion Optimization

For the environmental diffusion optimization, technical information based on pre-existing managerial accounts have been summarized through the OECD's nuclear safety measures reports, and are highlighted here in Table 7. In connection to this, there have been alternative theoretical explorations about environmental nuclear governance issues that refer to the integration of multi-level innovation systems. Complex managerial industry systems when operable, have often been associated with multilateral adaptation strategies, tackling the specific flow of local capacity provisions in view of enhancing mutual professional responsibilities as well as societal trusting levels.

The main state administrators performing in collaboration with social representative counterparts have been able to determine specific ethical guidelines on local environmental protection policies, which have been put into connection with the industrial production patterns, that have been basically identified according to diversified autonomous accesses to regional technological innovation capacities. For instance, the constructive governance models involving social adaptation factors, as well as, science and technology management affirmations, have been in place while reflecting environmental participatory models, which can refer to common normative collectivities mediating between the dividing lines of progressive institutional fronts. It has been highlighted that "societies should not be dictated by technologies instead policy should facilitate a mutual acknowledgement between social aims and technological predictions."[68]

The ontological reformulation of human environmental practices specified according to the changing technological innovating conditions, has been drafted through reflective considerations associated with inclusive cognitive inputs that can emerge in the case of responsive organizational frameworks cobbled by social local actors. Multi-level organizational criteria have been identified in conjunction with networking environmental practices forming through public regulatory systems. In turn, the common identification measures have led to environmental interdependent organized initiatives that have been developed in order to mediate on the effects of technological innovation transitions, conceived through a marked ethical configuration.

At a broader level, the national participative democracies have tended to redefine their respective civilian engaged status through a concrete collective planning based on the division of public-private structural capacities – associated to the rationalization of regulative goals while indirectly integrating the science knowledge transfers through local industrial management offers and multiple local innovation designs – stimulating at the same time the affirmation of social granting rules.[69] Basically, the societal democratic aims which have incorporated moral ethical meanings applied to local programmatic decisions and common

---

68 Katinka Waelbers, "Doing Good With Technologies. Taking Responsibility for the Social Role of Emerging Technologies," (Springer Science+Business Media B.V. 2011). p. 33.
69 See: Andrew Kakabadse, Nada K. Kakabadse, Alexander Kouzmin, "Reinventing the Democratic Governance Project Through Information Technology? A Growing Agenda for Debate," *Public Administration Review* 63.1/2003, pp. 44–60.

environmental actions, have been crucially interrelated with environmental and biological survival systems, through the distribution of public-private social provisions, which have involved incremental participation levels, as well as, locally upgraded implementation and monitoring conditions.[70]

In effect, the interdependency of subjective assimilations from substantive governance with mediation roles and assumed responsibilities has been put in connection with regional transitioning communities. Through practical operable models of common knowledge and learning systems, expanding within mass consumption patterns, the material evolution of multilateral regulatory domains was arranged, set internally according to comparative political dimensions; where mutual public-private PPP knowledge relations have been expressed according to comparative levels of effective compromises and local adaptation conditions seen as *modus operandi* in terms of national enhanced capacities.[71]

Moreover, the systematic combination of national and international regulatory approaches to science nuclear policy has been cultivated through a recurrent challenging maximization of contemporary power relations that influence comparative societal advantages and disadvantages, which have created connected transitional coordination spaces for alternative environmental opportunities. To add to this, process, the regional characterization of development and local industrial impacts has been therefore opened up along the lines of liberal, structural, and radical economies, which have produced temporary institutional arrangements, while becoming locally accessible in terms of sectoral innovation designs, and local assessment plans.

At the same time, the structural economic processes implemented through international public-private PPP approaches have increased a sense of respective territorial understanding of environmental local areas of participant communities. Looking at the progression of multi-layered renewable energy plans, the social environmental interactions have been linked to actual operational changes of industrial technical models, interrelated to local environmental and sustainability issues, which involve states' interdependent provisions as well. As an example, the progressive lack of national sponsoring tools for updating the regular collective assessments on *public attitudes toward science and technology* (PATSAT) in the U.S., has driven to the policy review of core quality

---

70 Waelbers, "Doing Good With Technologies," p. 33.
71 See: Chris Skelcher, Erik-Hans Klijn, Daniel Kubler, Eva Sorensen, Helen Sullivan, "Explaining the Democratic Anchorage of Governance Networks. Evidence from Four European Countries," *Administrative Theory & Praxis* 33.1/2011, pp. 7–38.

determinants, which have been typified through societal environment tendencies pointing at reciprocal changes based on public environmental concerns, when associated with transnational technological shifts.[72]

Considering the continuous transition of technological embedded networks, the organizational adaptation models have also been selected according to uncertain environmental factors. Essentially, societal beliefs systems have led toward the affirmation of public actual behaviors, which can become polarized in terms of environment collective interests, determining a parallel way for national reforms legally configured for the effective environmental acquisition goals of involved communities. Therefore, about the material environmental impacts related to nuclear power energy safety and security factors, the sectoral organization strategies have been developed according to collective adaptation policies, also based on precise recognition aspects of critical implementation grounds.

Then, as mentioned about *public attitudes toward science and technology* (PATSAT),[73] the use of a relational approach placed according to the adoption of governmental innovation programs is now commonplace, resulting in either directly or indirectly favoring multilateral trading exchanges of multi-level innovation systems; transnational criteria reflecting territorial participatory systems, have in turn evolved according to defined local identities and connected communitarian values.[74] Therefore, the geographical identification of environmental disaster areas associated with public risk perception levels, put in combination with environmental collective concerns, expressed by public organizational representatives, has provided the opportunity to develop a comparative civic knowledge dimension for critical reflections.

For instance, the expansion of global renewable energy markets with interrelated functional mechanisms, has been included in the production routing channels through decision-making propositions, which have been based on S&Ts (science and technology) initiatives, highlighting the determinant institutional implications in order to get a continuous reliability of multi-level technology information systems, including a compatible testing availability, which has been used for environmental research evidence, when the case.[75]

---

72 Chenyang Xiao, "Public Attitudes Toward Science and Technology and Concern for the Environment: Testing a Model of Indirect Feedback Effects," *Environment and Behaviour* 45.1/2011, pp. 113–137. p. 114.
73 See: Zygmunt J.B. Plater, "Law, Media, & Environmental Policy: A Fundamental Linkage in Sustainable Democratic Governance," *Boston College Environmental Affairs Law Review* 33.3/2006, pp. 511–549.
74 Xiao, "Public Attitudes Toward Science," pp. 113–137.
75 Xiao, "Public Attitudes Toward Science," pp. 113–137.

9.1.8 Systemic Transfer Orientations

Fundamentally, the common acknowledgement about the prevalence of optimistic as well as pessimistic tendencies referring to institutional transition behaviors has partly been stressed through regional networking relations. On such accounts, there have been transnational economic exchanges which have integrated the participatory enhancement of (S&T) programs, with systematic local knowledge transfers adapted for the operational dimension of trading innovation and consolidation. As result, at regional regulatory level have been defined a number of diversified institutional fragmentation processes, involving multilateral coordination approaches, which have been translated into concrete public orientations. Especially for the case of national energy production directions, which have been associated to an increasing involvement of national and local counterparts already spread across transnational producing regions, also interlinked with an environmental protection sphere, which can determine distributional effects for the embedded communities.[76]

Community development approaches have been critically undertaken in reference to nuclear power commercial facilities and local politics of renewable energy plans. In fact, a critical understanding maintained in social cognitive theories has been discussed according to the emergence of local regulation strategies, which have been tackled through national decision-making processes that have referred to the social participation for a comparative institutional engagement planned in the long-term. Public-private regulatory partners have had necessarily to deal with, among other issues, multi-level energy facilities management and assessments converging on: local legitimacy conditions, transparency development measures, and local procedural assessments, conducted according to multiple governance networks diffused across regional states, also active with the operational standardization of multilateral proceedings for correlated intervention mechanisms.

For instance, collective environmental orientations do matter, especially about local risk perception cases related to the nuclear power energy and nuclear waste management facilities operating in Japan; a country that has historically been marked by critical war events, which have characterized common governmental patterns that were established since the WWII post-war nuclear drama; as has been noted, "in real life as in television the first discussions of nuclear

---

76   See: Joseph Heath, Wayne Norman, "Stakeholder Theory, Corporate Governance and Public Management: What Can the History of State-Run Enterprises Teach Us in the Post-Enron Era?" *Journal of Business Ethics* 53/2004, pp. 247–265.

power at a policy level [was] in early 1950, the physicist Taketani Mitsuo argued that in principle, atomic power was a "wonderful thing" but acknowledged that the nuclear age today has a "tragic meaning."[77]

The Japanese civil society mobilization about nuclear power energy and radiological waste has involved state's regulation directions and the public affirmation of multilateral civic expressions, which have been aggregated in terms of mobilized civil oppositions maintaining a municipal base. During recent decades, especially after the 2011 Fukushima-Daiichi nuclear power plant industry disaster, the multiple follow-up disasters at commercial NPPs have led to investigative and interpretative legal measures, associated with constant reformulations of strategic and ethical management issues. When we try to redefine in a simple sentence the interrelated managerial technical issues of the post-Fukushima-Daiichi nuclear disaster crises, there is only one simple question that needs to be asked "How do local dynamics work?."[78]

For the direct alleviation of environmental proximate dangers related to nuclear structural exposures, the incoming incidental explanations that have been provided on local facilitation recovery plans and potential coordinating dynamics, have come closer to comparative degrees of understanding acquired in terms of public policy expectations and specific statutory undertakings. This coordination process can be supported according to the evolution of incidental environmental plans that have differently involved the public local perceptions that include common local reactions, and local practical considerations.

In combination with science and innovation policy strategies, the local municipalities have been able to establish country-based linkages, which have tentatively been put in line with energy production systems through efficiency requirements that have been integrated with the knowledge management process. As a result, there have been formal renewed assessments despite the critical supporting conditions regulated in multi-level environmental governance modes about the nuclear power energy programs. Nonetheless, local states' agencies put in association with key innovative leaderships have also reached a point of contradiction, in particular, about the enlargement of land-based environmental effects, which have been tackled for the systematization of local cooperation initiatives and participatory progressive relations, established within same communities.

---

77 Martin Dusinberre, and Daniel P. Aldrich, "Hatoko Comes Home: Civil Society and Nuclear Power in Japan," *The Journal of Asian Studies* 70.3/2011, pp. 683–705. p. 686.
78 Dusinberre, and Aldrich, "Hatoko Comes Home," pp. 683–705. p. 703.

For instance, in Canada's technology case the formal coordination levels that have been defined according to societal shared principles and common trust values, interrelated with individual ethical attachments, and public local commitments, have also been specified dynamically for different changing environments. Essentially, in the foreseeable future, the comparative functional environments can be affected by relational distributive changes, which can modify the evolution of country-based technological innovation programs, that can be shifted toward complex energy development designs.[79]

Basically, the public national or local communication kept about science innovation and science technology has been essential to everyone; but it has remained quite dependent on the material diffusion of institutional learning practices, as well as, on the implementation of local knowledge transfers intertwined with regional knowledge partnerships about science and innovation programs. Under such conditions, the social learning experiences, which have converged on a number of different cognitive fields, based on behavioral science disciplines, have also included a comparative understanding conceptually interrelated to interdisciplinary areas focused on e.g. business administration, environmental governance, sociology, anthropology, and the social sciences.

For the purpose of promoting multi-level participatory networking in local governance, the inclusion of public social dialogues and local motivational understanding has come as a helpful time window. The multilayered societal and behavioral elements, in turn, have been integrated according to the technological and socio-economic development stages, which have been tested through the availability of common value-based preservation systems, concurring on similar ethical ideals. In which case, the adoption of adaptive environmental governance that can form in conjunction with local participative approaches, has reflected the performing planning exercises that have been advanced through the experimental consensus-based models for local sustainability conditions.[80]

---

[79] Mary Roduta Roberts, Grace Reid, Meadow Schroeder, and Stephen P. Norris, "Causal or Spurious? The Relationship of Kowledge and Attitudes to Trust in Science and Technology," *Public Understanding of Science* 22.5/2011, pp. 624–641.

[80] Monika Kurath, and Priska Gisler, "Informing, Involving or Engaging? Science Communication, in the Ages of Atom-,Bio- and Nanotechnology," *Public Understanding of Science* 18/2009.

CHAPTER TEN

10.1 National Configurations

In the case of nuclear energy policy and technological industrial development, this combined dimension has inherently implied an evolution of domestic rulings on management innovation assets, that concur to the facilitation of standardized international maintenance programs. At managerial level, the possible integration of multilateral monitoring programs has certainly been aligned with public knowledge and learning exchanges, while specifying the formal regulatory measures able to tackle national programming objectives and local monitoring issues. A resulting structural interaction serving the purpose of establishing public local safety and security initiatives has been consequentially linked to respective institutional knowledge diffusion processes, relying on comparative learning domains.[1]

Moreover, environmental science and innovation in contemporary operational domains have redrawn the boundaries for nuclear energy power assets and transnational industrial programs. Let's consider, for example, the case of damage risk analyses concerning unexpected nuclear radioactive accidents like the 1999 Japan Tokai-mura[2] event; which have either directly or indirectly increased the systematic incentivization of territorial advocacy plans. For public local safety and decentralized security measures, established under the regional management of private local operators, engaged on the fuel reprocessing industrial activities, the centralized energy companies have also adhered to territorial transitioning stages, that are adapted through local functional plans.

In Japan's regional nuclear security context, we can highlight that public roles and national responsibilities have been shared exclusively by key stakeholders which have included: the ministerial agencies, the industrial trading actors, the international financial institutions, the regional scientific experts' commissions, among others. In fact, these functional inter-organizational representative agencies have been put in charge of leading the science technology programs, which have indirectly engaged national voluntary groups.

---

1 See: Justin Alger, Trevor Findlay, "Strengthening Global Nuclear Governance," *Issues in Science and Technology* 27.1/2010, pp. 73–79.
2 See: Mary Byrd Davis, "A 'Blue Flash' Hits Tokai-mura," *Earth Island Journal* 15.1/2000, p. 26.

Concerning the public-private partnerships PPP dimension, national planning efforts conducted on nuclear energy power plants NPPs have been comparatively driven toward the provision of systemic regulatory identifications, which have addressed concurrent levels of national local responsibilities, in order to facilitate and commercialize the nuclear fuel reprocessing technologies, also in view of future development strategies about the management of complex energy landing systems.[3]

In retrospective terms, it can be considered that the modern involvement of Japan in the nuclear power development industry has continuously surfaced across recent decades, also through long-term industrial cooperative commitments, which have been made for the reason of enhancing national comparative energy systems and territorial distribution capacities, while preserving independent orientations on public-private energy security safeguards.

However, these national energy distribution plans which have been expanded in Japan during the recent past, have consequently been adapted according to post-World War II interdependent contexts. The national political environment has been combined with the presence of international diplomatic allies, because of post-war conditional security cooperation frameworks, which were undertaken in agreement with the United States.

Despite the actual development of international supporting conditions, with security safeguards agreements provided in order to improve the comparative energy capacity systems and renewable technological acquisitions, the civic societal counterparts instead have focused more attention on local communities' environmental risk impacts, and public risk perceptions about nuclear implemented technologies, while also remaining attentive especially at municipal local level. Therefore, the delivery capacity process in Japan's case has, in fact, determined mounting preoccupations for communities involved, because of the increase of national plutonium stockpiles,[4] and nuclear power energy local maintenance problems, reported at NPP construction sites.[5]

---

3   Susan E. Pickett, "Japan's Nuclear Energy Policy: From Firm Commitment to Difficult Dilemma Addressing Growing Stocks of Plutonium, Program Delays, Domestic Opposition and International Pressure," *Energy Policy* 30/2002, pp. 1337–1355.
4   Note: In view of Japan's recycling programs, it is highlighted that "Japan has nearly 40 tons of plutonium in storage, fuelling international concerns over its potential for use in nuclear weapons." Junichiro Nagasaki, "Utilities Move to Reduce Plutonium Stockpiles Held Overseas," *The Asahi Shimbun*, National Report section, March 8, 2022. Web-page available at: https://www.asahi.com/ajw/articles/14556963
5   Pickett, "Japan's Nuclear Energy Policy," pp. 1337–1355.

In essence, some major difficulties about nuclear energy plans and territorial municipal arrangements that are undertaken for the local management of spent nuclear fuel storage activities, have been raised in connection to national risk communication responses, and effective environmental information made available for local residential communities. In critical emergency response cases, the local participant groups have formalized public demands for national relief assistance supports and local mediation activities run by governmental agencies, as well as, by transnational industry representatives. The formulation of relief assistance programs has basically been introduced in strict correlation with municipal decision-making processes, including public regulatory mandates requiring the application of specific industry infrastructural configurations, which have been progressively adapted for nuclear power plants and waste disposal facilities.[6]

At critical times, the nuclear power energy safety provisions for fuel cycles operated in local reprocessing facilities have drawn attention to the multilateral decisional process developed with respective legal counterparts that can interact across the mediation table, while analyzing the preparation of centralized public arrangements, also drafted by political practitioners in this field. The actual integration of national and local public representatives for open mediation activities and local discussions undertaken with regional public-private energy groups has been essentially reaffirmed because of national security issues of NPPs, which can foster in time public environmental pressures for common regulatory decisions, and for common understanding of environmental compromising efforts. Countervailing proposals have been carried out in order to promote the sustainability of nuclear science projects, interrelated with territorial technology policies legitimized through the acceptance models of multilateral governmental frameworks.

In fact, sustainable energy development has been supported in terms of a rational deployment of regulatory planning measures, which must be able to ensure the public interest on local safety prevention and protection measures, also in cases of critical incidents. In terms of resulting trials, there have been both commonalities and differences about sectoral organizational demands, that refer to: the specification of energy production coordination impacts, the status of nuclear power energy programs, the evaluation process of waste disposal systems, or the reduction of local proliferating risks posed under uncertain conditions.[7]

---

6   Pickett, "Japan's Nuclear Energy Policy," pp. 1337–1355.
7   See: Paul L. Joskow, John E. Parsons, "The Economic Future of Nuclear Power" *Daedalus* 138.4/2009, pp. 45–59.

However, the imposing variety of public-private PPP organizational roles intertwined with multi-level adjustment processes selecting economic, environmental, and political factors, for cooperative exploration areas, has similarly added complexity to the development of national regulatory characteristics; which should be able to match the public needs for accessible and cost-effective technologies. The transnational PPPs have been based on social diffusion models and consumption trading patterns[8] which have evolved according to the availability of national industrial resources, and transnational multi-level energy delivery capacities.[9]

In essence, the comparative degree of differentiation for renewable energy sectors and nuclear power industry assets, has become evident through regional states' economic configurations that capitalize on socio-economic interests, especially for the security and stabilization of independent regulative systems. Such type of national confrontation set between international, regional, and domestic representative entities has progressively amplified the structural interdependent needs for enhancing long-term sustainable energy plans, which have been established according to multilateral assessed conducts performed at the local, national, and transnational level.

In addition to this, under the influence of comparative geostrategic political options, the corresponding formation of decision-making activities has been the result of critical policy arrangements, which have been produced in parallel with national energy requirements, and local marketing exchanges put in connection with regional trading blocs, divided across the European Common Market, the International Monetary Union, or the Post-Soviet Republics in Central Asia, and Eastern Europe.[10]

Specifically, for the nuclear power industry production lines, the complex regulation environments associated with challenging local management issues have caused in the past strained relations across world regions, particularly for the case of nuclear facilities' accidents at the Three Mile Island,[11] the Chernobyl

---

8   See: Josep Vives-Rego, Serge Caschetto, Jordi Faraudo, Diego Prior, "Management Options for the Increasing Demand of Energy and Water: Is the Problem Soluble in Technosciences Only?" *Ambio* 37.2/2008, pp. 134–136.
9   Ioannis N. Kessides, "Nuclear Power and Sustainable Energy Policy: Promises and Perils," *The World Bank Research Observer* 25/2009, pp. 323–362. (Published by Oxford University Press on behalf of the International Bank for Reconstruction and Development).
10 Kessides, "Nuclear Power and Sustainable Energy Policy," pp. 323–362.
11 See: 1) Axel Roesler, "Lessons from Three Mile Island: The Design of Interactions in a High-Stakes Environment," *Visible Language* 43.2/3/2009, pp. 169–195; 2) NPR,

disaster,[12] and the accidental releases of radioactive material in the Russian Federation, Canada, or the United States.[13]

10.1.1 Centralized Operations

The common adherence to effective public-private PPP energy approaches has continued in line with the revalidation of alternative distribution transfers in the national energy market supply, with multiple associated uses of compatible advanced technologies. Essentially, due to the minimization of environmental energy crises, states' regulatory agencies need to establish comparative multi-risk development approaches, which also involve local environmental risk assessments, for the purpose of reducing the increase of potential environmental risk exposures; the main result has been to try to reach sustainable energy targets while enforcing national environmental measures with more stringent control mechanisms.[14]

Indeed, the industrial maintenance process of nuclear power energy plants NPP, with local disposal activities of nuclear contaminated waste has prompted the public and private energy operators to actualize the formation of territorial security options, also because of the national compliance plans formulated about sustainable storage measures, which have been carried out in different organizational environments at local, regional, or transnational level. Overall, despite substantial increases on transnational energy demands,[15] the diffusion of technical local barriers has influenced the growing intensification of political adaptation action-plans for either approvals or disapprovals about nuclear power energy development. In terms of local mitigation action plans, and direct

---

"Profile: How The Residents and Activists Surrounding Three Mile island Feel Regarding their Community 20 Years After the Nuclear Accident," *Weekend Edition. Saturday* 1/1/1999 (Washington, D.C.: National Public Radio); 3) John T. Scholz, "American Politics – Hostages of Each Other: The Transformation of the Nuclear Power Industry After Three Mile Island by Joseph V. Rees," *The American Political Science Review* 89.2/1995, p. 503.

12 See: Ismail Aktar, "A Comparison of the Effects of the Chernobyl and Three Mile Island Nuclear Accident on the U.S. Electric Utility Industry," *Sosyoekonomi* 2.11/2005, pp. 13-33.
13 Kessides, "Nuclear Power and Sustainable Energy Policy," pp. 323-362.
14 Kessides, "Nuclear Power and Sustainable Energy Policy," pp. 323-362.
15 See: Kenichi Matsui, "Global Demand Growth of Power Generation, Input Choices and Supply Security," *The Energy Journal* 19.2/1998, pp. 93-107.

or indirect community mediation effects,[16] the nuclear energy compliance policies have required constant adequacy and supporting innovative frameworks, in order to secure global energy availability sources, despite vulnerable risk transitions.

In brief, for science policy assessment models, the political equity accesses as well as structural public interventions have been launched in connection with the duality of regulatory incremental objectives pursued through PPPs. But in order to be able to regulate through different institutional and industrial programmatic orientations, the public-private evaluative commissions run by regional development field experts, have effectively reconsidered the societal conditions for possible redistributive innovation programs, also undertaken for the promotion of scientific knowledge aims identified in view of regional industrial assimilations. Similarly, there is a dividing line in governance for the inclusion of public acceptance models in the case of pro-nuclear or anti-nuclear civilian mobilization platforms;[17] which essentially have spontaneously emerged in relation to local recognized concerns over the technical nature of transnational nuclear industrial arrangements, extended for multi-level energy planning systems, involving the direct impact measures.[18]

For instance, during the last four decades the drastic industrial energy production impacts have been witnessed by anti-nuclear civilian alliance groups, which have opposed the localization of nuclear power facilities with waste burial proposals in countries such as the U.S., Germany, Italy, UK, and France. The reasons for significant civic oppositions have been related to the fragmentation of information available, provided through the reporting activity of multinational energy companies, with integrated technical analyses reclassified according to the local changing dynamics about nuclear power pollution aspects, as well as, the roles of management performing networks that operate in multi-level

---

16 See: Edward A. Parson, Karen Fisher-Vanden, "Integrated Assessment Models of Global Climate Change," *Annual Review of Energy and the Environment* 22/1997, pp. 589–628.
17 See: 1) Cy Gonick, "Are We Coming to the End of the Growth Era?" *Canadian Dimension* 46.2/2012, pp. 40–43. 2) Sherry Cable, "Political Processes and Institutions – Mobilizing Against Nuclear Energy: A Comparison of Germany and the United States by Christian Joppke," *Contemporary Sociology* 24.1/1995, p. 49.
18 Herbert P. Kitschelt, "Political Opportunity and Political Protest: Anti-Nuclear Movements in Four Democracies," *British Journal of Political Science* 16.1/1986, pp. 57–85.

complex energy domains;[19] in which case the main involved PPP actors have redefined comparative expectations through decision-making discussions laid out among multilateral or bilateral socio-economic partners.[20]

Moreover, the comparative review activity of international and national energy policy experts has indirectly had the effect of creating a correspondent managerial *compartmentation* linked to the constitution of knowledge information approaches. In both common spheres of available knowledge information and local knowledge management associated with public local regulations, the civilian environmental campaigning actions have grown according to specific cognitive environmental domains.[21] For the specialized cognitive notions focused on nuclear science policy and normative adaptation conditions, there have been concurrent public restrains and adjustments pursued through the political interrelated areas, in order to keep at distance societal local parties partly involved in cognitive disclosures about civil nuclear management programs, and future energy development strategies.[22]

At any rate, a dichotomous interplay of civilian local protests, and societal mobilizations across more recent decades (1990s-2000s) can be observed because of the increase of public environmental concerns. Retrospectively, we may observe that a crucial regulatory variation about multinational energy power maintenance systems and country-level municipal local responses was enhanced. Social environmental responsiveness has been the result of political differentiation elaborated through comparable alternatives, which have provided mutual confrontation grounds for the common restructuring of civilian opposing coalitions that have already manifested in the case of nuclear power energy objectives, and management of radioactive waste disposal sites, with some additional purposes of contributing to the revision of national energy reforms.

When in place, public-private energy reformed strategies within PPP, associated with contingent local planning stages have aimed at the revaluation of regional distributed environmental capacities, targeted according to different territorial needs of comparative localities. Fundamentally, the environmental planning reviews combined with local energy adaptation cycles have been

---

19 See: Nathalie Berny, "Europeanization as Organizational Learning: When French ENGOs Play the EU Multilevel Policy game," *French Politics* 11.3/2013, pp. 217–240.
20 Kitschelt, "Political Opportunity and Political Protest," pp. 57–85.
21 See: Erica Simmons, "Grievances Do Matter in Mobilization," *Theory and Society* 43.5/2014, pp. 513–546.
22 Kitschelt, "Political Opportunity and Political Protest," pp. 57–85.

followed up according to the nature of regional trading efforts of complex dynamic societies.

For ascribed terms that pertain to every distinct managerial energy sector, the engaged societies and local communities have been tied up in more closer and open relations for the facilitation of mutual functional relationships, which have affected the public understanding, as well as, the comparative legal dispositions about recurrent participatory policies which tend to focus on social environmental campaigns, e.g. about: how to better express nuclear energy policy today? Pro-active democratic states such as Sweden, France, U.S., and Germany have been confronted with similar political environmental backgrounds, developed through a combination of normative and social identity perspectives.[23]

Therefore, a possible societal knowledge intersection applied across the public regulatory fields referring to national renewable energy agreements has emerged, putting into evidence the distribution of single states' production capacities, and the different orientations adopted about the maintenance of national and local public dialogues. On a temporal scale, public local dialogues have been analyzed in association with the actual characterization of regulative political systems. But at the same time, both policy parties and state representatives have been faced with national local regulatory issues, including the corresponding levels of public acceptance and public participation that have been shaped through a common assimilation of civic state models for the arrangement of interdependent elements; which have encompassed regulative procedural strategies, while addressing the national formalization of public environmental assertive recognitions.[24]

To these objectives, an element of relativism about organizational industrial relationships has emerged, as they have been constantly modified in terms of centralized or decentralized development determinants. Depending on the level of public mobilization, the social determinism has been reflected into the flow of institutional dynamics, because of the integration of interrelated politics and local planning approaches. The composition of multi-level policy making processes has been made possible through the material correspondence of states, which have been reliant on natural local resources that have led to applicable managerial designs, and connected frameworks about the maintenance of multilateral energy innovation systems and specialized managerial networks.[25] In case that we decide to move a bit closer to focused conceptual meanings about the

---

23 Kitschelt, "Political Opportunity and Political Protest," pp. 57–85.
24 Kitschelt, "Political Opportunity and Political Protest," pp. 57–85.
25 See: David B. Spence, "Regulation, 'Republican Moments,' and Energy-Policy Reform," *Brigham Young University Law Review* 5/2011, pp. 1561–1623.

modern environmental governance and local strategic planning routes, several criteria involving correlated defining elements have been aggregated, such as:

1. Normative & structural specifications
2. Provisional quantitative-qualitative basis
3. Industrial management internalization
4. States descriptive and operable guidelines
5. Level of local context and comparative applicability

Source: (Powell, 2001)[26]

10.1.2 Mutual Communities and Critical Knowledge Intersections

The reformulation of environmental adaptive models has been studied in order to readdress not just the public choices dimension, put in relation with public ecological aims, but also to re-frame mutual PPP perspectives, for the possible industrial planned scenarios, which can be foreseen about nuclear power energy projects and local sustainability innovation programs.[27]

The relative convergence of structural PPP performances territorially defined in order to bring the states to common functional identifications, based on national or transnational plans, has required concerted formalized actions normatively accessible and, at the same time, technologically justified. Basically, the combination of dimensional local provisions related to production flexibility factors has additionally influenced the regional industrial states, focused on comparative evaluation reviews and technical classification methods, which tackle the implementation of modified innovation systems. In a way, the territorial facilitation of transnational market relations has corresponded to either direct or indirect institutional adaptative management with local operational impacts,[28] such as for the case of nuclear energy security agreements drafted for national commercial facilities.[29]

---

26 J. H. Powell, "Generating Networks for Strategic Planning by Successive Key Factor Modification," *The Journal of the Operational Research Society* 52.4/2001, pp. 369–382.
27 Powell, "Generating Networks for Strategic Planning," pp. 369–382. p. 370.
28 See: 1) Atsuyuki Suzuki, "Toward a Robust Nuclear Management System," *Daedalus* 139.1/2010, pp. 82–92. 2) Annika Beelitz, Doris M. Merkel-Davies, "Using Discourse to Restore Organizational Legitimacy: 'CEO-Speak' After an Incident in a German Nuclear Power Plant," *Journal of Business Ethics* 108.1/2012, pp. 101–120.
29 Powell, "Generating Networks for Strategic Planning," pp. 369–382. p. 370.

Besides there have been some retaining aspects concurring to the future projection of strategic business plans identified together with knowledge acquisition processes, which have been contextually required, for progressively redefining the national provisions with public consensus supported through comparative societal bases. The public acceptance model has been meant for local incorporation in terms of changing organizational transfers, completed through the modification of functional factors that have been developed in the case of local environmental innovation and local decision-making procedures. For instance, for the case of nuclear energy facilities' remediation activities, the environment process operated in the United States – for contextual nuclear power facilities located in Denver, Rocky Flats in Colorado[30] – has been framed with participatory collaborative platforms, and methodology preparation activities that have been assessed for long-term environmental remediation processes.

This example about the U.S. Colorado NPP decommissioning based-case[31] has offered the possibility to confront multiple socioeconomic issues[32] about ecological radioactive contamination and policy planning, referred to long-term managerial organization, which has been put in association with collective societal determinants, that recall attention upon the performing roles and responsibilities of local participatory officials, public health protection practitioners, and local environmental economic experts, mostly involved in the planning map for cooperative advanced projections.[33] This type of institutional understanding in regional and local municipal cases has been enhanced by the comprehensive inclusion of local participant groups who have collectively entered into the local

---

30 Seth Tuler, and Thomas Webler, "Competing Perspectives on a Process for Making Remediation and Stewardship Decisions at the Rocky Flats Environmental Technology Site," *Long-Term Management of Contaminated Sites Research in Social Problems and Public Policy* 13/2007, pp. 49–77. Elsevier Ltd.

31 See: Shiloh R. Krupar, "Transnatural Ethics: Revisiting the Nuclear Cleanup of Rocky Flats, CO, Through the Queer Ecology of Nuclia Waste," *Cultural Geographies* 19.3/2012, pp. 303–327.

32 See: IAEA 464 Report. "Fort St. Vrain NPP in Colorado was shut down in the early 1990s. As the plant had never performed satisfactorily, shutdown was not a great surprise. Nevertheless, an adverse social impact of shutdown was experienced as the site is in a somewhat remote area. It has more recently been rebuilt as a gas fired plant, which has ameliorated the situation." International Atomic Energy Agency, IAEA, Technical Reports Series No. 464. "Managing the Socioeconomic Impact of the Decommissioning of Nuclear Facilities." IAEA, Vienna, 2008. p. 44.

33 Tuler, Webler, "Competing Perspectives on a Process," pp. 49–77.

constitution of the 'Rocky Flats Coalition'[34] for instance, in order to facilitate the land management transition process for the public environmental reconversion of the same public area.[35]

However, there have been different considerations about the NPP nuclear clean-ups processes, taking into account the practical local limitations due to technical oversimplifications of systematic management measures, that involve the adoption of different multilateral perspectives, which determine variable provisions of multi-level transnational regulatory responses.

In the making of environmental remediation plans that address the NPP administrative conditions regarding the local hazardous risks and radioactive waste materials processed for the final disposal sites, the regulatory management mandates have been responsive to the supporting of sustainable critical modalities, which serve in terms of local pre-emptive safety actions, and technological adaptations for nuclear energy facilities, also reviewed in consideration of local modified designs.[36]

Moreover, the objective sustainability of governmental participatory approaches has been addressed through socio-technical functional options interrelated to local renewable energy supporting systems. The sustainable regulatory approaches have been reaffirmed in order to continue to serve the public for the common good. They have been developed in order to try to establish comprehensive knowledge dispositions, functionally based on constant needs for local technical renewals of NPP energy operating systems, which have been

---

34 Note: For the national legacy of Rocky Flats, the U.S. DOE confirmed the regulatory changes: "Cleanup [was] completed under the Comprehensive Environmental Response, Compensation and Recovery Act (CERCLA) and Colorado Hazardous Waste Act (CHWA) / Resource Conservation and Recovery Act (RCRA). Rocky Flats Cleanup Agreement (RFCA) is a legally binding agreement on cleanup decisions between DOE, U.S. Environmental Protection Agency, and Colorado Department of Public Health and Environment (CDPHE). Significant participation in cleanup decisions [was made] through participation of Rocky Flats Coalition of Local Governments, Rocky Flats Citizens Advisory Board, and numerous public meetings." Scott Surovchak, "Cleanup and Remedy Implementation at the Rocky Flats Site, Jefferson County, Colorado," *Department of Energy (DOE)*, Interstate Technology and Regulatory Council, Office of Legacy Management (LM), 2011, page 8.
35 Tuler, Webler, "Competing Perspectives on a Process," pp. 49–77.
36 Kevin M. Kostelnik, James H. Clarke, Jerry L. Harbour, Florence Sanchez, and Frank L. Parker, "A Sustainable Environmental Protection System for the Management of Residual Contaminants," *Long-term Management of Contaminated Sites Research in Social Problems and Public Policy* 13/2007, pp. 117–137.

put in line with the regulatory mandates of socioeconomic interacting environments, specifically indicating the common operational time-bound ground.

Additionally, it has been specified that "[the] institutional responsibilities can be viewed as a series of critical functions and activities that must be continually conducted to ensure that the remedial system performs as designed. The functions of a sustainable contaminant isolation system include: maintaining active remedial processes, maintaining the engineered barriers, maintaining institutional control"[37] Essentially, the problem of seismically isolated nuclear facilities is often recalled due to the need to *communicate* not only at the level of daily maintenance operations and configured performance systems, but also according to modified internal changes introduced at different stages to national governmental bodies, through programmatic relations and local managerial innovating plans.

10.1.3 Material Dualist Interplay

Multilateral public choices can basically lead to the identification of multiple constructive options that have been pursued according to specific time-frames. While, at the same time, multi-level managerial innovation decisions have endured conflictual positions over the negotiation of adaptive administration plans, which have been previously established for national innovation energy facilities, maintained for long-term periods.[38]

In addition, institutional decision-making can get very complex in the case of contractual factors reviewed for long-term monitoring and evaluation plans, which have been intertwined with local risk management practices.[39] Some authors have emphasized the aspect of codependency within PPP dualistic management ecosystems. Codependency theory referred to knowledge management innovation and centralized or decentralized public affairs has suggested the formation of composite dualist adaptations, including major organizational changes and practical industrial approaches that have tackled e.g.: the natural resources management and environmental protection rules; the service distribution networks and local capacity provisions; and the land environment remediation

---

37 Kostelnik, Clarke, Harbour, Sanchez, and Parker, "A Sustainable Environmental Protection System," pp. 117–137, p. 123.

38 Kostelnik, Clarke, Harbour, Sanchez, Parker, "A Sustainable Environmental Protection System," pp. 117–137.

39 See: Genevieve Fuji Johnson, "The Discourse of Democracy in Canadian Nuclear Waste Management Policy," *Policy Sciences* 40.2/2007, pp. 79–99.

practices; progressively the compatible regulatory systems have been reformed through local implementation practices and transitional environmental norms, periodically extended to local management schemes.[40]

We can also specify the fact that governmental ecological plans involving the local arrangements and centralized operable transitions – put into practice through the land management process and local environmental remediation activities – have remained interdependent for the applicability and effectiveness of governance practices, which have been established at multi-level territorial contexts.[41]

In essence, the territorial dynamics can be analyzed according to comparative PPP specialized knowledge areas, including the systematic reporting activities that refer to common local environment conditions on e.g.: the actual size of local population involved; the organizational functioning bodies and local representative bureaus; or the industrial management agencies. In terms of multi-level PPP cooperation and adaptation programs, public policy decisions and corresponding local inclusive proceedings have entailed regulative control mechanisms, which have had time-related limitations but also affirmative operational impacts, felt at the local agency level, for supporting either national or regional adaptation processes.[42]

For instance, in regard to U.S. Department of Energy (DOE) nuclear sites, it has been pointed out that "the US nuclear weapons complex comprises approximately 5000 facilities located at 16 major sites and more that 100 smaller sites. Some 113 of the DOE sites around the country contain chemical and radiological waste generated by the production of nuclear weapons"[43] This means that a radical combination of long-term NPP nuclear clean-ups that include local management approaches has been developed in consideration of local governmental performances.

---

40 Joanna Burger, Nellie Tsipoura, Michael Gochfeld, and Michael R. Greenberg, "Ecological Considerations for Evaluating Current Risk and Designing Long-Term Stewardship on Department of Energy Lands," *Long-term Management of Contaminated Sites Research in Social Problems and Public Policy* 13/2007, pp. 139–162.
41 See: Susan H. Bragdon, Marc Pallemaerts, Veerle Heyvaert, Iwona Rummel-Bulska, "International Hazard Management Other Than Nuclear," *Yearbook of International Environmental Law* 6.1/1996, pp. 275–292.
42 Burger, Tsipoura, Gochfeld, Greenberg, "Ecological Considerations for Evaluating Current Risk," pp. 139–162.
43 Burger, Tsipoura, Gochfeld, Greenberg, "Ecological Considerations for Evaluating Current Risk," pp. 139–162. p. 143.

Restorative environmental measures, targeting different nuclear energy industrial sites, have been integrated with public local planning activities, which have essentially been conducted through direct/indirect involvement of major mobilized civilian groups, which have been present in collaboration with: a. the government transitional bodies; b. the industry transnational corporations with open table discussions; c. as well as the participatory advocacy networks. As such, the cooperating civilian networks have become an integral part of local responsive and interactive processes that have been based in the U.S., but also elsewhere.[44]

Similarly, another major example is the local rehabilitation and resettlement provisions that have been planned for the Australia's islands in the western Pacific region, in which the institutional programming aims and practical effective results have been connected to: "[the] U.S. Department of Energy (USDOE) and the Australian Nuclear Science and Technology Organization (ANSTO) as well as [to] a host of other agencies, [for] over fifty years and hundreds of millions of dollars on the rehabilitation of former nuclear weapons test sites ... they [basically] hope to turn over to traditional landowners."[45]

In sum, the national nuclear energy adaptive configurations, which have resulted from local rehabilitation and resettlement policies, have been developed through the influence of national compromises affecting public trust issues. Because, both technical productions processes and societal participatory agreements have particularly put into evidence the local regulatory aspects of multilateral management plans about e.g. the removal and transportation of contaminated soils, in addition to the reduction of local environmental risks and the safety protection concerns about health and food consumption. As said, the safety reduction risk policies have been perceived in association with social preservation scopes, for the pursuance of traditional local practices performed in the industrial exposed locations of nuclear testing sites.[46]

Essentially, by taking into account comparative governance analyses related to the mutual process of local managerial adaptation in the case of long-term environmental stewardship programs, some authors like Bierly, 2007; Rosa,

---

44 Burger Tsipoura, Gochfeld, Greenberg, "Ecological Considerations for Evaluating Current Risk," pp. 139–162.
45 Anne Ballou Jennings, Amy M. Seward, and Thomas M. Leschine, "Living in a Nuclear Landscape: Rehabilitation and Resettlement of Proving Grounds in Australia and Islands of the Western Pacific," *Long-Term Management of Contaminated Sites Research in Social Problems and Public Policy* 13/2007, pp. 165–192. p. 166.
46 Jennings, Seward, Leschine, "Living in a Nuclear Landscape," pp. 165–192.

National Configurations 207

2007, Leschine 2007, have drawn attention to some emerging elements about environmental collaboration policies. There have been interrelated collaborative changing conditions at institutional level, which have determined different degrees of local programmatic actions, performed through comprehensive territorial targets requiring e.g..: the flexibility of adjustments; the national or local characterizations of specified policies; and the active environment exploration pursued despite the presence of managerial uncertainties.[47]

In fact, national regulatory levels of co-operation and effective integration for common normative reforming stages have been basically modified for the progressive alignment of long-term practices and local organizational approaches. These l regulative alignments have been meant to assimilate different contextual industrial aggregations, which correspond to the technological factors and knowledge innovation programs.

This substantiation of technical knowledge factors has also included the renovation of participatory supporting attitudes fostering public environmental awareness. Essentially, the public information agencies and local environmental networks can favor the affirmation of local developmental plans in view of long-term socio-economic regulative models, bringing to resulting PPP upgraded activities for e.g.: the land remediation programs, or the local environmental measures.[48] Moreover, the public regulation of ecological/environmental local policies has been directed toward the registration of comparative participatory assessments, while associated to the actual implementation of ecological management plans. Then, the supporting PPP on national industrial review schemes have been adapted according to structural ecological differences tested at specific energy industry's localities.

In distributed territorial municipalities which have adapted through local regulatory activities and transnational managerial programs (e.g. in the U.S. federal system), states have tended to favor the institutional local accesses to knowledge

---

47 See: David D. Briske, "Translational Science Partnerships: Key to Environmental Stewardship," *Bioscience* 62.5/2012, pp. 449–450.
48 Denise Bierly, "A 19-Year Perspective on Long-Term Care Issues," *Long-Term Management of Contaminated Sites Research in Social Problems and Public Policy* 13/2007, pp. 213–226. See also: Eugene A. Rosa, "Long-Term Stewardship and Risk Management: Analytic and Policy Challenges," *Long-Term Management of Contaminated Sites Research in Social Problems and Public Policy* 13/2007, pp. 227–255. See also: Thomas M. Leschine, "Introduction: Long-Term Management of Contaminated Sites," *Long-Term Management of Contaminated Sites Research in Social Problems and Public Policy* 13/2007, pp. 1–10.

innovation and societal integrated processes, while also reinforcing the collaboration of interacting responsible entities that have been based at federal, state, and province levels.[49]

The actual propagation of ecological collaborative networks operating within states' planning and organizational boundaries,[50] has been dependent on transnational characteristics that refer to PPP public-private adaptation schemes, which have developed constructive local planning initiatives for common environmental objectives. In fact, the regional environmental partnerships have been modified in time according to a progressive search of cultural and institutional arrangements that have been set in between private vs. public organizing spheres, which have been selected in connection with the formation of distinctive territorial responsibilities.[51]

The terms of social mutual representations developed according to environmental participatory programs, with public land classifications and governmental review assessments, which refer to local environmental protection issues, have been redefined for the establishment of national conservation practices validated in multilateral contexts, for the public utilization of local natural resources.

Participatory civil movements have been socially recognized through the emerging integration of environmental mutual alliances, which have been maintained both locally and nationally, when based on common ecological developmental aims.[52] In such conditions, are supporting cohesive levels of public discussions that have been conducted for the facilitation of civilian confrontation dialogues.

As such, public environmental objectives have also marked the formulation of national programmatic engagements that have been reviewed due to relevant local ecological issues. Through incremental organization processes, have been supported participatory tools for the knowledge management transfers and local

---

49 Christopher M. Weible, "Political-Administrative Relations in Collaborative Environmental Management," *International Journal of Public Administration* 34/2011, pp. 424–435.
50 See: 1) Joanna Burger, "Restoration, Stewardship, Environmental Health, and Policy: Understanding Stakeholders' Perceptions," *Environmental Management* 30.5/2002, pp. 631–640; 2) Charles T. Driscoll et al., "Science and Society: The Role of Long-Term Studies in Environmental Stewardship," *BioScience* 62.4/2012, pp. 354–366.
51 Weible, "Political-Administrative Relations," pp. 424–435.
52 See: Po-Hsin Lai, Michael G. Sorice, Sanjay K. Nepal, "Integrating Social Marketing Into Sustainable Resource Management at Padre Island National Seashore: An Attitude-Based Segmentation Approach," *Environmental Management* 43.6/2009, pp. 985–998.

networking activities, substantiating inputs provided through public environmental advocacy forums.[53] Eventually, follow-up practices and public policies combined with local environmental strategies have been formalized according to specified environmental targets, which involve a public revisionist scrutiny for reporting and reviewing acts.[54]

Therefore, the national configuration of decentralized ecological policies still matters in terms of public democratic exchanges, for the actual consolidation of socio-economic interests. The decentralized interconnection of local territorial domains has also engaged on multi-level comparative distributed responsibilities of formal representatives and local administration bureaus, or multilateral management groups, collaborating according to respective consultative PPP schemes within defined functional systems – which have promoted local environmental strategies, and local managerial cooperation, among other mutual commitments.[55]

In a way, the integration of local participatory knowledge has been shaped according to prospective public environmental interactions. Overall, the actual implementation of decentralized regulation linkages supporting the multi-stage adaptation process at federal, state, and municipal levels, has facilitated the affirmation of different environmental protection measures. Eventually, interdependent programmatic actions have been lined up according to different localities and regional constituencies, fixed within specified environmental domains; the corresponding local justice approaches and normative criteria, have been aggregated according to changing regulatory cases concerning, for instance, the hazardous chemical pollution found at transnational natural river basins.[56]

Moreover, a set of knowledge management determinants was explored in view of emerging trends on public ecological concerns. Governmental environmental

---

53 See: Ben S. Malayang III Ma., and Zita B. Toribio, (2002). "Networking, Environmental Management and Governance," *Produced by the Department of Environment and Natural Resources-United States Agency for International Development's (DENR-USAID) Philippine Environmental Governance (EcoGov) project through the assistance of the USAID under USAID PCE-1-00-99-00002-00.* Webpage available at: https://faspselib.denr.gov.ph/sites/default/files/Publication%20Files/Networking%20Environmental%20Mgmt%20and%20Governance.pdf?msclkid=c834552da91011ecb7eb95ff5f7d574f
54 Weible, "Political-Administrative Relations," pp. 424–435.
55 Weible, "Political-Administrative Relations," pp. 424–435.
56 Don Grant, Nell Trautner, Liam Downey, and Lisa Thirbaud, "Bringing the Polluters Back In: Environmental Inequality and the Organization of Chemical Production," *American Sociological Review* 75.4/2010, pp. 479–504.

determinants have been associated with national legal frameworks, and regional policy targets translated into formalized collaborative guidelines. This inclusive process has fundamentally implied local regulative changes, also about a renewed provision of public-private PPP mechanisms, combined with effectiveness criteria for local modified environments.

To add to this, depending on national environmental categorizations and different managerial relations, have been associated multiple levels of direct public interactions with national consultative bodies, in order to provide harmonized distributive policies, referring to, for instance, the local pollution restrictions of concurrent distributive bases of industrial compounds, for territorial *green* adjustments.[57] In a collaborative status, the changing ecological interdependent factors, that emerge within different energy production networks, internally arranged according to the territorial regulation plans, have led to the establishment of socio-economic reforms, which have been progressively advanced in order to favor the environmental integration policies, as well as, institutional local adaptation measures.

For instance, the focused knowledge-based practice approaches about ecological modernization spheres, have been taken into account for the national and local regulatory diffusion, which has been promoted within industrial innovation territorial bases; also the de-localization of organizational processes has been able to transform the regional production capacities, with resulting comparative energy systems.[58] In Table 8, we provide the NEA 2010 summary report through brief extracts, about public acceptance models conjugated with single countries' managerial approaches, for explorative material issues regarding the nuclear waste disposal programs, which have been commonly identified through elements as: *Voluntarism, Right to Veto and Collective Empowerment Criteria.*

---

57 Grant, Trautner, Downey, Thirbaud, 2010. "Bringing the Polluters Back In," pp. 479–504.
58 David Gibbs, "Prospects for an Environmental Economic Geography: Linking Ecological Modernization and Regulationist Approaches," *Economic Geography* 82.2/2006, pp. 193–215.

Table 8: Public Acceptance of Radioactive Disposal Programs

Country	*Voluntarism* Disposal site selection based on voluntarism	*Right to Veto* Formal veto rights granted to local or regional governments	*Empowerment* Community empowerment measures
Belgium	"ONDRAF/NIRAS invites volunteer municipalities from local communities as well as from interested communities for a new siting process." (p. 21)	"Municipal councils have the right to support or reject the development of a facility even though there is no legal basis for local veto right." (p. 22)	"Partnerships approach: local community directly involved in the process of facility design and socio-economic terms." (p. 21)
Canada	"Municipalities and the public have been engaged in the development of disposal facilities, in the environmental assessments as well as in long-term provisions." (p. 38)	"Each community has a veto power on the development of low-level nuclear waste (LLW) management facility for their area (p. 38). For low-intermediate level waste of deep geological repository (LILW DGR) community consultation advisory groups have been also formed for the regulatory process." (p. 39)	"Municipal engagement in the process with governmental funds for project-related costs. Expert consultants act as intermediaries for the municipalities that have hired them to assist with local citizens meetings, among others, and project proposal development." (p. 40)
Czech Republic	"A volunteer process in return for economic benefits is rarely (if ever) used in the industrial field of Czech Republic." (p. 48)	"Organized communities have strengthened their role for the siting process. Geological work requires the permission of the Ministry of the Environment. However local settlements do not have a right to veto. Despite local referendum existence overall state administrative issues will not be subjected to local referenda." (p. 48)	"Communication and mutual understanding approach: maintenance of good relations with the local population living nearby operating repositories. But only limited public support has been present at specific sites." (p. 49)

(Continued)

CHAPTER TEN

Table 8: Continued

Country	*Voluntarism* Disposal site selection based on voluntarism	*Right to Veto* Formal veto rights granted to local or regional governments	*Empowerment* Community empowerment measures
Finland	"Consultation process with affected communities for potential identification of 'volunteer sites' for fuel storage and final disposal." (p. 51)	"The country has adopted 'The Decision in Principle' (DiP) through which it is required by law a positive decision by the local municipality and regulators' support statements as preliminary conditions to build a repository in a specific area. It is the Municipal Council that has a right to veto and decision about the repository development process." (p. 51)	"Co-operation and follow-ups approach: involvement of municipality officials and elected officials including monthly assessments and regular information meetings. Companies' financial loans given to municipality agencies e.g. reconstruction plans." (p. 52)
France	"Economic and social benefits accorded to volunteer sites also for the process of radioactive waste management facilities." (p. 53)	"Local officials and members of the public from sites management are consulted for preliminary conditions before construction of disposal facility. However, communities have no veto power in France." (pp. 54–55)	"Local information and partnerships approach: work run with state representatives, in regional district and local governmental units, with Parliament members, Environmental groups and trade unions (p. 55) Local Communities get tax reimbursement for nuclear facilities projects in the form of an economic development tax and a technology diffusion tax. Budget from taxes finances local economy and employment." (p. 56)

Table 8: Continued

Country	Voluntarism Disposal site selection based on voluntarism	Right to Veto Formal veto rights granted to local or regional governments	Empowerment Community empowerment measures
Japan	"Volunteer site selection process run by NUMO a Japanese waste management organization. It invites volunteers from all municipalities to meet and confirm eventual satisfaction of geological conditions for the listing of volunteered areas." (p. 67)	"NUMO engages direct relations with the public for national discussions and round table meetings. The agency also maintains public forums, local residents' opinion exchanges and information networks about projects stages." (pp. 68–69)	"Socio-economic benefits for local municipalities going to education, welfare status and local industrial promotion with business opportunities." (p. 69)
South Korea	"Principles adopted for potential sites development of LILW include: • involvement of an independent site selection committee • Voluntary participation of local government • Democratic/open procedures for public acceptance • Financial support programs for local Communities." (p. 72)	"The suitability of the volunteer site is confirmed by the selection committee and their assessment is passed by the local government prior to a local vote in the area to determine local support for hosting the facility. This voting system concedes a local veto right." (p. 73)	"Hosting communities have been granted with annual disposal fees for the operation of the facility (with life duration of 50 years)" (p. 73)

Model adapted. **Source:** OECD/NEA, 2010. "Partnering for Long-term Management of Radioactive Waste. Evolution and Current Practice in Thirteen Countries." OECD/NEA No. 6823.

CHAPTER ELEVEN

11.1 Territorial Governance Connectivity

About intra-regional PPP innovation linkages, the related territorial research[1] focused on local constructive organizations and regional innovation systems (RIS),[2] has contributed to the diversification of knowledge innovation transfer approaches, in terms of local contextual adaptations. Comparable regional innovation approaches have been followed for the reason of proposing additional ways in which regional innovation organizations can be effectively connected to societal production activities.[3] Similarly, the actual development of institutional theories, in particular, has been explored through systematic management aspects of the local innovation cycles of industrial energy processes, which have been structurally linked with national local governance and regional industrial rooted domains that have characterized different populations.[4]

Targeted local populations have interacted collectively but according to different interdependent environment spheres related to socio-economic systems, which have been expressively formalized through states functional and structural linkages; societal mutual partners have envisaged, consequently, adaptive relationships for the materialization of local resource-related innovation processes.[5] As for comparative regional indicators about specific technology innovation measures, the influence played by innovative adjustment policies and institutional assimilation mechanisms can be taken into account,[6] as they have purposely facilitated the application of diversified transitional production

---

1 See: Abdelillah Hamdouch, Marc-Hubert Depret, Jean-Louis Monino, Christian Poncet, "Regional Policies, Key Levers of Regional Innovation Dynamics," *SSRN Working Paper Series* 2009, Copyright Social Science Research Network.
2 See: Thomas Cleff, Christoph Grimpe, Christian Rammer, "Demand-Oriented Innovation Strategy in the European Energy Production Sector," *International Journal of Energy Sector Management* 3.2/2009, pp. 108–130.
3 David Doloreux, "What We Should Know About Regional Systems of Innovation," *Technology in Society* 24/2002, pp. 243–263.
4 See: Helen Anne Curry, "Industrial Evolution. Mechanical and Biological Innovation at the General Electric Research Laboratory," *Technology and Culture* 54.4/2013, pp. 746–781.
5 Doloreux, "What We Should Know," pp. 243–263.
6 See: 1) Philip G. Robertson, Janet C. Broome, Elisabeth A. Chornesky, Jane R. Frankenberger; et al., "Rethinking the Vision for Environmental Research in US

schemes that have been amplified through the extensive interactions of territorial productive domains.

Following on these aspects, regional public and private technology partnerships PPP, rearranged for functional production assets and local technological activities, have been operated according to contextual local knowledge transfers, and through structural innovation components, mostly put in relation for instance to: *a. the infrastructural entrepreneurial bodies; b. the regional power networks for diffusion; c. and the state, local collectivities, which have been involved with the industrial innovative interactions;* comparable foundational elements have given public recognition status to institutional concerted accesses managed for the implementation of local information practices.[7]

Focusing on concurrent local governmental practices, the territorial development processes have been related with the combination of knowledge-based technical and social models, which have characterized the affirmation of regional innovation system*s* (RIS).[8] The material association of technical management issues aligned with social networking factors, has also been central to validate the common interactions and open learning paths; which have brought together the local environmental institutions and government representatives, in order to assess corresponding industrial cooperation conditions, and therefore reaching a compromise on common intra-regional environmental knowledge dimensions.

Essentially, the formal inclusions of both scientific research areas and technology innovation processes have involved specified intra-regional knowledge platforms connected to systematic national changing interactions, which can either favor or disfavor contemporary local adaptability contexts, that implement knowledge-based normative approaches, also in view of the recurrent technological management variations and different collaborative environments.[9] What has also been considered about the possible inclusion of regional innovation systems

---

Agriculture," *Bioscience* 54.1/2004, pp. 61–65; 2) Susan M. Walcott, "*Chinese Science and Technology Industrial Parks,*" (Aldershot, Hants: Ashgate, England). 2003.

7   Doloreux, "What We Should Know," pp. 243–263.
8   See: 1) Tai-Ming Ben, Kuan-Fei Wang, "Interaction Analysis Among Industrial Parks, Innovation Input, and Urban Production Efficiency," *Asian Social Science* 7.5/2011, pp. 56–71; 2) Hsieh-Sheng Chen, "The Relationship Between Technology Industrial Cluster and Innovation in Taiwan," *Asia Pacific Management Review* 16.3/2011, pp. 277–288.
9   David Doloreux, Saeed Parto, "Regional Innovation Systems: Current Discourse and Unresolved Issues," *Technology in Society* 27/2005, pp. 133–153.

Territorial Governance Connectivity 217

(RIS)[10] has been the common interplay of scientific regional interactions, with stabilized socio-economic relations, and distributive assimilation trends.

In which case, the introduction of critical theoretical ways to try to enhance science innovation specialized fields has been mediated through responsive regional industrial systems, whose performance objectives have gained comparable innovation value, when aggregated to the spatial growth and local territorial innovation of transnational production units.[11]

Moreover, the RIS organization and development approaches have been essentially formalized according to the evolution of different functional modalities of local governance. Respective public regional, national, or local, management arrangements have been dependent on comprehensive policy planning and material transmission capacities that have included the affirmation of local knowledge transfers.

But at the same time, the regional accumulation process of local inclusive innovation and technological extensive capacities has taken place according to the changing dynamics of local planning strategies and mutual institutional expressions, which have contributed to the economic development as well. In particular, the now-available type of systematic formulation of developmental innovation approaches, which have been contextually identified in Canada's case concerning the federated industrial concentration processes.[12]

In essence, the specific integration of multi-level regional approaches[13] has directly or indirectly led to the affirmation of innovation systems already aligned within transitional paths, while implemented in order to favor the regeneration of regional operating clusters for the production and research activities, particularly associated with local collective interests like in the northern-eastern regional EU-Asian clusters, which have been networked in view of regulatory stability confirmed for different territorial innovation systems.[14]

Similar considerations have remained instrumental to understand the comparative states' aggregation frameworks and political endurance strategies,

---

10 See: Trond Haga, "Action Research and Innovation in Networks, Dilemmas and Challenges: Two Cases," *AI & Society* 19/2005, pp. 362–383.
11 Doloreux, Parto, "Regional Innovation Systems," pp. 133–153.
12 Doloreux, Parto, "Regional Innovation Systems," pp. 133–153.
13 See: Katharina Rietig, "Reinforcement of Multilevel Governance Dynamics: Creating Momentum for Increasing Ambitions in International Climate Negotiations," *International Environmental Agreements: Politics, Law and Economics* 14.4/2014, pp. 371–389.
14 Doloreux, Parto, "Regional Innovation Systems," pp. 133–153.

adapted according to stratified managerial environments. Moreover, the progressive cohesiveness of national economic and societal spheres, determined through receptive policy strategies, has involved as well, modified learning inputs about local innovation networks and resulting configuration patterns; thus applicable distribution models have entailed common regulative goals, that relate to the industrial maintenance processes, addressing for instance: the local waste treatments and the waste disposal management process. Public management, in fact, can be arranged in coordination with functional responsive capacities interrelated with upcoming changing scenarios of local servicing activities, also based on demand and supply conditions, while taking into account the availability of local environmental information outlets.[15]

In addition to this, the systemic operability of complex environmental management capacities and local organizational systems expressed under uncertain conditions, has also required the diversification of national objectives and intensification of decisional scopes, formulated according to realistic managerial practices, which have included e.g. the case of municipal solid waste management.[16] In fact, it has been pointed out "for a waste management system to be sustainable, it needs to be environmentally effective, economically affordable, and socially acceptable."[17] For the actual development of similar managerial innovation processes have been prioritized the social and economic interests, which have been compatible with industrial revised plans, depending on technical affirmation stages put in conjunction with local regulatory criteria.

Essentially, the comprehensive governance models tackling local waste management programs have involved concerted local actions and public regulatory efforts, which have been undertaken in order to coordinate interlinked functional processes, supported by: 1.environmental protection agencies; 2.regional economic organizations; 3. and/or local networked interest associations; therefore all together have promoted the facilitation of decentralized recognition practices and stabilization cycles managed through PPP public-private negotiation architectures. Analogously, public environmental assessments and comparable sequential policies have been redefined in relation to local incremental

---

15 Goran Finnveden, Anna Bjorklund, Marcus Carlsson Reich, Ola Eriksson, Adrienne Sorbom, "Flexible and Robust Strategies for Waste Management in Sweden," *Waste Management* 27/2007, pp. S1–S8.
16 A.J. Morrissey, J. Browne, "Waste Management Models and Their Application to Sustainable Waste Management," *Waste Management* 24/2004, pp. 297–308.
17 Morrissey, Browne, "Waste Management Models," pp. 297–308. p. 298.

Territorial Governance Connectivity 219

innovative adaptations, also referred to the case of local waste disposal management alternatives.[18]

Moreover, the characterization of regional innovation systems for the case of waste management processes has often been associated to variable environmental and industrial resettlement and recovery choices. As such, the multilateral recovery plans have been undertaken according to functional regulative approaches that have been operated through public responsible agencies investing on regional common policies, technically elaborated in view of a full life cycle managerial process of urban waste disposals.[19]

In other words, the multi-functional and operational elements about public waste programs have, essentially, been identified in order to aggregate locally diversified approaches, which can substantiate the effectiveness of regional environmental provisions.[20] In comparative terms, for the case of controversial renewable energy plans developed in the United States, about the nuclear waste management practices (NWM), central significant shifts have come into place as well. Table 9 presents a brief summary about past political shifts related to regulatory environmental issues. This extract comes from the U.S. Congress D.C. 1982 document, based on public local controversies. The Table 9 has been divided into seven major waste management WM dimensions.

---

18 Morrissey, Browne, "Waste Management Models," pp. 297–308. p. 298.
19 Reinout Heijungs, Jeroen B. Guinée, "Allocation and 'What If' Scenarios in Life Cycle Assessment of Waste Management Systems," *Waste Management* 27/2007, pp. 997–1005.
20 Heijungs, Guinée, "Allocation and 'What If' Scenarios," pp. 997–1005.

Table 9: Political Controversial Issues for Radioactive Waste Management (WM) in the U.S.

WM Dimensions	Historical Follow-ups on WM
Waste Characterization	"The U.S. (AEC) Atomic Energy Commission, regulators and developers of nuclear power technology opted in the 1950s to 1970s for long-term solutions if possible given a low priority status of WM during previous decades." (p. 200)
Storage Process	"U.S. policy makers since the 1950s drove their actions about waste management storage facilities in virtue of a 'Technological Optimism' meant as a systematic perception of scientists and technical experts for defining possible solutions through RAT (readily available technologies). But for instance: The storage of military wastes at Hanford has been plagued with numerous problems. *Tanks expected to hold the liquid waste for 50 to 100 years have corroded and leaked after less than 25.14. Although the waste stored at the Savannah River facility have not leaked into the environment*" (p. 201)
Disposal Sites Location	"A lack of attention to non-technical issues of WM produced a divide between managers e.g. AEC and decision-makers who followed AEC technological optimism approach (1945 to 1975) [p. 203]. During the U.S. Reagan administration the away-from-reactor storage previously proposed as a high priority for decision on a geological disposal plan, was dismissed in favor of private sector development on the storage of spent reactor fuel." (p. 208)
Engineering Designs for NW facilities	"As result of low prioritization of WM with low budgets despite concerns about the adequacy of WM programs, organizational and technical infrastructures were not fully developed prior to the mid-1970s." (p. 203) "For high-level waste of commercial and naval facilities in the U.S. from early 1950s to late 1970s the concept of a repository structure underwent revisions and compromises between DOE and earth science agencies."(p. 212)
Radioactive Risk Exposure Levels	"High-level waste streams and spent reactor fuel can be put in different waste forms before disposal. In the U.S. since the 1950s work on radioactivity was concentrated on risk and safety limits to reduce migration of radioactive elements. However, WM organizational practices were reduced at major waste storage/disposal facilities at West Valley, Savannah River, Hanford and Idaho. In which prior to 1970s the economic and health costs associated to technological uncertainties about the uses of transformation of advanced waste forms overwhelmed any short-term advantages of waste processing." (p. 215 -216)

Table 9: Continued

WM Dimensions	Historical Follow-ups on WM
	"A policy shift in 1974 started the process for determining acceptability in waste management. EPA (Environmental Protection Agency) after years of bureaucratic battles established a firm stance in the domain of radiation protection standards. This agency issued radiation standards for the front-end of the nuclear fuel cycle, for reactor operations, and for reprocessing of spent nuclear fuel. At the same time the Energy Reorganization Act of 1974 abolished AEC and established ERDA and NRC in order to reduce conflict of interest for regulation and nuclear power development." (p. 219)
	"The NRC (Nuclear Regulatory Commission) was established as an independent review authority involved in a formalized process of regulation. ERDA and DOE since the 1974 Energy Act could select a site and design a disposal facility in line with NRC regulations that included EPA's environmental standards." (p. 219)
Governmental Partnerships	"In the U.S. the Federal Government has acted in partnership with States' jurisdictions for NW management with the mediating regulation activity of the NRC for agreements between States' activities of low-level waste burial grounds and Federal authorities. (p. 221). Federal/State relationships have evolved with the States being given more active role in waste disposal policymaking." (p. 231)
NW management and NP (Nuclear Power production)	"The logic of nuclear waste related to the commercial nuclear industry activities remains linked to its disposal." (p. 225)

Model adapted. **Source:** "Managing the Nation's Commercial High-Level Radioactive Waste." (Washington, DC: U.S. Congress, Office of Technology Assessment, OTA-O-171, March 1985).

CHAPTER TWELVE

12.1 Regional Specialized Networks (RIS)

In the specialized management area of international nuclear commissions such as the (NRC) Nuclear Regulatory Commission in the United States,[1] the public policy authorities have basically been involved on public allocation schemes promoting local technical monitoring tools, which in time have influenced the provisional distribution of functioning organizational models related to nuclear waste disposal management programs.[2]

Further decentralized monitoring protocols and national decision-making indications, have required comprehensive public-private managerial agreements, in order to cover the tangible but also intangible relevant aspects of competitive transnational energy markets, moving closer to multilateral energy management through local adaptive frameworks, operating within contingent public choice scenarios.[3] For instance, the prioritization of environmental regulative and evaluative processes has been pursued in view of advancing the national policy discourse about environmental protection goals that have been considered as indispensable.

Main environmental programmatic objectives and operable local guidelines have been assessed in terms of inclusive energy targets, structured in view of NPP life cycles and nuclear disposal management processes, which have been integrated into nuclear power diversified programs of industrialized interacting countries, such as: Sweden, France, Canada, and others.[4] In order to incorporate the so-called comparable know-how about local safety and security of procedural methodologies, the emerging states have laid out provisional technical statements, based on the evolution of renewable energy distribution networks,

---

1 See: US Fed News Service, Including US State News, "NRC Chairman Klein Addresses to American Nuclear Society at Raleigh," *HT Media Ltd.* 2008, Washington, D.C., U.S.
2 See: Katherine McIntire Peters, "Power player," *Government Executive* 39.12/2007, pp. 40–46.
3 See: Michael Gochfeld, Sandra Mohr, "Protecting Contract Workers: Case Study of the US Department of Energy's Nuclear and Chemical Waste Management," *American Journal of Public Health* 97.9/2007, pp. 1607–1613.
4 See: Magali Delmas, Bruce Heiman, "Government Credible Commitment to the French and American Nuclear Power Industries," *Journal of Policy Analysis and Management* 20.3/2001, pp. 433–456.

including the annual financial plans that are verified through the domestic taxation procedures.[5]

Similarly, the environmental impact assessment reviews (EIA) associated to the protection of affirmative objectives, have been structured according to the existing levels of industrial anti-pollution risks programs and local risk prevention plans, which have been upgraded through comparative participatory innovation designs. For such reasons, in order to favor the production of both qualitative and quantitative public-private renewable energy options, the standardized regulatory measures have to be established according to effective levels of national administrative capacities and public regulation bodies that need to activate local accountable stakeholders.[6]

In a long-term view, the industrial management of technology and knowledge transfer networks aligned to territorial planning processes, set in conjunction with the systematic identifications of regional environmental policies,[7] has allowed the application of regional innovation standards and technical environmental measures. At the same time, local knowledge innovation transfers have been progressively introduced according to the integration of comparative international and regional nuclear waste disposal management programs. Nuclear waste disposal activities are embedded together within local and national regulatory conditions, stabilized through local innovation management and comparable environmental adaptation, also based on recurrent organizational changes that remain still in progress.[8]

From another angle, it can be considered that context-related regional belief systems with mutual cultural characteristics belonging to single states, have emerged in parallel with the normative and legal frameworks, which have shaped concurrent policy determinants, interlinked with behavioral public attitudes, that result into different public orientations. Therefore, the regulatory provision of local safety and security programs has corresponded to societal risk awareness conditions, in which the integration of territorial information networks has been

---

5   Anna E. Bjorklund, Goran Finnveden, "Life Cycle Assessment of a National Policy Proposal – The case of a Swedish Waste Incineration Tax," *Waste Management* 27/ 2007, pp. 1046–1058.
6   Bjorklund, Finnveden, "Life Cycle Assessment," pp. 1046–1058.
7   See: Joy Parr, "A Working Knowledge of the Insensible? Radiation Protection in Nuclear Generating Stations, 1962–1992," *Comparative Studies in Society and History* 48.4/2006, pp. 820–851.
8   David L. Feldman, "The Future of Environmental Networks – Governance and Civil Society in a Global Context," *Futures* 44/2012, pp. 787–796.

sustained through the facilitation of local monitoring and local control measures promoted publicly, also with the inclusion of international climate planning directions, as well as, environmental management practices.[9]

For instance, the incentivization of local territorial strategies for managerial innovation and comparative knowledge acquisitions enhanced through public-private partnerships PPP, has been operated in the case of sectoral management for renewable energy units, that involve local interactions distributed among different administration bureau, which have also been responsible for the application of scientific and ethical principles, associated with theoretical reflections for the actual development of local environmental protection schemes.[10]

Fundamentally, the technology-based innovation schemes have been functional to the verification of local sharing opportunities. Such multi-level PPP plans and environmental protection schemes have been worked out in conjunction with the constitution of civil participatory groups, providing their support (when the case) about public environmental exchanges. Participatory public discussions based on comparable scientific and socio-economic issues, have also emerged in reference to local environmental commonalities tracked down for public protection plans, which have required support for the diffusion of local preservation and local conservation policies addressing, among other areas, global ecosystems diversity, common land-use patterns, and technical diversified procedures, envisioned for the environmental territorial practices.[11]

Therefore, the effective combination of comparable governmental responses, delineated through local protection policies, has converged toward corresponding local risk impact assessment processes, especially in the case of environmental development projects, which have been openly discussed by interested policy-makers, including the civilian representative local parties. This multi-purpose organizational dimension, incorporating the local regulatory distribution plans about eco-innovation activities, has been shared for the reason of preparing a coherent transition[12] about the energy

---

9   Feldman, "The Future of Environmental Networks," pp. 787–796.
10 See: Matthew Rimmer, "A Proposal for a Clean Technology Directive: European Patent Law and Climate Change," *Renewable Energy Law and Policy: RELP* 2.3/2011, pp. 195–204.
11 Feldman, "The Future of Environmental Networks," pp. 787–796.
12 See: Daniel K.N. Johnson, Kristina M. Lybecker, "Challenges to Technology Transfer: A Literature Review of the Constraints on Environmental Technology

diversification programs, which have included the promotion of public knowledge transfers.[13]

For example, in the European Union, in the United States, and in East Asia, the regional member states have transitioned to multi-level political adaptation plans, when inter-linked to the promotion of energy production networks. In such cases, the adoption of eco-innovation plans and decentralized energy strategies has been implemented through major decisional changes that have influenced, either directly or indirectly, the technical organizational conditions of transnational energy markets, opting for deregulation or liberalization of common trading practices, which have been fostered through the improvement of regional industrial transfers and local innovation designs.

Essentially, the government planning schemes PPP, the ecological management impacts, and the environmental adaptation mechanisms, have involved, in a comprehensive way, very large public distribution channels and intra-regional commercial areas, which have been politically intertwined with the regulation of sustainable delivery measures targeting, for example: the local water storage facilities and supply distribution chains; the industrial energy innovation parks and the local production cycles; the national renewable energy uses; and the technological innovation systems, explicated in terms of transitional competitive models.[14]

Overall, transnational ecological and commercial dispositions intertwined with territorial governance and regional trading trajectories have been perceived as critical research interest areas. As a result, territorial development applied through systematic leading orientations has been modified through strong/weak governmental supporting coalitions; formalized public coalitions in turn need to be able to determine eco-innovation policies for common emerging markets and local trading areas. Common integrated commercial areas have been explored in view of specific competition procedures of comparative industrial firms, which have basically facilitated the socio-economic transitions established under broader aims and uncertain grounds.[15]

In addition to this, it should be questioned the different types of regional industrial development and locally-based operating platforms; because the

---

Dissemination," *Colorado College Working Paper* No. 2009-07 – SSRN Working Paper Series, Aug. 2010.
13   Philip Cooke, "Transition Regions: Regional-National Eco-Innovation Systems and Strategies," *Progress in Planning* 76/2011, pp. 105–146.
14   Cooke, "Transition Regions," pp. 105–146.
15   Cooke, "Transition Regions," pp. 105–146.

knowledge innovation intensive capacities have been regularly standardized for coordination management plans, based on the eco-innovation policies and comparable industrial knowledge acquisitions. But for what reason? The indication is that the eco-innovation adaptation promoted according to the sustainable mobility of global production systems, has been interrelated to the political regime systems, which have distanced mutual expectations or otherwise modified co-evolution proceedings for the comparative adoption of socio-economic technological approaches. This long-term evolution process has been especially sustained for the reason of developing the modern traceability of technical and social interlinked cognitive trends, also reflecting the participatory terms reaffirmed about science development and local specialized knowledge.

From a governance perspective, the fragmentation of multi-level knowledge-systems within centralized/decentralized regulatory spheres, has essentially required a modern revision process about the local information provision and contextual adaptation on local technology practices. Such progressive reviews have been critically standardized, according to common socio-economic traits, in order to validate or not the specialized managerial innovation status of participant states, which have in time decided to reaffirm the amalgamation of multiple scientific and experimental methods, also applied to the hybrid designs of nuclear power energy models.[16]

At country level, for instance, the industrial mobility referring to an energy innovation transition has been intertwined with multiple service innovation choices, that also lead to a regional reconfiguration, in which respective governmental bodies such as the UK government agencies, and the Swedish ministerial entities,[17] have decided to put a major emphasis on an additional combination of local environmental strategies and local industrial distribution targets. Local capacity distribution has been included in national planning strategies that have reflected respective industrial intensification factors, which have been associated to socio-economic conditions, also interrelated to national territorial cohesion initiatives and to regional sustainability backgrounds.[18]

---

16   Bjorn Nykvist, Lorraine Whitmarsh, "A Multi-Level Analysis of Sustainable Mobility Transitions: Niche Development in the UK and Sweden," *Technological Forecasting & Social Change* 75/2008, pp. 1373–1387.
17   See: Ernst Ragnar Lofstedt, (1993). "The Swedish Energy Policy Dilemma and the Role of Behavioral Energy Conservation." PhD Dissertation submitted at Clark University, Worcester Massachusetts. The Graduate School of Geography, United States.
18   Nykvist, Whitmarsh, "A Multi-Level Analysis," pp. 1373–1387.

In a way, this means that the regulation of territorial local productions aggregated with industrial technical developed areas, has received political attention in terms of integration of specialized STS knowledge transfers, which have exposed the local intra-regional economies to highly sensitive operable activities, undertaken within regional innovation systems (RIS). However, the progressive evolutive patterns based on territorial innovation production zones, have been connected with spatial public policies and environmental associative interactions, that when possible have been expressed for facilitating common knowledge information platforms.

In addition, the ICT local transfer mechanisms and knowledge information systems have been able to communicate in multilateral environments through the specific dynamics of multiple innovation systems, which in turn have been able to favor the commercial policies of regional industrial economies that have operated within multi-dimensional collaborative spheres.[19]

Nonetheless, there have been controversial management issues of RIS in local governance models. In accordance with socio-interrelated exchanges, the interdependence of local transitional environments has included the local learning adaptability contexts, which have been put into association with international industrial innovation productions. Public knowledge management transfers have emerged through the local networking abilities of national territorial hubs such as: the EU regional memberships for the interdependent activities of industrial innovating states.[20]

This means in a sense that the incremental promotion of contextualized local innovation policies within comparative industrial production fields, has been technically addressed according to public learning activities, and PPP public-private knowledge transfer agreements. The mediated knowledge transfer agreements have also come into existence through government collaboration with interdependent sectoral agencies, which have pursued the common learning adaptation practices, by following mutual levels of national differentiation, based on national regulatory principles.

At the same time, the development of instrumental management tools and local regulation policies tackling regional innovation clusters' productions, has been reviewed through common coordination and cooperation mechanisms set up across policy sub-units, integrating the participation of key stakeholders

---

19 Bjorn T. Asheim, and Lars Coenen, "Knowledge Bases and Regional Innovation Systems: Comparing Nordic Clusters," *Research Policy* 34/2005, pp. 1173–1190.
20 Asheim, Coenen, "Knowledge Bases," pp. 1173–1190.

such as: the local industrial firms; the international/national service clients; the energy production associates; and the institutionally-driven interdependent representatives; which have provided the *RIS* with multilateral progressive knowledge, local technical information, and multi-level management capacity.[21] See Figure 2: The Multilateral Compromise, a summarized model of the multilateral mediating activities lined up through managerial integrated assets.

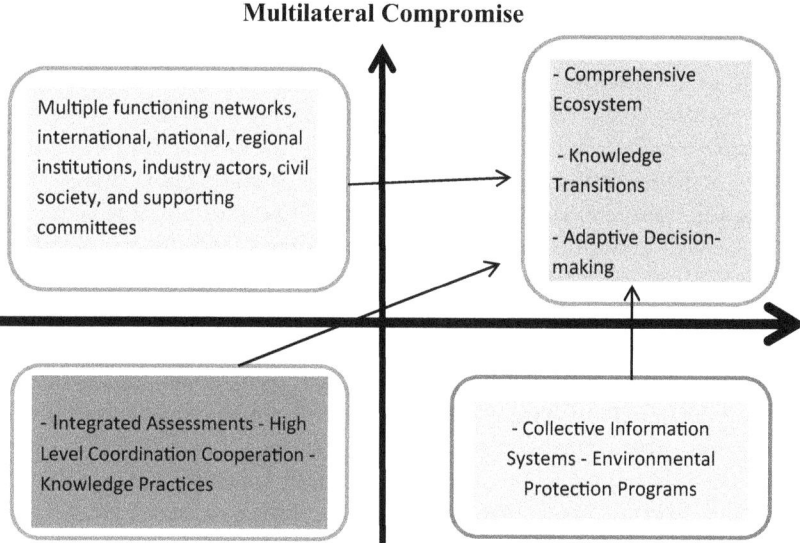

Model adapted. **Source:** OECD/NEA 2001. "Better Integration of Radiation Protection in Modern Society." Workshop Proceedings Villigen, Switzerland 23–25 January 2001. OECD 2002, Publications, Paris, France. P. 216. **Source:** EEA, 2008. "Annual Management Plan 2008." European Environment Agency, Copenhagen.

As a resulting, there have been deterministic values as well as separate visions which have referred to the integration of common renewable energy markets, through power consumption models and multilateral economic liberalization conditions that have characterized regional transitions, with direct/indirect local governance implications, also concerning the aspects of assimilation about

---

21   Franz Todtling, Michaela Trippl, "One Size Fits All? Toward a Differentiated Regional Innovation Policy Approach," *Research Policy* 34/2005, pp. 1203–1219.

different public organizational paths. In previous literature about knowledge innovation tracks, we have found some highlights converging about RIS development. It has been reaffirmed that the dynamic knowledge exposures which have been based on regional innovation systems, have been extended beyond the traditional PPP public-private coordination lines, for instance, through the regional governance approaches that have been public in nature, by operating according to diversified structural management grounds.

12.1.1 Local Ecological Policies

In terms of comparative ecological policies, understanding the mutual levels of international collaborative energy domains has implied concrete local steps undertaken within institutional learning processes. The introduction of common learning activities of the sectoral organizations, including public sub-systems tackling the transnational flow of local industrial productions, has been promoted through a national diffusion of corresponding knowledge transfer elements, classified according to their connection to the local, and national business cycles entailing instrumental planning tools.[22]

On such aspects, there are distinctive relationships emerging according to public organizational approaches and systematic progressive aggregation of cognitive envir0nments, interrelated to the development of common innovation systems; which then have resulted into regional public policies targeting, for instance, the eastern and northern European zones for industrial technology development. Such territorial processing zones have been monitored by involved representative stakeholders, through proportional regulatory conducts for regional production areas, which have been interlinked to multi-level territorial dimensions. Therefore, the comparative RIS integrating common learning dimensions have reflected national industrial economic capacities, which have been adapted according to specialized sectoral innovation processes.[23]

The regional governance process can effectively combine national organizational behaviors according to common political and normative measures by addressing for instance the growth of (RIS),[24] which have been continuously

---

22 Philip Cooke, Mikel Gomez Uranga, and Goio Etxebarria "Regional Innovation Systems: Institutional and Organisational Dimensions," *Research Policy* 26/1997, pp. 475–491.
23 Cooke, Uranga, Etxebarria, "Regional Innovation Systems," pp. 475–491.
24 See: Jon Birger Skjaerseth, "Governance by EU Emissions Trading: Resistance or Innovation in the Oil Industry?" *International Environmental Agreements: Politics, Law and Economics* 13.1/2013, pp. 31–48.

reaffirmed in parallel with the integration of local knowledge activities, retraced within national regulatory spheres. Nonetheless, the local collectivities involved in the process of regional systematic innovation – through the development of rational choices – have also come to terms with a comparative establishment of multi-level local environmental settings; which can indirectly operate toward the affirmation of positive/negative ideas supporting collective indications about the science innovation aims and complementary technology goals, validating a common political and democratic coexistence.[25]

When we look closer at different regulatory perspectives on RIS relationships, we can consider that there are still several unsolved dilemmas, about the effective formulation of regional environmental delivery provisions. The inclusions of complex trading processes of technological transformation have been interrelated to the presence of multi-level societal participants, supporting the integrative public acceptance models.

However, the actual causes of transformation and adaptation at societal levels have involved the polarization of local participatory dynamics, through the assessment of governing opposing aspects commonly interrelated to the changing environmental conditions and modified terms of institutional evaluation reviews, also meant as local visioning exercises for the long-term transition. In the case of common innovation management goals targeted on national, and local scales, through science networking systems, there have been common local practices and societal orientations, which have been diversified according to responsive environments, also for the reason of obtaining respective levels of internal mobility of regional innovation productions, through the inclusion of transnational innovation structures.[26]

Moreover, the separate technological reorganization of RIS made according to the corresponding spatial dimension has entailed critical managerial changes with local production shifts, which have been related to the parallel expansion of newly emerging developed regions. Such newly developed industrial regions have positioned themselves, through prospective production networks and planned growing sectors, by matching the specific degree of industrial innovation programs. In fact, the newly developed sectoral clusters have involved multi-dimensional knowledge industries such as: the car industry with its

---

25 Cooke, Uranga, Etxebarria, "Regional Innovation Systems," pp. 475–491.
26 Philip J. Vergragt, Halina Szejnwald Brown, "Sustainable Mobility: From Technological Innovation to Societal Learning," *Journal of Cleaner Production* 15/ 2007, pp. 1104–1115.

economic growth factors, corresponding to a search for alternative ecological projects destined to newly formed innovation hubs, like: the United States technological clustering models, China's internal specialized economic zones, and the European Union intra-regional science industry aggregations, also endorsing the petroleum-based production systems.[27]

Similarly, the establishment of territorial regulatory entities for instance based in the European Union members' zones has been faced with the progressive integration of science innovation policies. This local adaptation process elaborated through normative applications of countries that have 'adjusted' to internal coincident measures of national industrial hubs, has been focused on the organizational boundaries and facilitation mechanisms, which can mediate on an initial set of rules incorporating policy alternatives, particularly required in knowledge manufacturing processes and in industrial renewable energy fields.

To add to this, there is the contemporary association to the cultural beliefs and local practices established within common societal environments, which have been concurrently redefined through regional transition processes, mostly based on national industrial diversification contexts, as well as, on personal or collective materialistic explorations,[28] also involving the study of complex sociotechnical correspondences framed within regional performing dynamics.[29]

Within the social changing environments, there have been multiple orientations of local communities and multi-stakeholders' groups, which have had in essence to incorporate the programmatic nature of industrial innovation-based policies. Correspondingly, environmental policies have emerged in local and regional processes, which have been adapted according to the contextual evolution of participatory mechanisms; where for instance, the increasing regulatory pressures for the case of European Union bodies have led to local coordination arrangements, also for the preparation of regional legal dispositions, and for the affirmation of involved referential agencies, which have been classified in terms of supportive operational networks for e.g.: the environmental institutional programs and local waste management disposal activities.[30]

---

27 Vergragt, Brown, "Sustainable Mobility," pp. 1104–1115.
28 See: Bob Jessop, "Revisiting the Regulation Approach: Critical Reflection on the Contradictions, Dilemmas, Fixes and Crisis Dynamics of Growth Regimes," *Capital & Class* 37.1/2013, pp. 5–24.
29 Vergragt, Brown, "Sustainable Mobility," pp. 1104–1115.
30 Simin Davoudi, "Planning for Waste Management: Changing Discourses and Institutional Relationships," *Progress in Planning* 53/2000, pp. 165–216.

12.2.2 Local PPP Indications

During last three decades, 1990s, 2000s, 2010s[31] the territorial actualization of public and private partnerships PPPs[32] has been affected by environmental and regulative pressures, related to regional waste management sectors, with corresponding planning indications and local development targets, which influence the modified institutional relations.

In fact, the comparable negotiation procedures tackling local environmental waste strategies have been possibly followed up through the corresponding distribution of public roles with mutual local assignments and mutual internal responsibilities maintained, for instance, on the issue of centralized or decentralized municipal waste programs; as well as on the interdependence trading aspects of regional local configurations, interrelated to industrial innovative assets; and on the local networking status of environmental agencies. These negotiated arrangements have been partially able to respond to critical conditions on programmatic environmental assessments, where the local opposition parties have directly put into question the functional active roles about local environmental regulatory mandates.[33]

Another example is the case of China's contemporary urban development and the local planning context; this rapid urbanization process has involved territorial knowledge exchanges based on the transferable ecology principles, and local power energy configurations, as well as, sustainable managerial implementations. Through comprehensive understanding, the regional institutional interplay among PPP development counterparts has been formed according to practical local assessments of national urban ecosystems, still dependent on the local delivery of systemic planning approaches.

---

31 See: 1) Peter Newell, "The Political Economy of Global Environmental Governance," *Review of International Studies* 34/2008, pp. 507–529; 2) Ray Hudson, "Cultural Political Economy Meets Global Production Networks: A Productive Meetings?" *Journal of Economic Geography* 8.3/2008, pp. 421–440.

32 See: World Bank Report, "PPPs in the energy sector come in different shapes, sizes, and structures and are used mainly in generation and transmission. The methodology used varies, depending on the place, the government and the specifics of the operation; therefore each one is tailored to the needs and circumstances given at the time when the partnership is created." *World Bank, PPPLRC Public-Private-Partnership Legal Resource Center*, 2019, "Energy and Power PPPs."Available on website: https://ppp.worldbank.org/public-private-partnership/sector/energy

33 Davoudi, "Planning for Waste Management," pp. 165–216.

However, the continuous formalization of local urban planning applying technical concepts that emerge from regional ecosystems, has been identified in terms of characteristic cultural landscapes, entailing locally-defined governmental measures, which have been put in conjunction with communitarian approaches, reminding about common socio-ecological perceptions.[34] In essence, a regional sustainable developmental dimension arranged through different institutional planning practices, that address local regulatory terms about environmental productive resources, has remained interrelated to national territories and wider internal societies, which have constantly adapted to various types of institutional *rapprochement*, also at local level. More popular attention has been directed at local participatory mandates and life-cycles of administrative dynamics, where communities involved have interacted according to the available territorial resources, the science innovation provisions, and the local knowledge adaptation methods.

As previously discussed, the regional managerial environments involved with national and local renewable energy power systems, have been operated according to an integral access to local organizational components; which have been dependent on the level of technological implementation processes, added to public policy reforming programs, with spatial dynamic configurations, and local administrative convergences.

Whereas, the industrial mechanization systems associated with national research plans and progressive development activities, and structural open networks, have been analyzed in terms of comparative governance practices that point at the sustainable territorial capacities, and multilateral environment management systems, which essentially have been intertwined with proximate regional technology clusters and local facilities' management methodologies.[35] In a way, comparable institutional adaptability paths established in view of societal changing dynamics and knowledge management transfers, including systematic procedures of technological innovation networks, have been promoted according to progressive developmental stages, which reflect at the same time governmental temporal relations, interlinked to different territorial human boundaries.[36]

---

34 Jurgen Breuste, Salman Qureshi, Junxiang Li, "Applied Urban Ecology for Sustainable Urban Environment," *Urban Ecosystems* 16/2013, pp. 675–680.
35 Ron A. Boschma, Robert C. Kloosterman, *Learning from Clusters. A Critical Assessment from an Economic-Geographical Perspective*, (Published by Springer, The Netherlands, 2005).
36 Boschma, Kloosterman, "Learning from Clusters," pp.

Regional Specialized Networks (RIS) 235

Moreover, the formation of local development practices and local innovation agreements tackling technology clusters' requirements, has been combined according to diversified comparative assessments, scrutinized through the common collective norms as well as problematic territorial dilemmas; it has been highlighted that "for some participants, social problems are essentially territorial problems and must be tackled as such; for others, they have a directly territorial dimension but are not essentially territorial in character; for yet others, they are territorially conditioned or mediated but their effective causation is not itself territorial; and for others, social problems are neither territorial nor significantly territorially conditioned but that they have territorial effects"[37]

In other words, the formal need to meet on common grounds about territorial societal conditions, through governance collaborative transfers and political mutual dialogues, has acquired significant practical emphasis. In fact, for the case of nuclear environmental disputes, the public available solutions have involved societal cognitive assessments of the local environmental practices made in use, which have been put into relation with determinant hybrid innovation industrial models; so that relevant local innovation prospective propositions have been matched according to specific geographical production sites.[38]

For instance, about the territorial environmental contexts of Europe and eastern European Union regions, there have been emerging local competitive factors interrelated to entrepreneurial innovation systems. Local trading mechanisms have basically been embedded into local entrepreneurial dynamics and local economic growth impacts, similarly combined to regional innovation systems (RIS), which have offered cooperative production operations for supporting the industrial national formations.

Likewise, there are national urban-rural partnerships reaffirmed through the potential sustainable strategies comprising local economic profit assets and science innovation projects' initiatives, regionally connected to mutual societal affirmations. However, the common principles and normative presumptions about regional groups' participatory indications have been openly subjected to PPP public-private shifts, due to comparative changes which emerge in regard to various degrees of regional development actions expanded across neighboring

---

37 Bas Arts, Arnoud Lagendijl, and Henk van Houtum, *The Disoriented State: Shifts in Governmentality, Territoriality, and Governance*, (Springer Science + Business Media B.V. 2009). p. 84.
38 See: Mike Raco, "Competition, Collaboration and the New Industrial Districts: Examining the Institutional Turn in Local Economic Development," *Urban studies* 36.5/6/1999, pp. 951–968.

countries. In fact, the progressive implementation of science innovation and transitional industrial schemes, has particularly pointed at the need to strengthen comparative science and technological targets, which in any case have increased the level of material disparities among the urban-rural regional environments.[39]

Therefore, in order to be able to overcome constructively the regional governance shifts with evolving policy and industry arrangements, there are reformist countries such as Canada, Ireland, and the UK, which have directly approached their own communities, also for the purpose of developing self-regulated districts, through autonomy legal entities and the administrative devolution of powers.[40] By bringing up a conceptual flexibility on sub-national systems and institutional representations in governance, for the local planning purposes, the territorial dimension has emerged in its public capacity through respective levels of decentralization favoring local provincial activities, which have been directly operated for public regulatory objectives.[41]

The institutional common powers transfer with decentralized adaptation models has remained quite consistent with the constitution of decentralized public networks, which have favored a type of inclusive distributive options, particularly for the possible maintenance of transnational or local sectoral trading patterns, and the affirmation of autonomous management paths.[42] Fundamentally, the socio-political implications that characterize the local governance implementation issues have been conjugated to the aspect of institutional legitimacy, also referring to science innovation systems.

At the same time, the comparative environmental compatibility factors and local sustainable relationships have required committed structural implementation plans, able to renew common participatory arrangements. These community-based arrangements have essentially reflected the normative ordinary status of

---

39 Wendelin Strubelt, *Guiding Principles for Spatial Development in Germany* (Springer-Verlag Berlin Heidelberg 2009).
40 See: Gilbert Gagne, "Policy Diversity, State Autonomy, and the US-Canada Softwood Lumber Dispute: Philosophical and Normative Aspects," *Journal of World Trade* 41.4/2007, pp. 699–730.
41 Markku Suksi, *Sub-state Governance Through Territorial Autonomy. A Comparative Study in Constitutional Law of Powers, Procedures and Institutions* (Springer-Verlag Berlin Heidelberg 2011).
42 See: Werner Sengenberger, "Local Development and International Economic Competition," *International Labour Review* 132.3/1993, pp. 313–329.

public legitimacy frameworks, which can facilitate the affirmation of regional socio-economic innovation targets.[43]

From an environmental perspective, the public interactions developed by regional or local organizations and promoted through legitimacy conditions by public autonomy entities have been modelled according to normative preferential systems with the addition of regulative specialization. Similarly, the implementation of cooperative relations at regional level with mutual representative counterparts has involved programmatic functional conditions. Through which, it has been taken charge of a societal transmission in reference to political binding choices, which have characterized modern science projects and research coordination actions based on knowledge transfer principles.

At critical modern junctures, the regional autonomy paths for industrial innovation production zones and urban-rural ecosystems have involved the social participatory interactions with local learning groups, for mutual development and commercial purposes. Therefore, territorial industrial sectors have been continuously rearranged in view of technological and innovation changes leading up to public reorganization models, and competitive management strategies. However, the corresponding institutional regulation of local proceedings and social participative inclusions has not necessarily overcome the composite pressure of different localities. In fact, within comparative internal productive dynamics, the regional states have explored respective co-management assessment needs, and competitive distribution logics applied at local system levels.[44]

Overall, the major institutional determinants about the spatial integration of knowledge-based innovative productions have incorporated territorial governmental shifts, as well as, local environmental adaptation programs. The presence of knowledge-based local performing platforms has reflected the national degree of centralization and decentralization of societal operating processes. In which case, the national or local distribution of public sharing networks has had either direct or indirect impacts on the specific diffusion of science innovation particularly in local production environments.

---

43   Jurgen R. Grote, Bernanrd Gbikpi, *Participatory Governance. Political and Societal Implications* (Springer Fachmedien Wiesbaden GmbH 2002).
44   Frank Moulaert, and Jacques Nussbaumer, 2005. "Beyond the Learning Region: The dialectics of Innovation and Culture in Territorial Development," in: *Learning from Clusters: A Critical Assessment*, eds. R.A. Boschma and R.C. Kloosterman, (Springer. Printed in the Netherlands, 2005), pp. 89–109.

CHAPTER THIRTEEN

13.1 Local Territory and Local Identity

Looking at some descriptive environmental studies[1] debating over local identity politics and common environmental justice, there are interrelated issues concerning the emerging tendencies of national and local territorial municipalities, which have been associated with a dissimilar sectoral development that has been actualized through local territorial practices, and intra-regional political economy insights. In line with societal local affairs, territorial environmentalism has been approximated according to the evolution of environmental local approaches, where the state local agencies have provided concrete governmental feedbacks that have become intertwined with the distribution process of local economic resources and local planning appraisals prepared at field level.[2]

As a matter of fact, common structural linkages of local social interactions and local environmental risk-exposures factors have been analyzed in terms of the material diffusion of geographical vulnerability and locally-aggregated social disparities. The actual increase of local environmental disparities has basically determined the contextual reorganization of both political and normative dimensions, bringing to actual responses based on the development of national regulatory systems, which still need to address directly the local preparation of environmental protection measures, by focusing on balanced and restorative justice programs.[3]

In addition to this, although the adoption of social equity approaches[4] has remained a critical indicator reflecting the scale of sensitivity on

---

1 See: 1) Mick Hillman, "Environmental Justice: A Crucial Link Between Environmentalism and Community Development?" *Community Development Journal* 37.4/2002, pp. 349–360; 2) Heather McLeod-Kilmurray, Gavin Smith, "Unsustainable Development in Canada: Environmental Assessment, Cost-Benefit Analysis, and Environmental Justice in the Tar Sands," *Journal of Environmental Law and Practice* 21/2010, pp. 65–105.
2 Hilda E. Kurtz, "Scale Frames and Counter-Scale Frames: Constructing the Problem of Environmental Injustice," *Political Geography* 22/2003, pp. 887–916.
3 Kurtz, "Scale Frames and Counter-Scale Frames," pp. 887–916.
4 See: Deniz Zeynep Leuenberger, Michele Wakin, "Sustainable Development in Public Administration Planning: An Exploration of Social Justice, Equity, and Citizen Inclusion," *Administrative Theory & Praxis* 29.3/2007, p. 394-.

ecological anti-pollution programs, and on local implementation of environmental regulations; the social equity notions have also been relevant for the case of a multi-level differentiation aligned on comparable environmental spatial activities that have been socially renewed, in particular, at the local level. In fact, the qualitative organizational linkages that create multilateral public platforms for national, regional, and international, governmental assessment processes, have been integrated by environmental participatory interactions, in order to enhance local grassroots environmental initiatives, by also involving the civilian groups mobilized across the regional states' regulatory spectrum.[5]

In practice, the political relationships that have been shaped according to different emerging levels of local societal interactions have been associated with environmental decentralized policies, including a selection of technical evaluation undertakings and the formation of possible social distributive solutions, when especially referring to local socioeconomic conflict disputes – confronted through the formulation of available common practice tools explored whenever required.

In effect, the formal inclusion of multilateral processes comprising national and local environmental representative stakeholders, has reflected territorial differentiation based on local environmental criteria and transitional knowledge-transfers for socio-economic programs. Through a territorial differentiation process, the decision-making initiatives have been accordingly intensified with the incorporation of environmental justice issues, which have been reported for the field analysis of local disputed cases such as: the geographical distance of local pollution facilities.[6]

At the same time, communitarian environmental approaches have involved (but not necessarily in a systematic manner) the spread of ethnic identity models arising in combination with the national territorial representations,[7] which have been depicted through the preventive acceptance schemes, and diffusion of decentralized regulative strategies adopted by local public divisions at a comprehensive level. To add to this, environmental politics has resulted in multiple qualitative and quantitative analyses and in local research trials, undertaken in view of territorial national assessments and multilateral industrial innovation transfers, connected with local cluster development managed at transnational, national, and local level.

---

5   Kurtz, "Scale Frames and Counter-Scale Frames," pp. 887–916.
6   Kurtz, "Scale Frames and Counter-Scale Frames," pp. 887–916.
7   See: Linda Robyn, "Indigenous Knowledge and Technology: Creating Environmental Justice in the Twenty-First Century," *American Indian Quarterly* 26.2/2002, pp. 198–220.

According to territorial knowledge management and local innovation strategies, involving the local environmental awareness issues, the different increase of regional innovation productions and liberal cross-sectors economic interests has determined corresponding social adaptive practices, contouring relativism elements, which have moved publicly in favor or disfavor of related environmental justice concerns, also expressed on respective knowledge grounds, and eventually coming to a point of social open confrontations.

Interdependent environmental disputes have been characterized by contextual territorial inclusions planned for instance in the post-Soviet republics (Latvia), which have been operating through regional environmental policies and a direct formation of local protection measures. In such cases, the coordinated regulatory governmental interventions have also marked the tone about different societal exclusion factors, which have been addressed according to the presence of multiple ethnicities that in Latvia's case have mainly referred to Russian identity affirmations, which have resulted in marginalized right-based transitions compromising the access to natural resources and economic industrial exchanging dynamics.[8] More in general, through comparative projections about worldwide environmental protection schemes, the resulting convergent programs have entailed the provision of upgraded knowledge communication processes for the possible adoption of local participatory policy making that can frame ecological and conservation management approaches, through specific environmental participatory indications.

Essentially, public environmental protection rules have been associated with multi-disciplinary normative adoptions, which have been reflected across national jurisdictions, also through the comparative requirements of hierarchical systems operated within defined roles and social responsibilities. Therefore, public normative consensus has been solicited in terms of formal and informal public affirmative relationships for the purpose of maintaining comprehensive power distribution systems, perceived in symbiosis with local governmental functional roles; which have tentatively supported a level of environmental ethnic inclusion for the stabilization of socio-economic conditions, such as in the case of Latin America's ecological conservation policies, affected by deeper problems for local participatory practices, touching directly entire marginalized populations.[9]

---

8 Jane I. Dawson, "Latvia's Russian Minority: Balancing the Imperatives of Regional Development and Environmental Justice," *Political Geography* 20/2001, pp. 787–815.
9 Juanita Sundberg, "Conservation and Democratization: Constituting Citizenship in the Maya Biosphere Reserve, Guatemala," *Political Geography* 22/2003, pp. 715–740.

The incidental organization of instrumental normative justice tools set in preparation for civil society engagements, has been concurrently pursued in order to defuse social tensions and local inequalities. The social identification of local projected interests for the protection of national local ecosystems has recalled to the promotion of public-private implementation programs. The PPP implemented programs have also meant to create sustainable development and to diversify complex territorial issues. There have been in fact interrelated societal implications about local environmental policies and civilian actors' participatory engagements, which have involved the management of local environmental resources according to compatible ecosystems and regional development targets, such as the case of nuclear restricted zones[10] with resource protection projects, also managed in Latin America.[11]

Essentially, the public administration of participatory protection rules associated with adaptive decision-making processes has led to the comparative integration of aggregated local democratic patterns, and transnational diffused managerial arrangements, which have certainly varied based on country-to-country diplomacy relations. Looking at the local mobilized collectivities, public environmental efforts have suggested to commonly reaffirm a more critical understanding of the single country's territorial risk-exposures mechanisms, co-related to local structural changing systems. Similarly, localities involved have directly interacted with the official management bodies in order to reduce consistent inequalities, and promote environmental justice cooperation.

For instance, in the United States, common regulatory resolutions about the environmental confrontation issues, for the case of industrial hazardous facilities have been approached according to states' local intervention policies, which have been mediated particularly between sectoral spheres of local governmental divisions, able to deliver territorial programmatic alternatives, also through the national and local public interactions.[12] Because, the public pursuance of mediation activities, that addresses the preparation of community empowerment programs has been dependent on formal or informal environmental cognitive aims identified on a national scale.

As main point, societal environmental activism has been followed up according to specific territorial local administrations; in which, the responsible official

---

10 See: Danielle Nierenberg, "U.S. Environmental Policy: Where Is It Headed?" *World Watch* 14.4/2001, pp. 12–21.
11 Sundberg, "Conservation and Democratization," pp. 715–740.
12 Robert W. Williams, "Environmental Injustice in America and Its Politics of Scale." *Political Geography* 18/1999, pp. 49–73.

agencies operating within the networked governance systems, have been able to reform a systemic collaboration through multi-level development strategies, consolidated both nationally and regionally.[13]

13.1.1 Public Ecosystems Strategies

Broadly speaking, this consolidation process means that a constitutive application of collaborative public interests can be offered according to the multiple coexistence of public-private development practitioners, which have either directly or indirectly adhered to multi-level territorial production dynamics that result in parallel environmental assessments, locally constrained under uncertain managerial operating procedures.

As result, the contextual environmental civil groups have been engaged into major social campaigns addressing environmental conflict issues such as: the local minorities' empowerment processes, interrelated with local ruling decisions framed according to legal national frameworks, like the case of the U.S. policy measures with the Nuclear Waste Policy Act (NWPA) and Indian minorities' rights about land hazardous storage impacts.[14] Another type of deferential influence about the human industrial interactions and environmental protection of multiple ecosystems has involved ethical choice-making strategies, basically supported by western industrial counterparts based in the U.S.[15] and in Eastern Asian states, like the case of e.g.: Japan's territorial eco-responsibilities.[16]

---

13 Williams, "Environmental Injustice," pp. 49–73.
14 Williams, "Environmental Injustice," pp. 49–73.
15 As noted "almost all industrial activities (as well as many commercial, governmental, and institutional activities) produce some type of hazardous waste. The economic development of the United States during the twentieth century has been accompanied by and, to a significant extent, based on the rapidly rising use of technology, including synthetic organic chemicals. These synthetic chemicals often pose difficult problems because of their stability, resistance to natural degradation, and sometimes hazardous properties…While industry has been relatively successful in limiting accidents to workers during production or industrial use, the consequences of inadequate disposal of industrial hazardous waste has emerged as a critical national problem." John C. Holmes, et al., in: *"Technologies and Management Strategies for Hazardous Waste Control."* March 1983, U.S. Government Printing Office, Washington, D.C. 20402." Library of Congress Catalog Card Number 83-600706 OTA pp. 3–407. P. 44.
16 Midori Kagawa Fox, (2009). "The Ethics of Japan's Global Environmental Policy," The University of Adelaide. Unpublished PhD Dissertation for the Asian Studies School of Social Sciences, University of Adelaide, Australia.

This evolving assertive pattern about the environmental ethics and technological industrial changes has been seen across different transnational societies, which have employed a multi-level regulatory focus for a comparative identification of comprehensive local eco-politics, in order to justify local environmental interventions, that also support the reconfiguration of territorial recovery plans, for instance, which target the reduction of local land degradation, set together with the practical availability of natural resources involving local programmatic conditions.[17]

As highlighted, the public ethical conducts and public societal norms have involved comparative socio-economic confrontations, which have influenced the characteristic incidence of institutional negotiations when needed for the local environmental practices.[18] Similarly, the nuclear power energy innovation systems managed in view of local renewable energy options have been developed through territorial and political procedures reflecting either conservative or progressive eco-industrial technology approaches.

For instance, about the case of nuclear science policy adopted in Japan, the national orientations have fundamentally shown a wider combination of prevailing elements related to the: "1) National identity as a non-nuclear weapon state; 2) Commitment to global nuclear non-proliferation and disarmament; 3) Realist security calculations."[19] About socio-economic strategic issues, the national affirmation of rationalism principles identified in view of national security interests for the comparative environments has in time generated overlapping dimensions, with the introduction of hierarchical organizational confines. As result, the separation of environmental organizational roles, as well as, of technical local responsibilities for the case of governmental proceedings, has been functional to the values of commitment and restraint integrated to the conditional decision-making activities, which have been also marked by realism criteria as a matter of national pride.[20]

In fact, the formation of identitarian *raisons d'être* within international and national strategic power systems and interdependent political affairs has been balanced for the protection of the national interest. Because it has provided functional goals interrelated with local security measures and public confidence platforms, integrated by local distributive relations; which have corresponded to

---

17 Fox, "The Ethics of Japan's," 2009.
18 Fox, "The Ethics of Japan's," 2009.
19 Mike M. Mochizuki, "Japan Tests the Nuclear Taboo," *Nonproliferation Review* 14.2/2007, pp. 303–328. p. 304.
20 Mochizuki, "Japan Tests the Nuclear Taboo," pp. 303–328.

the concretization of regulative environmental agreements that have been stipulated according to the legal terms and conditions of management innovation programs, including reliable collaborative applications.

Within the national security backgrounds, the mutual implementation of regional organizational standards matching with public aggregated choices has been targeted for the reduction of local uncertainties in the nuclear energy field. Through a number of comparative policy studies, respective scholars have considered uncertainty as an environmental risk factor related to the industrial waste disposal management, and to the systematization of practices operated within the national monitoring and evaluation modes.

For the actual necessity of interacting with complex regulatory systems and international production hubs, there is a prevalence of concentrated efforts that have been undertaken for the necessity to rationalize multi-level PPP within political and corporate distributed options; which have been made available for both technical and administrative activities operated through nationally-based agencies. In multiple settings, for instance the U.S. has followed a national cooperation pattern in view of local management practices and the affirmation of common normative goals, which have been identified according to major critical regulative areas such as:

a.	Environmental protection and health reduction risks
b.	Reduction of potential risks from hazardous waste
c.	Monitoring programs improvement and mitigation measures
d.	Acknowledgement of different levels of national regulation
e.	Technology alternatives to land waste disposal in order to minimize releases of hazardous constituents
f.	Quality of data and information systems about hazardous waste, facilities, for reliable risks assessments to (EPA) environmental protection agency
g.	Reducing costs' increases at Federal and State level for administration and regulatory compliance by industrial sectors
h.	Long-term risks reduction and maintenance costs for future generations

Source: John C. Holmes, et al. "Technologies and Management Strategies for Hazardous Waste Control." March 1983. U.S. Government Printing Office, Washington, D.C. 20402. Library of Congress Catalogue Card Number 83-600706 OTA p. 52–66

This type of process of cooperative objective identification about the national policy strategies – as mentioned – has also left an open margin of discussion about the applicable nature of environmental justice norms at regulative level. To the field of nuclear power energy with integrated managerial systems and waste disposal facilities, there have been public interfaces for the inclusion of constitutional activism especially on environmental risk hazards, which have been brought to the attention of national, international and local campaigning fronts for the actual programming activities.[21]

Narrowing the path about common environmental participatory traits with common risk perception factors, which have been related to the critical expansion of nuclear energy technologies,[22] we need to specify that the development of local managerial plans has also been interrelated with the projection of normative formations based on environmental civic justice determinants. Public risk perceptions developed across intraregional population groups have been reported in terms of multiple cognitive changes that lead to constructive knowledge assessment domains, which refer to the respective societal degrees of environmental involvement in the local context.[23]

In fact, there is a dichotomous understanding of the energy development field with common innovative production systems and social responsive adaptations that emerge with changing degrees of participation in national or local collectivities. Social changes have been dependent in time on the different affirmation of spatial industrialized structures with local environmental knowledge accesses that have materialized according to the prevalence of collective cognitive aspects of the so-called *nuclear colonialism*.[24] The national promotion of civilian nuclear power programs, which has been done since the early stages of technology development in the last century, has indeed left a controversial social legacy.[25]

In fact, there are comparative environmental regulations and public organizational undertakings focused on the classification of environment-related assessed elements, which expose contrasting results with consequent polarizing patterns about the acceptance or rejection of the nuclear power energy facilities

---

21 Danielle Endres, "From Wasteland to Waste Site: The Role of Discourse in Nuclear Power's Environmental Injustices," *Local Environment* 14.10/2009, pp. 917–937.
22 See: Andrew Symon, "Southeast Asia's Nuclear power Thrust: Putting ASEAN's Effectiveness to the Test?" *Contemporary Southeast Asia* 30.1/2008, pp. 118–139.
23 Endres, "From Wasteland to Waste Site," pp. 917–937.
24 See: H-Holger Rogner, 2010. "Nuclear Power and Sustainable Development," *Journal of International Affairs* 64.1/2010, pp. 137–163.
25 Endres, "From Wasteland to Waste Site," pp. 917–937.

at territorial level. The management of civil nuclear power activities has been deepened through the practical inclusion of public democratic discourses and political local interactions undertaken within a domestic environment that reflect the multi-level governance distribution and communication channels.

Essentially, the supporting governmental networks have progressed along the lines of national identitarian conditions, which have adhered to the societal value-systems openly defined through the construction of national identitarian expressions, such as the case of Taiwan territorial representation R.O.C., with the population directly or indirectly involved with multilateral socio-economic interactions while maintaining the affirmation of national identitarian criteria that have driven toward specific directions, also in reference to nuclear waste disposal decisions and local environmental planning prospects.[26] Moreover, some constitutive national principles are considered which are through the material affirmation of e.g. the public interconnected information options; the multiple forms of electronic communication channels; and the societal knowledge networks established across the emerging migrant societies; which have been intertwined with national and local regulatory transitions due to socio-economic public transfers, and concerted diversified interests.

In Taiwan's case, the public environmental configurations have shown a propensity for self-determination and economic liberalization. This means that the national identity factor has been interrelated with the local economic status and military security choices, which have emerged in association to the search of political governmental representations.[27] In this specific case, the critical question formulated about the national cross-strait Taiwanese policy relations with mainland China, which from its side has opted for the *"one country two systems' policy,"* has been influenced by the public consensus and legitimacy processes developed according to industrial international partnering accounts.

For instance, at territorial level Taiwanese party politics with internal guiding lines has been assessed through social reforming relations developed by the KMT (Kuomintang) party groups, and the DPP (Democratic Party) groups, among others; where the governmental representative mandates and the societal affirmative principles have been applied in parallel with the public administration

---

26 Hung-Chieh Chang, (2012). "Equal Education, Unequal Identities: Children's Construction of Identities and Taiwanese Nationalism in Education," Unpublished PhD Dissertation: The University of Edinburgh. United Kingdom.
27 Shirley Syaru Lin, (2010). "National Identity, Economic Interest and Taiwan's Cross-Strait Economic Policy 1994–2009," PhD Dissertation unpublished at the University of Hong Kong.

and management business levels.[28] A result, the identitarian translation of strategic economic alliances has also favored the national status-quo principle, reaffirmed for the local decision-making activities.

However in practice, Taiwanese civilian environmental groups have manifested their level of collective dissent and public protests about the case of NPP risk construction plans, for the fourth nuclear power plant's addition promoted in more recent years, with the progressive alternation of local transition periods with frequent public crisis, while the polarization of societal and national interests has come in the form of multi-facets initiatives, differently perceived at the local level.[29]

Essentially, the regional environmental contexts connected to innovative energy differentiation paths have been responsive to changing economic and strategic interests, which have led to the adoption of functional or even dysfunctional national regulative approaches, that have contributed to the enhancement of national and local planning diversity. But at the same time, regional institutional approaches have created concrete boundaries due to the establishment of idealized progressive lines pointing at national technology and innovation development established according to respective territorial associations, configured through the historical affirmative relations.

Similar historical national determinants that create comparative societal collaborative conditions and socio-economic empowerment attributions for legitimacy bases in regional politics, have been shaped according to major territorial organizational shifts, which have also determined the type of local programmatic combination for the functional operating proceedings maintained in governance – which have also been redefined through concrete embedded elements such as: the autonomy of territorial development provisions, the territorial availability of local innovation capacities, the sovereignty of collective national affirmations, and/or the local identity expressions, when already established.[30]

Retrospectively, we can affirm that there are significant aspects of social autonomy principles that have grown according to each population cultural backgrounds, based on the shared democratic rights and self-determination practices. For instance, in comparative regional countries, the environmental evolutive dimension associated with societal ICT knowledge groups, interacting

---

28 Lin, "National Identity, Economic Interest," 2010.
29 Lin, "National Identity, Economic Interest," 2010.
30 Davut Ates, "Political Legitimacy of Nation State: Shifts Within the Global Context," PhD Dissertation unpublished at the Middle East Technical University, Graduate School of Social Sciences. Turkey, 2004.

in European member states, and in East Asian regions, has included commercial extraterritorial relations that have been defined within respective sectoral frontiers.

Multi-territorial policy relations have acquired a structural leading direction in terms of environmental conservation strategies and public mobilization platforms, also characterized by collective identitarian inputs pertaining e.g. to the European or East Asian dimensions. For instance, through the implementation of progressive legislative measures, the single European parties have been able to retrace regional political trajectories, that tackle cross-national socio-economic and environment interests, as for the case of the United Kingdom – still in the EU before the new Brexit phase voted in 2016 – France, or Italy.[31] It has been highlighted that "for environmental activism this process is important: it shapes the domestic and supranational political structures in which environmental policies are made, in which institutions and their actors operate, and in which new norms and practices become diffused between policy-makers and publics"[32]

Parallel to this, the effective ideological representation of supra-national or local environmental protest movements has entailed the propagation of distinctive interpretative models with respective levels of involvement for civilian mobilized parties, and for the organizational associations engaged into comparative local development activities. Through a contextual analysis, the literature studies that discuss about local collective identity's movements, normative institutional systems, and socio-industrial multi-level environmental processes, have been focused on the mutual increase of multiple modernist expressions of nationalism as well as local activism, which have evolved according to the possible momentum of social confrontations and presentation of self-expressions, also based on collective discourse traits. In fact, the common ethical discourse traits have been combined with principles of democracy and shared legalism retaining a level of local public activism and modern innovating practice.

The validation of public constructive ideas about cooperative relational models has been put in association to distinctive technical innovation knowledge fields; which for the case of science technology and environmental development schemes have been framed according to resulting knowledge transfers. In

---

31 Louise Maythorne, (2012). "The Europeanisation of Grassroots Greens: Mobilisation in France, Italy and the UK." PhD Dissertation at the University of Edinburgh, UK. PhD Politics.
32 Maythorne, "The Europeanisation of Grassroots Greens," 2012, p. 2.

addition to this, representative decision-making conducts have been formulated in view of national distributive attributions, and associated PPP for evaluative production asset conformations.

This means that social parties' sectoral divide created about identitarian marginalization issues, and local environmental justice debates has been analyzed in terms of progressive institutional cooperation and civic adaptive platforms; which have been promoted by participatory groups, by individuals, and by public-private organizations, for the simple purpose of approving resolute applicable agendas about: local or national *green* politics; public norms and local practices; and constitutive national formations; where public managed solutions should correspond to the projected efforts that have been envisioned through local participatory paths.[33]

---

33 Karen C. Knop, (1999). "The Making of Difference in International Law: Interpretation, Identity and Participation in the Discourse of Self-Determination." PhD Dissertation at the Graduate Faculty of Law, University of Toronto, Canada.

CHAPTER FOURTEEN

14.1 Transnationalism

Through a common degree of comparable governmental adaptability, the environmental reporting activity about industrial distributive networks has become more apparent, also because the transnationalism aspect of natural material resources still remains a key factor to leverage on the international supply productions, also related to the uranium mines productions. In this resource-based sector, there have been interrelated policy contractual relations, as well as, public local interactions of transnational environmental agencies, which have stimulated open discussions about the use of natural material resources, which are available particularly in e.g. uranium supply countries; where the regional supply areas have experienced a socioeconomic differentiation based on the specific distribution of natural local resources, employed in transnational processing capacities of ex.: African sub-regions.[1]

Basically, what lies behind the territorial and environmental rulings for local governance capacity adaptation has been interrelated to the construction of common participatory systems, shaped through local governance schemes and territorial organizing bodies, which have been put in association with the civilian national organizations. However, despite the integration of locally-based public-private organizations, in order to gather a participative mutual consensus, a systematic search of transnational adaptive development strategies has been put in place, which involves controversial local aid policies and local environmental transition issues in place. Such structural development issues have emerged in reference to detrimental industrial processes leading, for instance, to the regional environmental degradation of natural resources, with progressive socioeconomic impoverishments, and a material land-use with inefficient territorial dispositions; which have caused all together significant environmental impacts, associated with domestic/local managerial agencies' continuous practices.[2]

In the sectoral multi-party development partnerships as for the case of the uranium mining industry, the degree of intensified transnational cooperation

---

1 Rasmus Klocker Larsen, and Christiane Alzouma Mamosso, "Aid With Blinkers: Environmental Governance of Uranium Mining in Niger," *World Development* 56C/2014, pp. 62–76.
2 Larsen, Mamosso, "Aid With Blinkers," pp. 62–76.

has drawn to critical concerns about the instrumental production diversification that has been put in place for the promotion of national or transnational development projects. Essentially, the environmental coordination activities formulated across transnational corporate industries, domestic parties and local NGOs representatives, and public regulatory officials, have been widened through a direct or indirect integration of regional policy groups mediating territorial interests, across transnational nuclear industries and local reporting agencies.

At the national level, the implementation of industrial mining programs has been supported through the reception of international beneficiary aid funds channeled through international development agencies such as the World Bank.[3] In retrospective view, local environmental planning aims with diversified technical measures has brought to local programmatic changes; however it has been highlighted that "the basic premise of community struggle for environmental justice in the face of...mining venture is that of a highly unleveled playing field. The regulatory and institutional realities of African mining remain heavily influenced by the policy reforms of 1980s and 1990s, which aimed to provide the conditions to attract foreign high-risk capital within a neoliberal paradigm, in which the public regulatory framework was relegated to ensure the stabilizing conditions for investors."[4]

In such type of territorial disputes, the transnational regulations of the mining industries, for instance, located in Niger and reported in the study conducted by Larsen and Mamosso, have involved the maintenance of defined roles and specific responsibilities of corresponding domestic distributive powers that have reviewed regional programmatic decisions and local environmental reporting, including at the local population level. Instead, the MNCs have acknowledged respective corporate social responsibilities (CSR) in the transnational industrial regions involved, but have also left behind a disputed ground due to constant instability and uncertainty in terms of critical environmental factors, despite the development of local monitoring practices and internal management tools put already in place.[5]

In essence, the building-up of knowledge innovation processes assessed under critical economic industrial transitions and comparative regulative approaches has necessitated matching with the changing technological dynamics of standardized innovating systems, operated within respective local productive

---

3   Larsen, Mamosso, "Aid With Blinkers," pp. 62–76.
4   Larsen, Mamosso, "Aid With Blinkers," pp. 62–76. p. 63.
5   Larsen, Mamosso, "Aid With Blinkers," pp. 62–76.

dimensions; similarly, the transnational provisioning plans and economic distributive factors have maintained societal directorships and local responsive participation.[6]

From a different angle, the increasing nature of more coherent strategies about the socio-economic growing sectors has required a territorial knowledge-based ecological characterization, particularly aligned with the verticalization of local production systems and transnational distribution programs for multi-level delivery services, which have also been adapted to the current modes of industrial developed acquisitions and global human consumption patterns.[7]

Nonetheless, we need to consider that the modern automation of both industrial and agricultural processes has implied the adoption of technical management with locally integrated knowledge, which has been transferred across different sectoral production units, also according to labor prospective conditions.[8] In such terms, the exploitation of preferential industrial transnational networks has globally entered the phase of internationalization through the intensification of regional innovation systems RIS. The comparative networking systems in turn have acquired a position of social capital formation, with local participation and diffusion of horizontal or vertical linkages, enhanced for the affirmation of PPP public-private business interactions, pursued in local transnational environments.[9]

Fundamentally, the industrial materialization of transnational innovation production assets, within liberal production transitions has directly or indirectly engaged the different governmental officials and international or regional private cooperating partners on local converging interests, in order to operate for short term structural innovation programs; which have corresponded to different types of collective local undertakings and comparable exchanging adaptations. Furthermore, the societal transitional adaptation can also reflect distinctive cooperation spheres, and inclusive institutional mechanisms, which have been aggregated according to systematic guiding principles, despite the nature of pre-existing political scenarios.[10]

About transnational coordinating aspects of mutual PPPs, the comparative transition processes have determined the practical reconfiguration of local

---

6   Carlo Sessa, Andrea Ricci, "The World in 2050 and the New Welfare Scenario," *Futures* 58/2014, pp. 77–90.
7   Sessa, Ricci, "The World in 2050," pp. 77–90.
8   Sessa, Ricci, "The World in 2050," pp. 77–90.
9   Sessa, Ricci, "The World in 2050," pp. 77–90.
10 Sessa, Ricci, "The World in 2050," pp. 77–90.

innovation systems. So that, the industrial sectoral environments have addressed comparative advanced levels of managerial skills of national, as well as, transnational industrial trading units, which have encompassed the diversification of national development programs.

For instance, the energy innovation sectors have been associated with the structural implementation of specified sustainable programs, however within distinctive local territorial boundaries. For which, the levels of business integrated organizations and the involvement of regulatory representatives placed e.g. in the European Region, have been identified according to single states' transitional renewable energy paths, undertaken through open governance approaches. Such type of transnational cooperative transition carries with it structural dynamic factors related to different territorial conditions such as: the local geography and local landscapes, territoriality and public identifications, and long-term migration issues related to the human mobility as well.[11]

About comparable energy research and local development (R&D), the governance relational implications have been analyzed in terms of territorial changes and local dynamic environments. Moreover, the consolidated knowledge innovation processes have been interrelated to specific geolocations, as well as, to sustainable conditions for the application of multi-level technical implementation measures, which have been differently targeted for science programs and technical operable systems, generally perceived according to mutual scopes of a spatial economic regionalization.[12]

Consequentially, main interaction elements about the spatial economic dimensions have been defined for the purpose of understanding technological progressive changes experienced in regional transition processes. In which, the available local solutions have been proposed in line with the possible dimensional explorations of prospective knowledge trading domains in renewable energy scenarios.

At the same time, conceptual approaches have been drawn for the type of 'locational proximity' definition in order to relate respective dynamics according to the measure of large-scale societies, which have shifted in size within local urbanities. The actual demand for compatible strategies about renewable energy designs[13] has come in the form of prioritized regulatory goals, also because of

---

11 Gavin Bridge, Stefan Bouzarovski, Michael Bradshaw, Nick Eyre, "Geographies of Energy Transition: Space, Place and the Low-Carbon Economy," *Energy Policy* 53/ 2013, pp. 331–340.
12 Bridge, Bouzarovski, Bradshaw, Eyre, "Geographies of Energy Transition," pp. 331–340.
13 Bridge, Bouzarovski, Bradshaw, Eyre, "Geographies of Energy Transition," pp. 331–340.

Transnationalism 255

spatial organizational elements intertwined with the diffusion of public developmental agendas. This simultaneous affirmation of liberalized economic directions at the national and local level, has also emerged through territorial expansion plans of technological diversification systems, corresponding to internal energy productions.[14]

However, the centralized decisional processes of convergent industrial sectoral transitions have essentially involved a number of collective societal dilemmas, that refer to the degree of local implementation managed in parallel with local regulative reforms, while combined with the national capacity of infrastructural available components. Public decentralized resources and corresponding local material components have in fact been interrelated with dynamic programmatic changes about e.g. the land-use disposition policies for urban waste disposals, or for the housing sprawls' extensions, locally clustered for the provision of long-term territorial housing services.[15]

To add to this, there have been cohesive societal identifications that favor local territorial groups and regional representative parties, which have promoted public open debates, for possible participatory steps connecting the divided local administrations with multi-level regulatory sectors, which have been identified also for the preparation of social emergence programs tackling, for instance, the local environmental marginalization effects,[16] such as the case of NIMBY syndrome about NPP local waste management.

From national and local political perspectives, the particular integration of cohesive societal approaches, has been granted through the mobilization of selected local parties commonly endorsing the engaged groups from the rural and urban communities. This process has been developed also for the purpose of translating respective local civilian alliances into constitutional parties, including formal or informal local environmental representations, with the possibility of establishing the territorial coordinates for collective emancipation aims running through the federated local networks.[17]

These collaborative environmental programs that evolve into social mobilization and political participation, have encouraged the formation of comparative knowledge-based information platforms, shared for public communication

---

14 Bridge, Bouzarovski, Bradshaw, Eyre, "Geographies of Energy Transition," pp. 331–340.
15 Michael Woods, "Deconstructing Rural Protest: The Emergence of a New Social Movement," *Journal of Rural Studies* 19/2003, pp. 309–325.
16 Woods, "Deconstructing Rural Protest," pp. 309–325.
17 Woods, "Deconstructing Rural Protest," pp. 309–325.

and participatory environmental debates. Under public prospective terms, both the rural and urban contexts have been affected by the regional distribution of centralized and decentralized regulatory powers, with respective performing functions, which have been explored only within comparative intra-regional dimensions of mobile societies.

CHAPTER FIFTEEN

15.1 Local Mobilization Sectors

Across time, the regional societies have become in a way elite sources of dynamic forces in compounded innovating systems, linked to the application of territorial schemes, which have openly required the proliferation of transparency, as well as, legitimacy[1] mechanisms. In essence, we can observe that territorial social groups have been the mobilized parties for transformation, leading to progressive societal initiatives, often contradictory, but which have been interrelated with local environmental causes, which touch upon intra-sectoral interests for the common land rights, the rural water conflicts, and the diversification of regional industry productions, assessed in regions such as Europe, Asia, Latin America, United States, and Africa.[2]

On the other hand, there is evidence of major conflicting points about the logic industrial intersections that have been aligned to social mobility factors, which have been either expanded or reduced according to public preparatory formulations supporting local distributive capacities, validated through national collective engagements as well as local integrative plans. The possibility of fostering environmental social inclusions and comparable practical participation undertaken at territorial level, has been aimed at specific territorial industrial transitions followed by local economic recognitions.[3]

The conceptual exploration about the regional participatory environments that can merge into local social movements, testifies the aggregation of multilateral groups that have operated in terms of categorical uncertain patterns, which reflect temporal industrial proceedings. Essentially, social environmental alliances have moved closer to regional multi-party societies by developing common external linkages that have been reaffirmed among local mobilization groups.[4]

For the direct implementation of local ecology and environmental initiatives, there are some pending questions about the societal effective integration associated with the formation of responsive environments, which can be adapted

---

1   Woods, "Deconstructing Rural Protest," pp. 309–325.
2   Michael Woods, "Social Movements and Rural Politics," Guest Editorial, *Journal of Rural Studies* 24/2008, pp. 129–137.
3   Woods, "Social Movements," pp. 129–137.
4   Woods, "Social Movements," pp. 129–137.

through different territorial groups. Because, the affirmation of comparative socioeconomic interests formulated within regional states has been substantiated on the basis of common eco-innovation projects, that include development transfer processes, taking place according to different levels of mutual cooperative engagements, undertaken through societal communication networks, which influence local public interactions on concerning issues referred to the reduction of environmental health risks and environmental local risk hazards.[5]

In view of public/private adaptive schemes combined with territorial industry policies addressing, for instance, the renewal of NPP nuclear power energy facilities productions, and consistent monitoring energy plans, there are interrelated aspects at the very base of regulatory national prospects. Such additional aspects have included the public orientations and diffused local interactions that emerge from social ecological movements, because they have been very significant for the societal integrative processes aggregated with effective regional production growing areas.

However, we need to consider that the resulting institutional framework about socio-economic energy distribution plans, verified at industrial level, has imposed a set of limits on the achievement of merged socio-economic environmental transfers related to the field of renewable energy expansion; these structural open limits have depended on differentiated country-level energy policies, which have been applied according to contextual regulatory dimensions that have been found in developed changing areas such as in Australia and the United States.[6]

Moreover, in the case of transnational renewable energy resources with incorporation of the uranium mining industry, the comparable review strategies have been intertwined with local governmental agencies for the development of local monitoring measures and internationally-based energy trading networks. Commercial trading networks focused on mineral extraction activities have faced social resistance with public endorsing campaigns, pursued by involved environmental groups and local mobilized communities that have collectively addressed mutual societal concerns about the common protection of life, and the common management of natural ecosystems.[7]

---

5   Marta Conde, Giorgos Kallis, "The Global Uranium Rush and Its Africa Frontier. Effects, Reactions and Social Movements in Namibia," Global Environmental Change 22/2012, pp. 596–610.
6   Conde, Kallis, "The Global Uranium Rush," pp. 596–610.
7   Conde, Kallis, "The Global Uranium Rush," pp. 596–610.

The conjunction of environmental groups initiatives engaged in climate globalism with cultural identifications patterns has been expanded through comparative regional strategic recognitions about the protection of the living human dimension of different adaptive societies. This is to say that a territorial accent of social movements has been given with a spatial characterization of local social dynamics and comparative deterministic values. Local environmental dynamics when shifted to the field of cognitive mutuality have been evidenced through reactive options and management mechanisms applied to the corporate reorganizations and industrial trading institutions, which have also marked the flow of productive and political strategic models for the participatory orientations of stratified communities.[8]

The similar interpretation given about national participatory approaches concerning the states lands' rules with practical local identifications and statutory organizational aims, has been influenced by respective geographical boundaries marked together according to cultural environmental commitments, which have indirectly favored national attachments to the spatial transmission of conservationist traditions.[9]

As underlined before, the multiple directions of industrial socio-economic relations have evolved according to the level of local engagements for dialectical past exchanges designed in association with environmental local activist parties. In which case, the transnational governance and transnational regulative policies have come to redefine comparative environmental impacts appraised against the authoritarian affirmation of divergent management scopes.

At various levels the local and transnational environmental groups located e.g. in South-east Asia have been historically intertwined with social activism initiatives that have gone beyond the local established roots due to correlated ecological policy networks.[10] Transnational collective interactions of environmental groups have taken place through the regional confrontation of nationalized economies, which have also entailed the material coordination of local governmental actors and cooperation of private industrial groups, as well as, the integration of emerging environmental interest parties.

---

8   Arturo Escobar, "Culture Sits in Places: Reflections on Globalism and Subaltern Strategies of Localization," *Political Geography* 20/2001, pp. 139–174.
9   Escobar, "Culture Sits in Places," pp. 139–174.
10 See: Adam Simpson, (2009). "Transnational Energy Projects and Green Politics in Thailand and Burma. A Critical Approach to Activism and Security," PhD Dissertation, School of History and Politics. The University of Adelaide.

On such paradigms, the combined levels of public governance interactions with transnational environmental activism currents and business managerial conducts have led to the recognition of mutual respective limitations, as in the case of transnational renewable energy projects. There is, in fact, a distinctive formation of public organizational impacts and local environmental aims that have been identified through the aggregation of socio-economic initiatives for the national knowledge innovation programs, which maintain common transmissible transfers under political cognitive reviews.

In addition, the parallel inclusion of local environmentalism through progressive cognitive patterns that involve the local participation of societal groups – through mobilization networks – has also applied to the newly developed countries, which have transitioned to the employment of comparable renewable energy systems, that include the nuclear power energy seen as alternative energy option such as in New Zealand's case.[11] In this instance, the political identification of nuclear power development plans has been publicly confrontational, with the alternation of manifest local civil interests, which have coalesced through the social networking exchanges of involved civil society organizations.[12]

Within institutional environmental frames, the national operating dynamics of local social movements have followed in time and in place directly with the people who have been confronted with the application of distinctive liberal economic growth models, and respective local cultural interpretations. In some specific cases, the public environmental campaigns, with aggregated levels of political consensus about the nuclear power energy programs, have been influenced by the formation of local opposing groups brought together through transnational alliance coalitions; which have provided supporting or critical links about the specific regulatory issues involved, while similarly launching comparative participatory models for the public environmental debates.

In essence, a cooperative environmentalism has provided cognitive inputs for specific civilian interests interrelated with national ecological research matters such as: the natural resources conservation plans; the land appropriation rules; and the water distribution policy.[13] In fact, since the 1970s the divergent positions held by regional, national, and local collective groups have led to

---

11 John Francis Hamilton Wilson, (2000). "Turf Wars in Environmentalism: Competing Discourses in Hydroelectric and Nuclear Power Campaigns in New Zealand." PhD Dissertation. Political Studies, The University of Auckland. New Zealand.
12 Hamilton Wilson, "Turf wars in environmentalism," 2000.
13 Hamilton Wilson, "Turf wars in environmentalism," 2000.

mobilized ecology associationism[14] approaches, maintained through multilateral aggregative participatory relations, which have been focused on common local environmental interests.

In addition, the fragmentation of cultural transitional movements and regional environment protesting groups has been polarized particularly about the environmental conservation policies, which have been adapted within internal – external demonstrative conditions, that lead to different types of local environmental collective practices, also undertaken in view of major shifts about global environmentalism. To add to this, the complementary legitimacy of the normative process drafted about the public-private PPP energy planning goals, that promote regional managerial agendas, has become critical because of comparable changing agreements which have reflected local environmental conditions, as well as, innovative production systems dependent on transparent accountability issues, for instance, about local knowledge adaptative transitions undertaken by European states, the United States region, and across the East Asian countries.

International environmental programs have been eventually clarified since the 1992 Rio de Janeiro conference, in which the related United Nations international environmental development conventions have been established.[15] Since the U.N. Rio de Janeiro conference held in 1992, the increasing orientations of transnational regulatory authorities and political environmental movements focusing about global climate patterns and technical environmental changes, have been influenced by international practical interventions, assessed by transnational local community associations that have been committed on collaborative defining paths, which involve legitimate political eco-innovation objectives.[16]

Nonetheless, despite this transnational inclusion of community-based collaborative programs that refer to national territorial preservation policies and local diffusion of environmental protection measures, other comparative social aspects interrelated to broader topics have also crystallized, such as environmental gender values, ecology ethics and local ethnicity relations, as well as labor changing conditions; which have converged toward eventual inclusions

---

14 Hamilton Wilson, "Turf wars in environmentalism," 2000.
15 D. Wastl-Walter, "Social Movements: Environmental Movements," *International Encyclopedia of the Social & Behavioural Sciences*, Elsevier Science Ltd, 2001, p. 14352–14357. https://doi.org/10.1016/B0-08-043076-7/04191-7.
16 D. Wastl-Walter, "Social Movements," p. 14352–14357.

of comparative participatory discourses based on local environmental politics and local institutional adaptations intertwined with performing governmental agencies.[17]

Moreover, the different employment of national strategies about common ecological factors, considering the societal opposing views, has indirectly shaped the concurrent patterns of different ideological affirmations, which have not necessarily led to the association of formalized local environmental practices. But for what reasons? Societal affirmation of both the science and technology perspectives should be kept in mind, as it combines decisional learning practices with national operable discussions (e.g., India's nuclear energy policy programs), and results in the inter-crossing of the national established powers' structures and local emerging transition systems, where the common supporting dispositions have characterized common ecological concerns and evolving critical patterns.[18]

To consider also the fact that due to alternative aspects of the environmental civilian opposing positions, have been drawn together particularly the voices of transnational *green* movements. Transnational environmental recognitions based on common motivational issues of the land partition rules and territorial conflicts that involve the maintenance of local NPP nuclear power plants and related operational infrastructures, have eventually spread at national and international levels through intra-governmental activities, which reflect public information and communication campaigns, also intertwined with different open controlling domains.[19]

In India's case, the technological adaptive dimension of nuclear power energy programs has been assumed across the last five decades, with the innovation knowledge process incorporating strategic national security proceedings. At the same time, the country's respective flow of systematic knowledge innovation policies formulated through incremental districts' development with related programming activities, has been operated in association with concurrent local democratic coalitions for political representations, which include local domestic parties adopting critical comparative logics.

In any case, within corresponding boundaries there has been an implicit presence of cultural determinants that emerge in connection with technology local risk factors and environmental safety issues, which have attracted critical

---

17   D. Wastl-Walter, "Social Movements," p. 14352–14357.
18   Monamie Bhadra, "Fighting Nuclear Energy, Fighting for India's Democracy," Science as Culture 22.2/2013, pp. 238–246.
19   Bhadra, "Fighting Nuclear Energy," pp. 238–246.

attention of transnational science energy experts, local ethnic communities, and large urban citizens' groups. Public attention has also led toward the consolidation of multiple national analyses and political modified orientations, related with cultural environmental protests and the diffusion of knowledge information exchanges, formed through local active dialogues, which have been eventually assimilated into the resulting local networking processes.

However, crucially, there is the fact that "yet, with reports of police brutality, charges of sedition levelled against non-violent protesters, and refusals by nuclear experts to acknowledge ethical and social concerns, underscoring all of these tensions are anxieties about the meaning and practice of Indian democracy"[20] In essence, the materialization of Indian nuclear science and technology development programs[21] in conjunction with regionally-based industrial energy policies, has commonly adhered to the structural governance regulations established within international evolving cycles of regional newly-constituted energy markets, as well as, through the liberalization of corresponding regulatory measures, which reflect specific socio-economic transitions.

Besides the relative causal effects of societal development changes associated to the industrialized field of nuclear power energy and its future development – have left an open ground for the implementation of local environmental debates concerning, for instance, the comparative nuclear mining activities and regional monitoring of complex built-up systems of national nuclear energy management facilities exploited, as in the India's case.[22] Regarding the participatory mandates for national and transnational democratic environmental debates, the comparative evaluation of either internal or external normative dimensions has been addressed, intertwined with national policy-making and local adaptation processes. These focused adjustment processes have also been the result of different transitional monitoring practices, lined up with governmental legitimacy mechanisms that have been put into place; where previous functional rules and local designs of regional institutional regimes have been established according to technical knowledge sharing procedures and local information patterns.[23]

---

20 Bhadra, "Fighting Nuclear Energy," pp. 238–246. p. 239.
21 See: Lydia Powell, Akhilesh Sati, and Vinod Kumar Tomar, "Nuclear Energy in India: Small May Not Be Beautiful." Observer Research foundation (ORF), War Fare, Feb. 03 2022. Website available at: https://www.orfonline.org/expert-speak/nuclear-energy-in-india/
22 Bhadra, "Fighting Nuclear Energy," pp. 238–246.
23 Bhadra, "Fighting Nuclear Energy," pp. 238–246.

On such aspects, the emergence of knowledge sharing approaches has influenced the science and technology (STS) analogous fields despite structural technical divides. In view of the mobilization of civil participatory groups, the states' empowerment routes can favor the consolidation of policy functional models meant as experimental legislative associations that favor local adaptative directives; especially when reasons involved with social related compatibility and local structural adaptability can be also considered in view of sustainable implementation schemes.[24]

This means that a multi-level interaction among the public agencies and transnational market actors has become essential for the possibility of reshaping the advancement of energy production systems, also in collaboration with non-governmental groups. In essence, public-private industrial operators have become more accountable about the domestic propositions produced on the alternative 'dynamic' solutions able to readdress the issues directly or indirectly referring to public coordination and integration measures, when based on the inclusive or exclusive determinants of alternative renewable energy strategies.[25]

Furthermore, the reformulation of collective environmental industrial arrangements has been interrelated to the affirmation of public-private PPP energy management systems. Such PPPs supported with organizational integrative functions have been defined through determinant elements linked to contemporary socioeconomic liberalization approaches for commercial trading relations, extended across competitive regional societies. Transnational commerce networks have operated through progressive adjustments with technological and knowledge innovation measures, extended to industrial and entrepreneurial processes.

The analytical scrutiny of local innovation factors, regional manufacturing technologies, and local governance adaptations, set in line with human labor mobility, and political mediating environments, either internal or external, have led to corresponding critical management steps, which have evolved according to comparable institutional learning performances and local trading adjustments.

In particular, the institutional adaptative settings have facilitated local collaborative paths reaffirmed through environmental functional compromises. Local environmental mediation processes, in essence, have been designed according

---

24 David J. Hess, "Transitions in Energy Systems: The Mitigation – Adaptation Relationship.," *Science as Culture* 22.2/2013, pp. 197–203.

25 Alastair Iles, "Choosing Our Mobile Future: The Degrees of Just Sustainability in Technological Alternatives," *Science as Culture* 22.2/2013, pp. 164–171.

Local Mobilization Sectors 265

to regional and transnational production strategies tackling, for instance, the consolidation of national renewable energy capacities with local attached industries. In order to achieve a comparative affirmation of direct/indirect convergent applications required – for technological industrial changes and societal participative integration – the local governance adaptability conditions have been analyzed in parallel with the multi-level management of domestic innovation activities, wherever possible.

In fact, the establishment of national renewable energy policies has basically been the result of both economic and social regulatory transformations, which have concurrently affected not only the internalization factors of production systems, but also the legitimation status of states' cooperation programs developed according to the sectoral development. For instance, public risk factors and local safety rules can be taken into account in association with the industrial energy's maintenance goals, for the common knowledge management programs undertaken within regional economic hubs in countries such as the United States, South Korea, and Germany. Moreover, the practical and operational outcomes have been politically re-discussed according to the adoption of distinctive environmental policy models related to local social protection conditions and public-private knowledge distribution models, which also validate the allocation of industrial structural funding assets.[26]

Likewise, the national interplay of multiple regional stakeholders over the last four decades has provided very distinctive patterns about states' governmental networks, that favor prioritized territorial initiatives associated with science innovation schemes, and related local environmental accountability.[27] Through conciliatory tones, the environmental national scholars as Jones 2013,[28] Laird 2013,[29] and Mulvaney 2013[30] have put forward comparative policy debates about the infrastructural energy development taking place for the transformation and adaptation of renewable energy projects (ex. Canada's case). A process, which is

---

26 Sheila Jasanoff, Sang-Hyun Kim, "Sociotechnical Imaginaries and National Energy Policies," *Science as Culture* 22.2/2013, pp. 189–196.
27 Jasanoff, Kim, "Sociotechnical Imaginaries," pp. 189–196.
28 Christopher F. Jones, "Building More Just Energy Infrastructure: Lessons From the Past," *Science as Culture* 22.2/2013, pp. 157–163.
29 Frank N. Laird, "Against Transitions? Uncovering Conflicts in Changing Energy Systems," *Science as Culture* 22.2/2013, pp. 149–156.
30 Dustin Mulvaney, "Opening the Black Box of Solar Energy Technologies: Exploring Tensions Between Innovation and Environmental Justice," *Science as Culture* 22.2/2013, pp. 230–237.

also undertaken in order to specify the variant range of policy and economic issues, which have applied to different transnational changing localities.

In fact, the transnational renewable energy schemes that involve the changing status of comparative technologies have particularly allowed the spatial consolidation of regional clustered industries – including spatial local factors regarding: 1. progressive national energy shifts; 2. national and local industrial production dynamics; and 3. the growing influence of public regulatory divisions on multilateral energy systems and independent management organizations performing for comparative structured ecosystems.

Moreover, the participatory interdependent approaches for regional energy systems operated with local alternative technologies (ex. wind – solar power) have been introduced against potential controversial issues of interrelated local health risks factors, which have been caused by local environmental pollution. Essentially, the local connecting communities have integrated the concurrent development of science and knowledge innovation networks in a systematic manner, in order to establish socio-economic and normative adaptation plans, also linked to local environmental risks reduction policies, and comparative human health effects of multiple industrial operating plants.

Over time, despite this convergence of independent public monitoring directions followed by scientific technical experts, an emerging divide has arisen, created between local environmental groups and regional commissioned practitioners, because of the lack of consensus about local environmental claims and technical environmental evaluations, related to direct monitoring actions.[31] In terms of governmental participatory processes, the progressive inclusion of local shared practices and mutual collaboration activities, for instance, in the case of industrial health risk hazards' expositions, has been subjected to the technical requirements and decision-making field strategies assessed between local environmental associations and public mandatory agents.[32]

This type of transfusion of objective programmatic purposes, within territorial development plans and societal intersecting programs, has determined the over-imposition of multilevel political environments that have evolved through centralized or decentralized core functions, including multiple public choices and internal constitutive options. In fact, these alternative public choice developments have been scrutinized in order to avoid selective transnational

---

31 Gwen Ottinger, "The Winds of Change: Environmental Justice in Energy Transitions," *Science as Culture* 22.2/2013, pp. 222–229.
32 Ottinger, "The Winds of Change," pp. 222–229.

productions imposing unequal terms for local environmental dynamic acquisitions.[33]

From an environmental justice point, the unbalanced science policy relations created between local environmental association groups, national governmental agencies, and centralized or decentralized power elites, including military agents, have in a way conditioned the structural transition linkages established within representative policy institutions and regional innovation production networks, available across states interrelated with multi-level organizations. At the same time, comparable states have configured different types of centralized or decentralized rational assertive distributions, relying on comparative industrial networking logics, referred to international corporate institutions and the federal public agencies.[34]

For the case of uranium mining pollution activities and human health exposure effects, the characteristic neighboring communities have suffered from health risk hazards related to local water contamination and local uranium extraction activities. Also, there is a possible lack of research information despite the availability of public discussion forums, local know-how science platforms, and technical management appraisals, connected to the cases of the United States toxic treadmills.[35]

As consequence, the marginalization of collective local environmental processes undertaken in connection with the nuclear mining industrial corporations has led to both past and present public legitimacy conflicts, with the formation of responsive policy facilitation mechanisms, which still need to take into account local sustainability goals, as well as, the territorial adaptability plans operated at times of critical environmental crises.[36] As a matter of practice, the regional policy experts have suggested the adoption of dialectical approaches which can be initiated between the national and local regulatory agencies, and the international multilateral energy corporations, through the formulation of voluntary environmental governance arrangements (VEGA), which can meet technical governance standards about the established development goals, also reviewed according to local environmental and cultural contexts.[37]

---

33 Joshua Sbicca, "Elite and Marginalised Actors in Toxic Treadmills: Challenging the Power of the State, Military, and Economy," *Environmental Politics* 21.3/2012, pp. 467–485.
34 Sbicca, "Elite and Marginalised," pp. 467–485.
35 Sbicca, "Elite and Marginalised," pp. 467–485.
36 Sbicca, "Elite and Marginalised," pp. 467–485.
37 Jeroen van der Heijden, "Voluntary Environmental Governance Arrangements," *Environmental Politics*, 21.3/2012, pp. 486–509.

However, the specific voluntarism character of common environmental arrangements (VEGA) has also been a concern due the actual implementation of transnational firms' conditional renewable energy deals, which have required the support of public implementing agencies, in order to obtain the correspondent environmental performance levels that have been voluntarily established.[38] This type of mediation activity, for the environmental voluntary process that includes the transnational energy firms, the non-governmental agencies, and the legislative local representatives, has been addressed according to problematic and competitive public-private interactions of key regulatory officials, also driven by profit-related conditions, as well as, by constitutive sets of rules for the local compliance platforms.[39]

Therefore, the regional knowledge mediation processes have been associated with the structural accountability and mutual understanding of comparative multilevel technical and policy officers, that favor or otherwise disfavor the organizational and practical dimension[40] of local knowledge and societal transfers. At the same time, the normative and scientific integration of governmental development approaches, delivered according to different environmental scopes and actual participatory conditions, have been expanded through the mutual alternation of national or transnational regulatory orientations, which have emerged both in theory and practice.[41]

As such, the operable notions about territorial social adaptations and socio-economic cohesive motivations put in line with respective institutional configurations, have reflected complex and dynamic elements of comparative associated regimes organized under different ruling systems and decision-making practices, which correspond to multi-level mutable policy dimensions.[42] Once again, the contextual transnational science approaches intertwined with local political institutions and corresponding organizational entities, have shown the correlation with environmental determinism values.

Essentially, the socio-economic confrontation boundaries about the governance knowledge dimension have differently emerged in newly formed liberal democracies, in which the national and transnational cooperation linkages have

---

38   Heijden, "Voluntary Environmental," pp. 486–509.
39   Heijden, "Voluntary Environmental," pp. 486–509.
40   Daniel Barben, "Changing Regimes of Science and Politics: Comparative and Transnational Perspectives for a World in Transition," *Science and Public Policy* 34.1/ 2007, pp. 55–69.
41   Barben, "Changing Regimes of Science and Politics," pp. 55–69.
42   Barben, "Changing Regimes of Science and Politics," pp. 55–69.

remained very critical.[43] For the case of nuclear energy technology systems, the respective operating states have shown common elements of political centralism combined with the process of constant nuclear energy security maintenance. The comparative regional states have basically re-assessed mutual policy orientations about the national interdependent mutual security schemes, and public environmental risks' awareness measures, programmed in cases of critical ecological expositions. Therefore, public energy policies and compatible local governmental schemes have evolved according to a progressive rise on the preparation of common alert response and security systems that can provide the immediacy of local safeguard actions.

Complex environmental preparedness action-plans have also been the result of organizational and regulative requirements predisposed in view of local practical aims and compatible developmental prospects, which have reflected the single countries' status about managerial knowledge operations, and contextual supporting practices related to local environmental sustainability plans.[44] In addition, the national cognitive perspectives about comparative transitions in common science policy directions, have been formed through the consolidation of local cooperative elements for decision-making strategies, which however do not necessarily reflect similar environmental justice goals, promoted across different industrial knowledge regions, which incorporate multi-layered functional directing agencies.

Essentially, the classified regulatory modalities of collective engagements to achieve environmental justice aims, through systematic political mediation activities have been diffused among industrial dynamic regions interlinked to local knowledge and practice organizations.[45] More in general, the international, transnational, and national, environmental justice movements have testified the regional potential of societal aggregations, which have retraced traditional demarcation lines about respective cognitive boundaries, in order to re-value subjective/objective notions according to intra-local and national governance processes.[46] Similarly, the combination of structural trading conditions and

---

43 Barben, "Changing Regimes of Science and Politics," pp. 55–69.
44 Barben, "Changing Regimes of Science and Politics," pp. 55–69.
45 Maria Isabel Casas-Cortes, Michal Osterweil, and Dana E. Powell, "Blurring Boundaries: Recognizing Knowledge-Practices in the Study of Social Movements. Meaning-Making in Social Movements," *Anthropological Quarterly* 81.1/2008, pp. 17–56.
46 Casas-Cortes, Osterweil, Powell, "Blurring Boundaries," pp. 17–56.

political assessed implications has suggested to direct more attention at cognitive collective identifications, while having in place realist concerns.

Nonetheless, the different cultural practices can fundamentally constrain public interests which have also been primarily focused on technocratic spatial impacts of nuclear power energy facilities NPP. For instance, the post-colonialist territories of Eastern regions have been in essence politically fragmented, while addressing the short-term and long-term economic and political priorities, which have been interrelated to states' planning aims and managerial systems performances, operated through local accountable practices destined for national collectivities.[47]

In connection to this, the multi-level managerial roles and corresponding responsibility aspects of national or local regulative environments, have been arranged in view of the democratic and communitarian practices that have been offered across sectoral development processes. The collective discussions about local environmental schemes and societal adaptation determinants have fundamentally evolved according to the sharing of common political beliefs, and the constitution of integrated cognitive dimensions.

In terms of regional policy assessments of governmental adaptive landscapes put in relation with economic managerial resources, as well as, technical innovation systems, the corresponding societal interactions and collective decisional processes about local adaptation rules, have been traditionally substantiated through a variety of dominant public-private organizational approaches PPPs, having merged into symbiotic transfers of local knowledge practices and redistributed developmental changes.[48]

As consequence, the environmental politics and social confrontations have involved both subjective and collective cognitive processes which have referred to climate environmental impacts and related local adaptation approaches, which often imply constrained levels of interest-based and multi-aggregative knowledge-practice relations.

At the same time, the historical configuration has been reasserted through major political party's directions, capturing the conservative and progressive preservation planning aims of democratic constitutive groups for the pursuance of communitarian programs. What then can we expect from the political

---

[47] Richard Falk, "A Radical World Order Challenge: Addressing Global Climate Change and the Threat of Nuclear Weapons," *Globalizations* 7.1–2/2010, pp. 137–155.

[48] Jedediah Purdy, "The Politics of Nature: Climate Change, Environmental Law, and Democracy." *The Yale Law Journal* 119/2010, pp. 1122–1207.

territorial ideologies that have prompted societal independent aggregations for a comparable development of science and knowledge innovation programs interlinked with typified local ecological approaches?

Essentially, it can be carefully weighed that knowledge policy transfers with multi-level approaches have been a central matter of action based on the geographic consistency of security initiatives, that reflect the respective population mobility patterns and institutional power dynamics, referring to the spatial economic transitions directed at the affirmation of energy productive systems.[49] About local environmental politics, the collective practice of supportive local actions associated to voluntary behavioral shifts has been combined with the technical and theoretical dimensions of different aligned societies, which show a changing set of priorities and goals about mutual environmental governance rules.[50]

However, the specific regional impacts of so-called socio-technical innovation transitions of spatial distributive networks have been limited in terms of participatory knowledge management activities and local environmental policy relations.[51] In essence, major evolving classifications about modern environmentalism and global transnationalism that refer to regional politics, science and technology, and business economics, have been pictured in association with the nationally-based industrial and political orientations, which have not necessarily corresponded to the states' territorial power diffusion networks[52] of comparative regulative environments.[53]

Fundamentally, what stays engaged in public local protests does not necessarily acquire the organizational stability for environmental protection plans. Since the 1980s, the regional and transnational networks have been acknowledged with associated environmental prism of collective manifestations. The theoretical diffusion of ideas in post-Washington security policy consensus has been transnational in nature, alongside national limitations of state parties

---

49 Mary Lawhon, and James T. Murphy, "Socio-Technical Regimes and Sustainability Transitions: Insights From Political Ecology," *Progress in Human Geography* 36.3/2011, pp. 354–378.
50 Lawhon, Murphy, "Socio-Technical Regimes," pp. 354–378.
51 Lawhon, Murphy, "Socio-Technical Regimes," pp. 354–378.
52 Brian Doherty, and Timothy Doyle, "Beyond Borders: Transnational Politics, Social Movements and Modern Environmentalism," *Environmental Politics* 15.5/2007, pp. 697–712.
53 Emily Brownell, "Negotiating the New Economic Order of Waste," *Environmental History* 16/2011, pp. 262–289. p. 262.

reflecting collective identities of comparative societies engaged in multi-level industrial innovation domains.

Corresponding important questions that remain are about the transnational flow of multilateral assertive processes versus the transitional governmental construction of multi-level strategies intertwined with local environmental cases. To add to this, the integrated development initiatives have been addressed in terms of national convergent practices and modified industrial trading relations, in particular, when put in association with territorial regulatory plans operated in parallel with technological local transfers, and societal mobility patterns.[54]

Finally, national environmental politics emerging across stratified localities has been driven according to contingent levels of local development and critical environmental affectations. In such context, the industrial technology configurations have transcended mutual borderlines by inheriting transnational political disputes of involved local collectivities, including the environmental cases of nuclear final repositories for waste disposals.[55] In Table 10, and Table 11, nuclear disposal regulation pattern defined at country-level are highlighted. These two tables briefly summarize the nuclear waste disposal regulations and confrontational regulative disposal issues, which have been emerging for the future of implementation schemes and local planning activities.

---

54 Su-Ming Khoo, and Henrike Rau, "Movements, Mobilities and the Politics of Hazardous Waste," *Environmental Politics* 18.6/2009, pp. 960–980.
55 Khoo, Rau, "Movements, Mobilities," pp. 960–980.

CHAPTER SIXTEEN

16.1 Conclusion

Overall, it can be can reported that through this research plan focused on STS innovation schemes and nuclear energy policy development discussed in terms of a comparative activity by searching aggregated national reports, scientific books, scientific articles, and public guiding documents regarding the changing patterns of science and technology with comparative environmental dimensions that involve the nuclear industry in specific national settings, and the nuclear power plants NPP with waste disposal management procedures, a critical point of analysis is finally reached. Because, in more definitive terms, due to the lack of convergent studies about complex governance knowledge transfers connected with science technology and societies, intertwined with transformative components of STS organizational domains, the contemporary international science policy characterizations remain underreported.

As a concluding remark, there is a noticeable prevalence of the formation of ethical modelling approaches which continue to be underestimated in terms of local rehabilitation processes and local environmental restoration projects. For such reasons, the environmental governance evaluative positions about the regional STS development areas, reaffirmed across multiple territories, depending on case-based situations, have evolved according to public policy approaches and comparative knowledge innovation management domains remaining open to compatible adaptive reforms for a common future.

Table 10 Nuclear Waste Disposal Periodic Regulation Frameworks

Risk Assessments	Integrative Decision-making Process	Long-term Compliance Process	Public Participation	Confidence & Trust	Timing Disposal Implementation
In view of nuclear waste cleanup in the United States, the U.S. Government Accountability Office (GAO) reports that "(Environmental Management) EM identifies cleanup-related R&D needs internally and through input from entities across DOE, including the sites themselves, national laboratories, and other DOE offices….	For the case of European nations, the 2019 OECD reports: "Directive 2003/4/EC of the European Parliament and of the Council of 28 January 2003 on public access to environmental information and repealing Council Directive 90/313/EEC, OJ L 041 (14 February 2003) … Amended Safety Directive requires EU member states to 'ensure that necessary information in relation to the nuclear safety of nuclear installations and its regulation is made available to workers and the general public, with specific consideration to local authorities, population and stakeholders in the vicinity of a nuclear installation.'	At periodic review stage the 2008 OECD workshop reports: "During the last two decades the community of nuclear waste management has acknowledged the need for more transparency, stakeholder participation and local community involvement in the decision-making processes. … We have learnt that radioactive waste management, due to its long-term nature, uncertainties, and range of societal impacts, is not the exclusive domain of technical expertise. … The decision-making process must be open, transparent, fair and participatory. The programmes have also become more communicative by requirements of Environmental Impact Assessment (EIA) at project level and Strategic Environmental Assessment (SEA) at the planning and programme implementation levels."	As a historical case study about Sweden nuclear energy and community, OECD reports: "In Sweden, the Oskarshamn community is one of the two communities chosen for further technical investigations in the site selection process for a high-level waste final repository. The successful basis for work with the municipality is the so-called 'Oskarshamn Model', which includes…: 1.Openness and participation…; 2. The use of the Environmental Impact Assessment process as a tool: this should constitute a joint basis for a decision by all parties (the industry, the authorities, the county, and the municipality with its citizens).	"The study analysis[ing] the results of the Eurobarometer 2005 and its 2006 update data concludes that public concern with respect to radioactive waste disposal is a key factor in reducing public support for nuclear energy in general." (OECD/NEA, 2008. "Timing of High-level Waste Disposal." NEA No. 6244. Page 10)	"Most countries have well developed waste management programs with time schedules for disposal implementation. However, experience has shown that in practice the time schedules envisaged proved to be ambitious." "Factors such as technical acceptability of a chosen sites and governance choice making influence public acceptability under time constraints to reach a final outcome."

Risk Assessments	Integrative Decision-making Process	Long-term Compliance Process	Public Participation	Confidence & Trust	Timing Disposal Implementation
In addition, the Technology Development Office identifies R&D needs–such as those relevant to multiple sites… EM officials and contractors at EM's sites identify project-specific needs, including needs that arise in the course of each site's cleanup operations. Sites often address such R&D needs by engaging the national laboratories or adapting commercially available technologies.	… In addition, some responding countries reported that their legal frameworks specific to nuclear energy establish specific bodies, either at a national or local level (for example, local information committees), tasked with providing information to the public on the safety of nuclear power reactors," (OECD/NEA, 2019. "Legal framework for long-term operation of nuclear power reactors." Legal Affairs, 2019. NEA No. 7504. Page 36–37). "A majority of responding countries reported that their legal framework for LTO (long term operation), include requirements regarding public participation.	(OECD Radioactive Waste Management report, 2008. "Regulating the long-term safety of geological disposal of radioactive waste: Practical issues and challenges." Nuclear Energy Agency, NEA, Workshop Proceedings, Paris, France, 28–30 November 2006. Page 93). "Current national programmes vary considerably the degree to which an extended open period prior to the complete backfilling and closure of a repository is foreseen. The ethical principle that future generations should be allowed flexibility in their decision-making favors assigning to future generations the decisions regarding backfilling and closure. … Another concern, particularly for repositories in saturated environments, is that detrimental changes to the system may occur or events take place during the open	3. The community council is a reference group: the competent elected officials are responsible to and on behalf of the voters. Public participation takes place within the framework of representative democracy. 4. the public is a resource… the 'public' is the real expert on many relevant issues. 5. the environmental groups are also a resource: … they have views that can help 'stretch' the industry. 6. 'Stretching' the implementer (SKB) to provide clear answers… 7. The competent authorities are the public's experts: the authorities must be visible throughout the process. The municipality decision on siting must come after statements by the competent authorities."	In the European region, the Eurobarometer survey 2010 about EU citizens and nuclear safety reports "Many Europeans are still afraid of nuclear power plants but a substantial percentage do not consider them to be a risk to them and their families. The major risks are considered to be lack of security against terrorist attacks in NPPs, and misuse of radioactive materials and the disposal of radioactive waste."	"The clear commitment and support of successive governments towards a national radioactive waste management programme will help its timely implementation and are important factors in reaching a publicly acceptable disposal solution." "Clear legislation and well-defined roles of actors in the decision-making process at the local, regional and national levels are key factors in a successful and timely HLW disposal programme."

(Continued)

276 CHAPTER SIXTEEN

Table 10 Continued

Risk Assessments	Integrative Decision-making Process	Long-term Compliance Process	Public Participation	Confidence & Trust	Timing Disposal Implementation
Officials at the Oak Ridge site in Tennessee identified the need to remove mercury vapor from the air during facility deactivation and decommissioning activities. Site officials engaged Savannah River National Laboratory to develop technologies to reduce mercury vapor and debris in the building to limit worker exposure." (GAO, United States Government accountability Office, 2021. "Nuclear Waste Cleanup. DOE needs to better coordinate and	Such requirements typically rest with the nuclear regulatory body or on another decision-making authority (e.g. public authority in charge of environmental protection or local authority) and may entail public hearings, comments in writing and/or the dissemination of draft decisions for public consultation, as well as requirements for the decision-making authority to take into account the comments received when reaching its final decision. ... Overall, eight responding countries indicated that their legal frameworks do not include any requirements for either the decision-making authority or the	period, and that the severity of these changes or events will increase the duration of the open period. ... It is ... recognised that such technical considerations need to be balanced against other factors, such as policies on monitoring and retrievability, which may require a more prolonged open period, or the views of the local community. Monitoring of a wide range of parameters within and around a repository is likely to be carried out prior to repository closure, and some monitoring may take place in the post-closure period. ... A cautious approach is generally applied in which no credit is taken for such measures in averting or reducing the likelihood of human intrusion beyond a few hundred years. This is because of the potential for societal changes and our inability to predict priorities of future generations."	(page 63, OECD/NEA, 2008. "Timing of High-level Waste Disposal." NEA No. 6244. Page 63). "It is confirmed in 2022 that "The government approves SKB's final repository system. The 27th of January 2022, the Swedish Government decided to allow SKB in Osthammar Municipality and an encapsulation plant in Oskarshamn. It is a historical decision that enables SKB to dispose of the nuclear waste that our generation has produced. This decision is met with open arms. We are now looking forward to implementing Sweden's largest environmental protection project"	(Special Eurobarometer 324. "Europeans and Nuclear Safety. Report." European Commission, Special Eurobarometer 324 /Wave 72.2 – TNS Opinion & Social. Publication on March 2010. Page 118). "Europeans have a moderate level of knowledge of nuclear issues: though few respondents knew that the European Union has the largest number of nuclear power plants in the world, they were not aware that nuclear waste is not exclusively	"The structure and transparency of the decision-making process and the level of and possibility for public participation are key issues for achieving public acceptance." (OECD/NEA, 2008. "Timing of High-level Waste Disposal." NEA No. 6244. Page 11) "Research and Development R&D on new technologies has the expected potential of significantly reducing the quantities of long-lived radioactive waste resulting in reduced volumes for disposal in a repository.

Conclusion

Risk Assessments	Integrative Decision-making Process	Long-term Compliance Process	Public Participation	Confidence & Trust	Timing Disposal Implementation
prioritize its research and development efforts." GAO@100 Highlights of GAO-22-104490, a report to the Committee on Science, Space, and Technology, House of Representatives. October 2021. Page 11–12) "EM uses a variety of mechanisms to coordinate R&D, but its efforts do not fully align with certain leading collaboration practices. EM uses both formal and informal coordination mechanisms throughout the complex, including the national laboratory network and working groups.	licensee to solicit public participation as part of the LTO-approval process." (OECD/NEA, 2019. "Legal framework for long-term operation of nuclear power reactors." Page 39). "Public participation is allowed during the continued operation authorisation process. While the decision-making authority does not have a legal duty to solicit public participation, the operator does not have such a duty. This duty is found in Article 103, 'Gathering of Residents Opinion', in the Nuclear Safety Act. During the operator's preparation of the Radiation Environmental Report, the operator shall gather opinions from residents within the scope determined by	(OECD Radioactive Waste Management report, 2008. "Regulating the long-term safety of geological disposal of radioactive waste: Practical issues and challenges." Nuclear Energy Agency, NEA, Workshop Proceedings, Paris, France, 28–30 November 2006. Page 58).	(Swedish Nuclear Fuel and Waste Management Company (Svensk Kärnbränslehantering Aktiebolag, SKB), 2022. "We take care of Swedish radioactive waste." SKB web portal available at: https://www.skb.com/)	produced by nuclear power plants." "Similarly, Europeans continue to be unfamiliar with safety issues related to nuclear power plants. Only a quarter of citizens feel 'very well or 'fairly well' informed, compared with three in four who feel 'not very well,' or 'not at all' informed about the safety of nuclear power plants"	It also holds appeal to people who are unconvinced by current proposals for deep geological disposal … R&D into partitioning and transmutation is not simply a response to public concern. It is part of a responsible and ethical approach towards good resource management, i.e. sorting recovery, recycling and therefore resource saving. However these technologies need significant development and time before they are deployable at a commercial scale."

(*Continued*)

Table 10 Continued

Risk Assessments	Integrative Decision-making Process	Long-term Compliance Process	Public Participation	Confidence & Trust	Timing Disposal Implementation
The agency also follows certain leading practices for collaboration – such as clarifying roles and responsibilities and including relevant participants. However, EM does not fully follow others, which affects its ability to identify, track, and evaluate the effectiveness of R&D efforts." (GAO, United States Government accountability Office, 2021. "Nuclear Waste Cleanup. DOE needs to better coordinate and prioritize its research and development efforts." GAO@100 Highlights of GAO-22-104490. Page 15)	the NSSC, upon making the draft Radiation Environmental Report available to the public for inspection or by holding a public hearing. The operator must describe, in general, each resident's opinion in the final Radiation Environmental Report, whether its opinion will be reflected in the assessment and, if not, why not. In such cases, a public hearing shall be held if there is a request from the head of a local government having jurisdiction over the area in which residents' opinions are to be gathered, or from the residents within the scope prescribed by Presidential Decree." (OECD/NEA, 2019. "Legal framework for long-term operation of nuclear power reactors." Page 93).			"Europeans are critical of the information offered in the media about energy in general and nuclear energy in particular: almost two thirds of the interviewees said it is insufficient. Large majorities in almost all of the countries surveyed mention television as the main source of information on nuclear energy." "Radioactive waste management and environmental monitoring procedures are the main aspects citizens would like to know more about. Scientists,	(OECD/NEA, 2008. "Timing of High-level Waste Disposal." NEA No. 6244. Page 12.)

Risk Assessments	Integrative Decision-making Process	Long-term Compliance Process	Public Participation	Confidence & Trust	Timing Disposal Implementation
"EM's Technology Development Office has not taken a comprehensive approach to prioritizing R&D. In the absence of a comprehensive approach, individual EM sites and DOE laboratories have developed their own approaches for making R&D prioritization decisions, according to site and laboratory officials. These individual approaches to prioritizing R&D differ, including in the extent to which they consider complex-wide issues."	In the case of Sweden, "members of the public may access specific information regarding nuclear safety through the Local Liaison Safety Committees, which are established in every region where an NPP is located." (OECD/NEA, 2019. "Legal framework for long-term operation of nuclear power reactors." Page 131). In the case of the UK, "nuclear sites are also regulated under other licenses, permits and authorisations issued by regulators, such as a generation license issued by the Office of Gas and Electricity Markets (OFGEM), the economic regulator, under the Electricity Act 1989. These documents include conditions and requirements that			followed at a distance by national nuclear safety authorities and international organizations working on uses of nuclear technology, are the three most trusted sources of information." "Only around one in four Europeans would like to be directly consulted in the decision-making process regarding the developing and updating of energy strategies. An identical proportion would prefer to leave the responsible authorities to decide exclusively on this matter	

(*Continued*)

280   CHAPTER SIXTEEN

Table 10 Continued

Risk Assessments	Integrative Decision-making Process	Long-term Compliance Process	Public Participation	Confidence & Trust	Timing Disposal Implementation
(GAO, United States Government accountability Office, 2021. "Nuclear Waste Cleanup. DOE needs to better coordinate and prioritize its research and development efforts." GAO@100 Highlights of GAO-22-104490. Page 28). "R&D plays an essential role in efforts by DOE's EM program to clean up contamination at 16 sites around the country remaining from 75 years of federal nuclear weapons production and energy research.	must be complied with throughout the life cycle of the site. The national nuclear regulatory body, the Office of Nuclear Regulation (ONR) has issued a list of Standard License Conditions attached to Nuclear Site Licenses. Other government bodies are responsible for environmental			and an additional fifth would prefer the national Parliament to be consulted and to participate in the decision-making process" (Special Eurobarometer 324. "Europeans and Nuclear Safety. Report." European Commission, Special	

Risk Assessments	Integrative Decision-making Process	Long-term Compliance Process	Public Participation	Confidence & Trust	Timing Disposal Implementation
Studies have found that investing in R&D may help EM identify safer, more effective, and cost-efficient cleanup approaches – especially needed because the federal government's environmental liability associated with cleaning up radioactive and hazardous waste is now over $400 billion and growing." (GAO, United States Government accountability Office, 2021. "Nuclear Waste Cleanup. DOE needs to better coordinate and prioritize its research and development efforts." GAO@100 Highlights of GAO-22-104490. Page 35).	and urban planning-related aspects of such operation, namely the Environment Agency, Natural Resources Wales or the Scottish Environmental Agency – depending on the location of the concerned reactor – and the relevant local planning authorities." (OECD/NEA, 2019. "Legal framework for long-term operation of nuclear power reactors." Page 146).			Eurobarometer 324 /Wave 72.2 – TNS Opinion & Social. Publication on March 2010. Page 119).	

Table 11 Comparative Nuclear Disposal and Safety Regulation Framework

Germany: "the HLW disposal has been regulated by three Ministries – the Ministry of Economics and Technology (BMWi) also in charge of R&D under the 5th Energy Research Programme "Innovation and New Technology." The Federal Ministry for the Environment, Nature Conservation and Nuclear Safety (BMU) and the Ministry of Education and Research (BMBF)." (OECD/NEA, 2008. "Timing of High-level Waste Disposal." NEA No. 6244. Page 34).
NUCLEAR SAFETY
In 2018, "The Federal Environment Ministry (BMUB) worked in partnership with the federal states, the Federal Office for Radiation Protection (BfS) and the Federal Office for the Safety of Nuclear Waste Management (BfE) to create a … knowledge platform … on nuclear safety. Previously, interested citizens had to scour multiple federal and state authority websites to find information on nuclear facilities, licensing procedures, the regulatory bodies of Germany and Europe, and emergency preparedness and response. The joint platform streamlines this knowledge in five subject areas and thus allows easy access." (BMU) Federal Ministry for the Environment, Nature Conservation, Nuclear Safety and Consumer Protection: 16.02.2018 – Press release No.035 – Nuclear Safety. Available at Website: https://www.bmuv.de/en/pressrelease/a-new-portal-compiles-information-on-nuclear-safety
Japan: "The responsibility for designing and constructing facilities for the geological disposal of high-level waste in Japan lies with the Nuclear Waste Management Organisation (NUMO)." (OECD/NEA, 2008. "Timing of High-level Waste Disposal." NEA No. 6244. Page 35).
NUCLEAR SAFETY
In 2018, (NUMO) published its "ACTION PLAN for REFORMING COMMUNICATION ACTIVITIES:" "Key points: 1. Plan various communication activities which are interactive and include participants' opinions (NUMO) will no longer utilize the support of external organizations. 2. Visit and deliver presentations in places capable of attracting people from a wide range of different backgrounds, improve approaches for engaging students and make greater use of the NUMO website and 3. Regularly review these efforts through staff discussion and reflection."
I. [About] "The Improvement in past approaches to seminars and information exchange meetings:
Current situation and challenges: … In response to participant feedback, we will discontinue the conventional nationwide symposia and undertake the organization of smaller meetings that include face-to-face discussions and interactive dialogue… / For the benefit of participants, NUMO will be more flexible when conducting small group discussions… In addition, we will regularly try out different styles/approaches to meetings in order to continually improve and enhance participant experience."

"Efforts for broader discussion: – Widely disseminate information by making live videos of meetings available on the NUMO website; – Request participation of representative stakeholders at meeting and encourage expressions of opinions; – hold panel discussions with experts or specialist in a range of fields related to geological disposal"

"Improvements to the NUMO website: – NUMO staff will make appearances on the NUMO website and also social networking services (e.g. Facebook and Instagram) to explain various aspects of geological disposal … – Share participant opinions a 'NUMO Test' that will allow interested people to check their knowledge on geological disposal – Make available on the website videos of the panel discussions with experts and specialists that work in various field related to geological disposal." Nuclear Waste Management Organization of Japan, 2018. Topics, About NUMO, "Action plan for reforming communication activities." April 13, 2018. Available on Website: https://www.numo.or.jp/en/about_numo/new_eng_tab08.html

Republic of Korea: "The Ministry of Trade, Industry and Energy (MOTIE) has the responsibility of establishing basic policies and project implementation plans for the storage, treatment and disposal of radioactive waste in the Republic of Korea." (OECD/NEA, 2008. "Timing of High-level Waste Disposal." NEA No. 6244. Page 35).

NUCLEAR SAFETY

In 2009 it was established KORAD, Korea Radioactive Waste Agency as a Quasi-Governmental Agency, put in charge of: "medium and low-level radioactive waste management, High-level radioactive waste management and promotion of radioactive waste." "KORAD was established to safely and transparently manage radioactive wastes … KORAD has been stepping on the center of the energy transition era in Korea. In the circumstance, safe management organization of high-level radioactive wastes can be regarded as a pioneer to open a new era for energy transition. Under the era of social responsibilities for safe management of radioactive wastes with expert knowledge, various communication, job creation and mutual development for the society, KORAD will become a 'Public Platform' to make people satisfied." SMEs Energy Technology Market, "KOMIPO KORAD at a Glance." December 31, 2020. Available at Website: https://www.energytechmarket.or.kr/svc/itd/enOrgIntro.do?org=KORAD

Sweden: "The responsibility for the management of spent fuel in Sweden lies with the Swedish Nuclear Fuel and Waste Management Company (SKB) which is jointly owned by the four nuclear utilities. SKB is regulated by two government agencies – the Swedish Nuclear Power Inspectorate (SKI) and the Swedish Radiation Protection Institute (SSI). The Swedish National Council for Nuclear Waste (KASAM) is an independent advisor to the environment ministry and regulators on all issues concerning nuclear waste management." (OECD/NEA, 2008. "Timing of High-level Waste Disposal." NEA No. 6244. Page 35).

(Continued)

Table 11 Continued

NUCLEAR SAFETY
In 2018, the Swedish Radiation Safety Authority reports specific requirements for public information and participation: Regarding the Licensing process at NPPs, it is pointed out that "There are several procedures that serve the purpose to involve the public in the siting of new spent nuclear fuel and nuclear waste facilities.… An application submitted to the Authority is sent on referral to a large number of stakeholders, e.g. other authorities, the municipality concerned, county administrative boards, universities and NGOs….   - SSM publishes all its significant decisions on the SSM website. Through an e-register on the website, the general public can view the documents sent from the Authority or submitted to it. The Constitution gives everyone the right to access the documents held by the Authority.". - As part of the preparation of an Environmental Impact Assessment (EIA), an applicant must, before the application documents are submitted, consult with the county administrative board, relevant authorities, the potential host municipality, other stakeholders, the public and NGOs…. The Authority also participates in consultation meetings primarily intended for the municipality and other stakeholders concerned.…. - According to the Act (2006: 647) and Ordinance (2017: 1179) on Financing of Management of Residual Products from Nuclear Activities, the municipalities that might host a spent nuclear fuel or nuclear waste facility, including a disposal facility, are reimbursed for information activities aimed at citizens. Municipalities have been reimbursed for their information activities since the mid-1990s.…. - Since 2005 non-profit non-governmental organisations may be reimbursed for costs incurred in relation to their engagement in consultations related to disposal of spent nuclear fuel and radioactive waste. Decisions concerning reimbursement to municipalities and non-profit organisations are made by SSM." Eva Gimholt, SSM2018-2869, Document number: 2018-2869-5. "Swedens second National Report on Implementation of Council Directive 2011/70/Euratom." Rapport of the Swedish Radiation Safety Authority, pages 35–36, 2018. Available at Website: https://www.stralsakerhetsmyndigheten.se/
United Kingdom: "Spent fuel is transferred … to the Sellafield plant in Cumbria for reprocessing or long-term storage. … Since the formation of NIREX (Nuclear Industry Radioactive Waste Executive) in the early 1980s the main focus on geological disposal in the United Kingdom has been in relation to intermediate and long-lived low level wastes (LLW). … A [national] Committee on Radioactive Waste Management (CoRWM) was established to examine a list of options … and following a process of extensive public consultations" (OECD/NEA, 2008. "Timing of High-level Waste Disposal." NEA No. 6244. Page 36).

NUCLEAR SAFEY In 2015 guiding lines published by the UK Nuclear Decommissioning Authority (NDA) about involving participant societies specifies that "Site Stakeholders Groups (SSGs), or their equivalents, are the interface between us [NDA], the local communities near our sites, and the site operator. They meet regularly and provide opportunities to 1. ask questions; 2. review, comment on and influence strategies, plans and achievements.". UK Government, Emergency preparation, response and recovery section, 2015. "Guidance. Engaging with Nuclear Decommissioning Authority: how to get involved." Nuclear Decommissioning Authority (NDA), 2015. Available at Website: https://www.gov.uk/guidance/engaging-with-nuclear-decommissioning-authority-how-to-get-involved
NDA Guidance Report, 2015. "The primary objectives of each SSG are: 1. to provide an opportunity for questioning the operators, NDA and regulators on behalf of the community; 2. to receive and comment on progress reports and forward plans for the sites; 3. to represent the views of the local community through the provision of timely access advice to NDA, operators and regulators. … SSGs are not decision-making bodies. The objectives above do not remove the accountability of relevant bodies for decision-making in the sites." NDA, Nuclear Decommissioning Authority, 2015. "NDA Guidance for site stakeholder groups." Ref. LAR3.0, page 1.
United States: "For many years, civilian and defense-related activities have produced spent nuclear fuel and high-level radioactive waste. These materials have accumulated and continue to accumulate, at 72 commercial and 4 U.S. Department of Energy (DOE) sites across the United States. Because these materials are highly radioactive, they must be isolated from the accessible environment. … The Congress adopted the *Nuclear Waste Policy Act of 1982* (NWPA) which created a comprehensive national program for the safe, permanent disposal of highly radioactive waste in a geologic repository. This program included the identification, characterization, and approval of a site for permanent geologic repository, and for its licensing by the U.S. Nuclear Regulatory Commission (NRC). The NWPA assigned lead responsibility to the Secretary of Energy and created the Office of Civilian Radioactive Waste Management (OCRWM) to develop and manage a federal system for disposing of commercial spent nuclear fuel and defense high-level radioactive waste" (OECD/NEA, 2008. "Timing of High-level Waste Disposal." NEA No. 6244. Pages 36–37).
NUCLEAR SAFETY At the Department of Energy (DOE), the Office of Environmental Management […] "EM has been charged with the responsibility of cleaning up 107 sites across the country [US] […]. Under the "SITE NAME and TYPE(s) OF CLEANUP WORK PERFORMED [the cleanup work at the Savannah River Site includes] – Transuranic and Solid Waste Disposition; –Tank Waste; – Special Nuclear Materials and Spent Nuclear Fuel; – Facility Deactivation and Decommissioning; – Soil and Groundwater Remediation." Cleanup Sites, Office of Environmental Management, U.S. Department of Energy, accessed on April, 2022. Available at Website: https://www.energy.gov/em/cleanup-sites

(*Continued*)

Table 11 Continued

Based on this case study, the local citizens involvement relates to: "The Savannah River Site (SRS) – Citizens Advisory Board (CAB) [which] is a part of the Environmental Management Site-Specific Advisory Board (EMSSAB), a stakeholder board that provides the Assistant Secretary for Environmental Management and designees with advice, information, and recommendations on issues affecting the EM program at various sites. Among those issues are clean-up standards and environmental restoration; waste management and disposition; stabilization and disposition of non-stockpile nuclear materials; excess facilities; future land use and long-term stewardship; risk assessment and management; and clean-up science and technology activities." U.S. Department of Energy, Citizens Advisory Board (CAB), 2022. "GA & SC citizens working together for a better tomorrow at SRS." SRS, CAB page, Website Available at: https://cab.srs.gov/srs-cab.html

Similarly, "Through the CAB, particularly the Facilities Disposition and Site Remediation (FD&SR), the Strategic and Legacy Management (S&LM), the Nuclear Materials (NM), and the Waste Management (WM) Committees, the parties are able to discuss their concerns and better understand the competing needs and requirements of the government and local citizens. The CAB also broadens the scope of decision making to account for local stakeholder issues in addition to consideration of technical data required under the Comprehensive Environmental Response, Compensation, and Liability Act (CERCLA) public comment rules. CAB Combined Committee Meetings, Full Board Meetings, and Committee Meetings are held bimonthly. DOE, the Environmental Protection Agency (EPA), and South Carolina Department of Health and Environmental Control (SCDHEC) work actively with the CAB members on the various committees." (ARF 17491, U.S. Department of Energy, Savannah River Site (SRS), 2011. "Savannah river site federal facility agreement community involvement plan (U)." WSRC-RP-96–120, Revision 7, February 2011, page 5).

"Key Priorities and Strategic Vision 2022–2032: – EM's mission is to complete the safe cleanup of the environmental legacy brought about from decades of nuclear weapons development and government-sponsored nuclear energy research." (U.S. DOE, Office of Environmental Management, 2022. "EM Strategic Vision: 2022–2032." EM Report, page 7. Website available at: https://www.energy.gov/sites/default/files/2022-03/DOE-EM-Strategic-Vision-2022-Final-3-8-22.pdf).

"The EM Program Plan will be informed by the latest knowledge in technologies and environmental management, safety and health (ES&H) and programmatic risks; and incorporates current site plans to complete work." (U.S. DOE, Office of Environmental Management, 2022. "EM Strategic Vision: 2022–2032."page 9).

Overall nuclear regulatory systems will comparatively differ in terms of costs density programs as well as progressive commitments toward integrated waste operations and local exchanges.

Major Source: OECD/NEA, 2008. "Timing of High-level Waste Disposal." NEA No. 6244

BIBLIOGRAPHY

Achillas, Ch., Ch. Vlachokostas, N. Moussiopoulos, G. Banias, G. Kafetzopoulos, A. Karagiannidis. "Social Acceptance for the Development of a Waste-to-Energy Plant in an Urban Area." *Resources, Conservation and Recycling* 55/ 2011, pp. 857–863.

Adamantiades, A., I. Kessides. "Nuclear Power for Sustainable Development: Current Status and Future Prospects." *Energy Policy* 37/2009, pp. 5149–5166.

Alexander, Catherine, Joshua O. Reno. "From Biopower to Energopolitics in England's Modern Waste Technology." *Anthropological Quarterly* 87.2/2014, pp. 335–358.

Al-Rodhan, N. R. F. "The Politics of Emerging Strategic Technologies. Implications for Geopolitics, Human Enhancement and Human Destiny." Palgrave Macmillan. 2011.

Alston, Margaret, Kerri Whittenbury. "Climate Change and Water Policy in Australia's Irrigation Areas: A Lost Opportunity for a Partnership Model of Governance." *Environmental Politics* 20.6/2011, pp. 899–917.

(d') Aquino, Patrick. "Empowerment and Participation: How Could the Wide Range of Social Effects of Participatory Approaches Be Better Elicited and Compared." *CIRAD (Centre de Coopération Internationale en Recherche Agronomique pour le Développement)*, 2007, France.

Aronson, James, Florian Claeys, Vanja Westerberg, Philippe Picon, Guillaume Bernard, Jean-Michel Bocognano, and Rudolf de Groot. "Steps Towards Sustainability and Tools for Restoring Natural Capital: Etang de Berre (Southern France) Case Study." In: *Sustainability Science: The Emerging Paradigm and the Urban Environment*, eds. M.P. Weinstein and R.E. Turner, (Springer Media, LLC, 2012), pp. 111–138.

Arts, Bas, Arnoud Lagendijl, and Henk van Houtum. *The Disoriented State: Shifts in Governmentality, Territoriality, and Governance*, (Springer Science + Business Media B.V. 2009).

Asheim, Bjorn T., and Lars Coenen. "Knowledge Bases and Regional Innovation Systems: Comparing Nordic Clusters." *Research Policy* 34/2005, pp. 1173–1190.

Ates, Davut. "Political Legitimacy of Nation State: Shifts within the Global Context." PhD Dissertation unpublished at the Middle East Technical University, Graduate School of Social Sciences. Turkey, 2004.

Barben, Daniel. "Changing Regimes of Science and Politics: Comparative and Transnational Perspectives for a World in Transition." *Science and Public Policy* 34.1/2007, pp. 55–69.

Beken, Tom Vander, Nicholas Dorn, Stijn Van Daele. "Security Risks in Nuclear Waste Management: Exceptionalism, Opaqueness, and Vulnerability." *Journal of Environmental Management* 91/2010, pp. 940-948.

Bhadra, Monamie. "Fighting Nuclear Energy, Fighting for India's Democracy." *Science as Culture* 22.2/2013, pp. 238-246.

Bierly, Denise. "A 19-Year Perspective on Long-Term Care Issues." *Long-Term Management of Contaminated Sites Research in Social Problems and Public Policy* 13/2007, pp. 213-226.

Bjorklund, Anna E., Goran Finnveden. "Life Cycle Assessment of a National Policy Proposal – The case of a Swedish Waste Incineration Tax." *Waste Management* 27/2007, pp. 1046-1058.

Blowers, Andrew. "Power, Participation and Partnership. The Limits of Co-operative Environmental Management." in: *Co-operative Environmental Governance*, eds. P. Glasbergen, (Kluwer Academic Publishers, 1998). pp. 229-249.

Bluth, Christoph. "Correspondence, Civilian Nuclear Cooperation and the Proliferation of Nuclear Weapons." *International Security* 35.1/2010, pp. 184-200.

Boschma, Ron A., Robert C. Kloosterman. *Learning from Clusters. A Critical Assessment from an Economic-Geographical Perspective*, (Published by Springer, The Netherlands, 2005).

Bozeman, Barry, Daniel Sarewitz. "Valuing S&T Activities. Public Values and Public Failure in US Science Policy." *Science and Public Policy* 32.2/2005, pp. 119-136.

Breuste, Jurgen, Salman Qureshi, Junxiang Li. "Applied Urban Ecology for Sustainable Urban Environment." *Urban Ecosystems* 16/2013, pp. 675-680.

Bridge, Gavin, Stefan Bouzarovski, Michael Bradshaw, Nick Eyre. "Geographies of Energy Transition: Space, Place and the Low-Carbon Economy." *Energy Policy* 53/2013, pp. 331-340.

Brownell, Emily. "Negotiating the New Economic Order of Waste." *Environmental History* 16/2011, pp. 262-289.

Burchell, Jon, Joanne Cook. "Banging on Open Doors? Stakeholder Dialogue and the Challenge of Business Engagement for UK NGOs." *Environmental Politics* 20.6/2011, pp. 918-937.

Burchell, Jon, Joanne Cook. "Sleeping with the Enemy? Strategic Transformations in Business-NGO Relationships Through Stakeholder Dialogue." *Journal of Business Ethics* 113/2013, pp. 505-518.

Burger, Joanna, Michael Gochfeld, David S. Kosson, et al. "Science, Policy, and Stakeholders: Developing a Consensus for Amchitka Island, Aleutians, Alaska." *Environmental Management* 35. 5/2005, pp. 557-568.

Burger, Joanna, Nellie Tsipoura, Michael Gochfeld, and Michael R. Greenberg. "Ecological Considerations for Evaluating Current Risk and Designing Long-Term Stewardship on Department of Energy Lands." *Long-term Management of Contaminated Sites Research in Social Problems and Public Policy* 13/2007, pp. 139–162.

Butler, Catherine, Karen A. Parkhill, and Nicholas F. Pidgeon. "Nuclear Power after Japan: the Social Dimensions." *Environment Magazine* 53.6/2011, pp. 3–14.

Buuren, Arwin van, Jean-Marie Buijs, Geert Teisman. "Program Management and the Creative Art of Cooperation: Dealing With Potential Tensions and Synergies between Spatial Development Projects." *International Journal of Project Management* 28/2010, pp. 672–682.

Buuren, Arwin van, Peter Driessen, Geert Teisman, Marleen van Rijswick. "Toward Legitimate Governance Strategies for Climate Adaptation in the Netherlands: Combining Insights From a Legal, Planning, and Network Perspective." *Regional Environmental Change* 2013. DOI 10.1007/s10113-013-0448-0. (Springer-Verlag Berlin Heidelberg).

Chan, Gabrielle. "We Have Lost Control:" NSW Farmers Battle Private Irrigation Companies for Water." 2019. The Guardian Australia Edition. https://www.theguardian.com/australia-news/2019/apr/30/we-have-lost-control-nsw-farmers-battle-private-irrigation-companies-for-water (16 Aug. 2022).

Chang, Hung-Chieh, (2012). "Equal Education, Unequal Identities: Children's Construction of Identities and Taiwanese Nationalism in Education." Unpublished PhD Dissertation: The University of Edinburgh. United Kingdom.

Chen, Liang-Chih. "The Governance and the Evolution of Local Production Networks in a Cluster: the Case of Taiwan's Machine Tool Industry." *GeoJournal* 76/2011, pp. 605–622. DOI 10.1007/s10708-009-9317-2.

Chung, Ji Bum, Hong-Kew Kim, and Sam Kew Rho. "Analysis of Local Acceptance of a Radioactive Waste Disposal Facility." *Risk Analysis* 28.4/2008, pp. 1021–1032.

Ciarli, Tommaso, Karolina Safarzynska. "Sustainability and Industrial Challenge: The Hindering Role of Complexity." *SPRU Working Paper Series*, SPRU Science Policy Research Unit, University of Sussex Business School, 2020.

Clark, Simon M. "Public Participation in Decisions Relating to the Environmental Management Ministry of Defence Sites." in: *Defence and the Environment: Effective Scientific Communication*, eds. K. Mahutova et al., (Kluwer Academic Publishers. Printed in the Netherlands, 2004), pp. 65–70.

Coenen, Lars, Paul Benneworth, Bernhard Truffer. "Toward a Spatial Perspective on Sustainability Transitions." *Research Policy* 41/2012, pp. 968–979.

Conde, Marta, Kallis. "The Global Uranium Rush and Its Africa Frontier. Effects, Reactions and Social Movements in Namibia." *Global Environmental Change* 22/2012, pp. 596–610.

Cooke, Philip, Mikel Gomez Uranga, and Goio Etxebarria. "Regional Innovation Systems: Institutional and Organisational Dimensions." *Research Policy* 26/1997, pp. 475–491.

Cooke, Philip. "Transition Regions: Regional-National Eco-Innovation Systems and Strategies." *Progress in Planning* 76/2011, pp. 105–146.

Cortes, Maria Isabel Casas, Michal Osterweil, and Dana E. Powell. "Blurring Boundaries: Recognizing Knowledge-Practices in the Study of Social Movements. Meaning-Making in Social Movements." *Anthropological Quarterly* 81.1/2008, pp. 17–56.

Davies, Anna R. "Clean and Green? A Governance Analysis of Waste Management in New Zealand." *Journal of Environmental Planning and Management* 52.2/2009, pp. 157–176.

Davoudi, Simin. "Planning for Waste Management: Changing Discourses and Institutional Relationships." *Progress in Planning* 53/2000, pp. 165–216.

Davoudi, Simin. "Governing Waste: Introduction to the Special Issue." *Journal of Environmental Planning and Management* 52.2/2009, pp. 131–136.

Dawson, Jane I. "Latvia's Russian Minority: Balancing the Imperatives of Regional Development and Environmental Justice." *Political Geography* 20/2001, pp. 787–815.

Dawson, Jane I., Robert G. Darst. "Meeting the Challenge of Permanent Nuclear Waste Disposal in an Expanding Europe: Transparency, Trust and Democracy." *Environmental Politics* 15.4/2006, pp. 610–627.

Dennis, Michael A. "Scientific and Technical Knowledge and the Making of Political Order." *History and Technology: An International Journal* 28.4/2013, pp. 415–421.

Deutz, Pauline, and David Gibbs. "Industrial Ecology and Regional Development: Eco-Industrial Development as Cluster Policy." *Regional Studies* 42.10/2008, pp. 1313–1328.

Doherty, Brian, Timothy Doyle. "Beyond Borders: Transnational Politics, Social Movements and Modern Environmentalism." *Environmental Politics* 15.5/2007, pp. 697–712.

Doloreux, David. "What We Should Know About Regional Systems of Innovation." *Technology in Society* 24/2002, pp. 243–263.

Doloreux, David, Saeed Parto. "Regional Innovation Systems: Current Discourse and Unresolved Issues." *Technology in Society* 27/2005, pp. 133–153.

Durant, Darrin. "Responsible Action and Nuclear Waste Disposal." *Technology in Society* 31/2009, pp. 150–157.

Dusinberre, Martin, Daniel P. Aldrich. "Hatoko Comes Home: Civil Society and Nuclear Power in Japan." *The Journal of Asian Studies* 70.3/2011, pp. 683–705.

Edelenbos, Jurian, Nienke van Schie, Lasse Gerrits. "Organizing Interfaces Between Government Institutions and Interactive Governance." *Policy Science* 43/2010, pp. 73–94.

Edwards, Arthur. "Tensions and New Connections Between Participatory and Representative Democracy in Local Governance." in: *Renewal in European Local Democracies*, eds. L. Schaap, H. Daemen (VS Verlag fur Sozilwissenschaften, Springer Fachmedien Wiesbaden, 2012).

EEA Report 2006. "Energy and Environment in the European Union. Tracking Progress towards Integration." European Environment Agency EEA Report No. 8/2006.

EEA, 2008. "Annual Management Plan 2008." European Environment Agency, Copenhagen.

EEA 2012. "Agriculture and the Green Economy." European Environment Agency TH-31-12-196-EN-N.

Eizaguirre, Santiago, Marc Pradel, Albert Terrones, Xavier Martinez-Celorrio, and Marisol Garcia. "Multilevel Governance and Social Cohesion: Bringing Back Conflict in Citizenship Practices." *Urban Studies* 49. 9/2012, pp. 1999–2016.

Endres, Danielle. "From Wasteland to Waste Site: The Role of Discourse in Nuclear Power's Environmental Injustices," *Local Environment* 14.10/2009, pp. 917–937.

EPA 2008. "309 Reviewers Guidance for New Nuclear Power Plant Environmental Impact Statements." EPA Publication 315-X-08-001 (U.S. Environmental Protection Agency).

Escobar, Arturo. "Culture Sits in Places: Reflections on Globalism and Subaltern Strategies of Localization." *Political Geography* 20/2001, pp. 139–174.

E.U. Special Eurobarometer 324. "Europeans and Nuclear Safety. Report." European Commission, Special Eurobarometer 324 /Wave 72.2 TNS Opinion & Social. Publication on March 2010.

Falk, Richard. "A Radical World Order Challenge: Addressing Global Climate Change and the Threat of Nuclear Weapons." *Globalizations* 7.1–2/2010, pp. 137–155.

Farla, Jacco, Jochen Markard, Rob Raven, Lars Coenen. "Sustainability Transitions in the Making: A Closer Look at Actors, Strategies and Resources." *Technological Forecasting & Social Change* 79/2012, pp. 991–998.

Federal Ministry for the Environment, Nature Conservation, Nuclear Safety and Consumer Protection (BMU): 16.02.2018 Press release No.035 Nuclear Safety. https://www.bmuv.de/en/pressrelease/a-new-portal-compiles-information-on-nuclear-safety (16 Aug. 2022).

Feldman, David L. "The Future of Environmental Networks – Governance and Civil Society in a Global Context." *Futures* 44/2012, pp. 787–796.

Finnveden, Goran, Anna Bjorklund, Marcus Carlsson Reich, Ola Eriksson, Adrienne Sorbom. "Flexible and Robust Strategies for Waste Management in Sweden." *Waste Management* 27/2007, pp. S1–S8.

Fox, Midori Kagawa. (2009). "The Ethics of Japan's Global Environmental Policy." The University of Adelaide. Unpublished PhD Dissertation for the Asian Studies School of Social Sciences, University of Adelaide, Australia.

Frey, B.S. "Public Governance and Private Governance: Exchanging Ideas." In: *Multidisciplinary Economics*, eds. P. de Gijsel and H. Schenk (Springer, printed in Netherlands, 2005), pp. 167–186.

GAO, United States Government accountability Office, 2021. "Nuclear Waste Cleanup. DOE Needs to Better Coordinate and Prioritize Its Research and Development Efforts." GAO@100 Highlights of GAO-22-104490, a report to the Committee on Science, Space, and Technology, House of Representatives. October 2021.

Gee, G.W., P. D. Meyer, and A.L. Ward. "Nuclear Waste Disposal." *Reference Module in Earth Systems and Environmental Sciences* 2005, pp. 56–63. From Encyclopedia of Soils in the Environment, Elsevier.

Gibbs, David. "Prospects for an Environmental Economic Geography: Linking Ecological Modernization and Regulationist Approaches." *Economic Geography* 82.2/2006, pp. 193–215.

Gibson, Kevin. "Stakeholders and Sustainability: An Evolving Theory." *Journal of Business Ethics* 109/2012, pp. 15–25.

Gimholt, Eva. SSM2018-2869, Document number: 2018-2869-5. "Swedens second National Report on Implementation of Council Directive 2011/70/Euratom." Rapport of the Swedish Radiation Safety Authority, pages 35–36, 2018. https://www.stralsakerhetsmyndigheten.se/ (16 Aug. 2022).

Gouldson, Andrew, Peter Hills, Richard Welford. "Ecological Modernisation and Policy Learning in Hong Kong." *Geoforum* 39/2008, pp. 319–330.

Grant, Don, Nell Trautner, Liam Downey, and Lisa Thirbaud. "Bringing the Polluters Back In: Environmental Inequality and the Organization of Chemical Production." *American Sociological Review* 75.4/2010, pp. 479–504.

Greenberg, Michael, David Lewis, Michael Frish. "Local and Interregional Economic Analysis of Large US Department of Energy Waste Management Projects." *Waste Management* 22/2002, pp. 643–655.

Greenberg, Michael R. "NIMBY, CLAMP, and the Location of New Nuclear-Related Facilities: U.S. National and 11 Site-Specific Surveys." *Risk Analysis* 29. 9/2009, pp. 1242–1254.

Grossmann, Matt. "Environmental Advocacy in Washington: A Comparison with Other Interest Groups." *Environmental Politics* 15.4/2006, pp. 628–638.

Grote, Jurgen R., Bernanrd Gbikpi. *Participatory Governance. Political and Societal Implications* (Springer Fachmedien Wiesbaden GmbH 2002).

(van der) Heijden, Jeroen. "Voluntary Environmental Governance Arrangements." *Environmental Politics* 21.3/2012, pp. 486–509.

Heijungs, Reinout, Jeroen B. Guinée. "Allocation and 'What If' Scenarios in Life Cycle Assessment of Waste Management Systems." *Waste Management* 27/2007, pp. 997–1005.

Hess, David J. "Transitions in Energy Systems: The Mitigation – Adaptation Relationship." *Science as Culture* 22.2/2013, pp. 197–203.

Hisschemoller, Matthijs. "Participation as Knowledge Production and the Limits of Democracy." in: *Democratization of Expertise? Exploring Novel Forms of Scientific Advice in Political Decision-Making* eds. Sabine Maasen and Peter Weingart, (Sociology of the Sciences 24/2005, Springer. Printed in the Netherlands), pp. 189–208.

Hisschemoller, Matthijs, Jan Eberg, Anita Engels, and Konrad von Moltke. "Environmental Institutions and Learning: Perspectives from the Policy Sciences." In: *Principles of Environmental Sciences*, eds. Jan J. Boersema and Lucas Reijinders (Springer Media B.V, 2009), pp. 281–303.

Hollingsworth, Rogers J. 1998. "Territoriality in Modern Societies: The Spatial and Institutional Nestedness of National Economies." In: *Territoriality in the Global Society*, eds. S. Immerfall (Springer-Verlag Berlin Heidelberg, 1998).

Holmes, John C., et al. In: *"Technologies and Management Strategies for Hazardous Waste Control."* March 1983, U.S. Government Printing Office, Washington, D.C. 20402. Library of Congress Catalog Card Number 83-600706 OTA pp. 3–407.

Hogselius, Per. "Spent Nuclear Fuel Policies in Historical Perspective: An International Comparison." *Energy Policy* 37/2009, pp. 254–263.

Honneland, Geir, Anne-Kristin Jorgensen. "Implementing International Agreements in Russia: Lessons from Fisheries Management, Nuclear Safety and Air Pollution Control." *Global Environmental Politics* 3.1/2003 by the MIT (Massachusetts Institute of Technology). pp. 72–98.

Hulbert, Margot, Kathleen McNutt, Jeremy Rayner. "Pathways to Power: Policy Transitions and the Reappearance of the Nuclear Power Option in Saskatchewan." *Energy Policy* 39/2011, pp. 3182–3190.

Hung, Hung-Chih, Tzu-Wen Wang. "Determinants and Mapping of Collective Perceptions of Technological Risk: The Case of the Second Nuclear Power Plant in Taiwan." *Risk Analysis* 31.4/2011, pp. 668–683.

IAEA, International Atomic Energy Agency, 2004. "Status of the Decommissioning of Nuclear Facilities around the world." IAEA, Vienna, Austria. STI/PUB/1201. Annex I "Nuclear Power Plants."

Iles, Alastair. "Choosing Our Mobile Future: The Degrees of Just Sustainability in Technological Alternatives." *Science as Culture* 22.2/2013, pp. 164–171.

Jasanoff, Sheila, Sang-Hyun Kim. "Sociotechnical Imaginaries and National Energy Policies." *Science as Culture* 22.2/2013, pp. 189–196.

Jenkins-Smith, Hank C., Carol L. Silva, Matthew C. Nowlin, and Grant de Lozier. "Reversing Nuclear Opposition: Evolving Public Acceptance of a Permanent Nuclear Waste Disposal Facility." *Risk Analysis* 31.4/2011, pp. 629–644.

Jennings, Anne Ballou, Amy M. Seward, and Thomas M. Leschine. "Living in a Nuclear Landscape: Rehabilitation and Resettlement of Proving Grounds in Australia and Islands of the Western Pacific." *Long-Term Management of Contaminated Sites Research in Social Problems and Public Policy* 13/2007, pp. 165–192.

Jessop, Bob. "Governance and Meta-Governance in the Face of Complexity: On the Roles of Requisite Variety, Reflexive Observation, and Romantic Irony in Participatory Governance." In: *Participatory Governance in Multi-Level context*, eds. H. Heinelt et al. (Springer Fachmedien Wiesbaden 2002), (2).

Johnson, Genevieve Fuji. "The Discourse of Democracy in Canadian Nuclear Waste Management Policy." *Policy Science* 40/2007, pp. 79–99.

Jones, Christopher F. "Building More Just Energy Infrastructure: Lessons from the Past." *Science as Culture* 22.2/2013, pp. 157–163.

Jorgensen, Ulrik. "Mapping and Navigating Transitions – The Multi-Level Perspective Compared with Arenas of Development." *Research Policy* 41/2012, pp. 996–1010.

Kennedy, Emily Huddart. "Rethinking Ecological Citizenship: The Role of Neighbourhood Networks in Cultural Change." *Environmental Politics* 20.6/2011, pp. 843–860.

Kessides, Ioannis N. "Nuclear Power and Sustainable Energy Policy: Promises and Perils." *The World Bank Research Observer* 25/2009, pp. 323–362. (Published by Oxford University Press on behalf of the International Bank for Reconstruction and Development).

Khoo, Su-Ming, Henrike Rau. "Movements, Mobilities and the Politics of Hazardous Waste." *Environmental Politics* 18.6/2009, pp. 960–980.

Kitschelt, Herbert P. "Political Opportunity and Political Protest: Anti-Nuclear Movements in Four Democracies." *British Journal of Political Science* 16.1/1986, pp. 57–85.

Knop, Karen C. (1999). "The Making of Difference in International Law: Interpretation, Identity and Participation in the Discourse of Self-Determination." PhD Dissertation at the Graduate Faculty of Law, University of Toronto, Canada.

Kostelnik, Kevin M., James H. Clarke, Jerry L. Harbour, Florence Sanchez, and Frank L. Parker. "A Sustainable Environmental Protection System for the Management of Residual Contaminants." *Long-term Management of Contaminated Sites Research in Social Problems and Public Policy* 13/2007, pp. 117–137.

Krupar, Shiloh R. "Transnatural Ethics: Revisiting the Nuclear Cleanup of Rocky Flats, CO, Through the Queer Ecology of Nuclear Waste." *Cultural Geographies* 19.3/2012, pp. 303–327.

Kunreuther, Howard, Douglas Easterling, William Desvousges, and Paul Slovic. "Public Attitudes toward Siting a High-Level Nuclear Waste Repository in Nevada." *Risk Analysis* 10.4/1990, pp. 469–484.

Kurath, Monika, Priska Gisler. "Informing, Involving or Engaging? Science Communication, in the Ages of Atom-, Bio - and Nanotechnology." *Public Understanding of Science* 18/2009.

Kurtz, Hilda E. "Scale Frames and Counter-Scale Frames: Constructing the Problem of Environmental Injustice." *Political Geography* 22/2003, pp. 887–916.

Lagendijk, Arnoud, Frans Boekema. "The Territoriality of Spatial-Economic Governance in Historical Perspective: The Case of the Netherlands." In: *The Disoriented State: Shifts in Governmentality, Territoriality and Governance*, eds. B. Arts et al., (Springer Science-Business Media B.V, 2009), pp. 121–140.

Laird, Frank N. "Against Transitions? Uncovering Conflicts in Changing Energy Systems." *Science as Culture* 22.2/2013, pp. 149–156.

Larsen, Rasmus Klocker, Christiane Alzouma Mamosso. "Aid with Blinkers: Environmental Governance of Uranium Mining in Niger." *World Development* 56C/2014, pp. 62–76.

Lawhon, Mary, James T. Murphy. "Socio-Technical Regimes and Sustainability Transitions: Insights from Political Ecology." *Progress in Human Geography* 36.3/2011, pp. 354–378.

Leong, Ching Ching, Darryl Jarvis, Michael Howlett, Andrea Migone. "Controversial Science-Based Technology Public Attitude Formation and Regulation in Comparative Perspective: The State Construction of Policy Alternatives in Asia." *Technology in Society* 33/2011, pp. 128–136.

Lesbirel, S. Hayden. "Project Siting and the Concept of Community." *Environmental Politics* 20.6/2011, pp. 826–842.

Leschine, Thomas M. "Introduction: Long-Term Management of Contaminated Sites." *Long-Term Management of Contaminated Sites Research in Social Problems and Public Policy* 13/2007, pp. 1–10.

Lin, Shirley Syaru (2010). "National Identity, Economic Interest and Taiwan's Cross-Strait Economic Policy 1994–2009." PhD Dissertation unpublished at the University of Hong Kong.

Linkov, I., A. Varghese, S. Jamil, T.P. Seager, G. Kiker, T. Bridges. "Multi-Criteria Decision Analysis: A Framework for Structuring Remedial Decisions at Contaminated Sites." In: *Comparative Risk Assessment and Environmental Decision Making*, eds. I. Linkov and A. Bakr Ramadan, (Kluwer Academic Publishers. Printed in the Netherlands, 2004), pp. 15–54.

Loveland, P.J., P.H. Bellamy. "Environmental Monitoring." *Reference Module in Earth Systems and Environmental Sciences* 2005, pp. 441–448, from Elsevier Encyclopedia of Soils in the Environment.

Markard, Jochen, Rob Raven, Bernhard Truffer. "Sustainability Transitions: An Emerging Field of Research and Its Prospects." *Research Policy* 41/2012, pp. 955–967.

Marshall, Alan. "The Social and Ethical Aspects of Nuclear Waste." *Electronic Green Journal* 1.21. 4/2005.

Maythorne, Louise (2012). "The Europeanisation of Grassroots Greens: Mobilisation in France, Italy and the UK." PhD Dissertation at the University of Edinburgh, UK. PhD Politics.

McIntyre, Janet. "Part 1: Working and Re-Working the Conceptual and Geographic Boundaries of Governance and International Relations." *Systemic Practice and Action Research* 18.2/2005, p. 173220.

Messer, Chris M., Alison E. Adams, and Thomas E. Shriver. "Corporate Frame Failure and the Erosion of Elite Legitimacy." *The Sociological Quarterly* 53/2012, pp. 475–499. (Midwest Sociological Society).

Miller, Clark A. "Science and Democracy in a Globalizing World: Challenges for American Foreign Policy." *Science and Public Policy* 32.3/2005, pp. 174–186.

Mochizuki, Mike M. "Japan Tests the Nuclear Taboo." *Nonproliferation Review* 14.2/2007, pp. 303–328.

Morrissey, A.J., J. Browne. "Waste Management Models and Their Application to Sustainable Waste Management." *Waste Management* 24/2004, pp. 297–308.

Moulaert, Frank, Jacques Nussbaumer, 2005. "Beyond the Learning Region: The dialectics of Innovation and Culture in Territorial Development." In: *Learning from Clusters: A Critical Assessment*, eds. R.A. Boschma and R.C. Kloosterman, (Springer. Printed in the Netherlands, 2005), pp. 89–109.

Mulvaney, Dustin. "Opening the Black Box of Solar Energy Technologies: Exploring Tensions between Innovation and Environmental Justice." *Science as Culture* 22.2/2013, pp. 230–237.

Nichols, Rodney W. "Innovation, Change, and Order: Reflections on Science and Technology in India, China, and the United States." *Technology in Society* 30/2008, pp. 437–450.

NDA Guidance Report, 2015. NDA, Nuclear Decommissioning Authority, 2015. "NDA Guidance for Site Stakeholder Groups." Ref. LAR3.0.

NDA Nuclear Decommissioning Authority, 2016. "Nuclear Decommissioning Authority Independent Research Board Terms of Reference." NDA Research Board Terms of Reference, Issue 3, November 2016. EDRMS No. 24761455.

Nielsen, Henry, Henrik Knudsen. "The Troublesome Life of a Peaceful Atoms in Denmark." *History and Technology* 26.2/2010, pp. 91–118.

Nuclear Waste Management Organization of Japan, 2018. Topics, About NUMO, "Action Plan for Reforming Communication Activities." April 13, 2018. https://www.numo.or.jp/en/about_numo/new_eng_tab08.html

Nykvist, Bjorn, Lorraine Whitmarsh. "A Multi-Level Analysis of Sustainable Mobility Transitions: Niche Development in the UK and Sweden." *Technological Forecasting & Social Change* 75/2008, pp. 1373–1387.

OECD/NEA 2000. "Stakeholder Confidence and Radioactive Waste Disposal." *Inauguration, First Workshop and Meeting of the NEA Forum on Stakeholder Confidence in the Area of Radioactive Waste Management*. Paris, France. Nuclear Energy Agency and Organization for Economic Co-operation and Development.

OECD/NEA 2001. "Better Integration of Radiation Protection in Modern Society." Workshop Proceedings Villigen, Switzerland 23–25 January 2001. OECD 2002, Publications, Paris, France.

OECD/NEA, Radioactive Waste Management Committee, 2007. "Approaches and Practices in Decommissioning of Facilities and Management of Radioactive Waste from Non-Nuclear Fuel Cycle Related Activities." *Proceedings of the Topical Session of the RWMC 40th Meeting at NEA Offices in France 14th March 2007*. (NEA/RWM 2007/9).

OECD/NEA, 2008. "Timing of High-level Waste Disposal." NEA No. 6244.

OECD Radioactive Waste Management report, 2008. "Regulating the long-term safety of geological disposal of radioactive waste: Practical issues and challenges." Nuclear Energy Agency, NEA, Workshop Proceedings, Paris, France, 28–30 November 2006.

OECD/NEA, 2010. "Partnering for Long-term Management of Radioactive Waste. Evolution and Current Practice in Thirteen Countries." OECD/NEA No. 6823.

Report OECD/NEA, 2011. "Road Map for Crisis Communication of Nuclear Regulatory Organisations – National Aspects." (NEA/CNRA/R (2011)11).

OECD/NEA Report. Radioactive Waste Management 2012. "The Evolving Role and Image of the Regulator in Radioactive Waste Management: Trends Over Two Decades." OECD 2012 (Organisation for Economic Co-operation and Development)/NEA (Nuclear Energy Agency). No.7083.

OECD 2012. "Reversibility and Retrievability in Planning for Geological Disposal of Radioactive Waste." *Proceedings of the "R&R" International Conference and Dialogue 14–17 December 2010 Reims, France*. NEA No. 6993 (Nuclear Energy Agency) – Organization for Economic Co-operation and Development (OECD Secretary-General).

Report OECD/NEA, 2012. "Good Practice in Effluent Management for Nuclear Power Plant New Build. A report from the CRPPH Expert Group on BAT (EGBAT)." NEA/CRPPH/R (2012)3.

Report OECD/NEA, 2012. "Geological Disposal of Radioactive Wastes: National Commitment, Local and Regional Involvement." NEA/RWM (2011)16.

OECD/NEA Nuclear Development Report 2014. "Managing Environmental and Health Impacts of Uranium Mining." Nuclear Energy Agency NEA No. 7062 & Organisation for Economic Co-operation and Development OECD.

Report OECD/NEA, 2014. "CFD for Nuclear Reactor Safety Applications (CFD4NRS-4) Workshop Proceedings." NEA/CSNI/R(2014)4.

OECD/NEA, 2019. "Legal Framework for Long-Term Operation of Nuclear Power Reactors." *Legal Affairs*, 2019. NEA No. 7504.

Ottinger, Gwen. "The Winds of Change: Environmental Justice in Energy Transitions." *Science as Culture* 22.2/2013, pp. 222–229.

Paraskevopolou, Evita. "Non-Technological Regulatory Effects: Implications for Innovation and Innovation Policy." *Research Policy* 41/2012, pp. 1058–1071.

Parthasarathy, Shobita. "Breaking the Expertise Barrier: Understanding Activist Strategies in Science and Technology Policy Domains." *Science and Public Policy* 37.5/2010, pp. 355–367.

Pennington, Charles W. "Comparative Population Dose Risks from Nuclear Fuel Cycle Closure and Renewal of the Commercial Nuclear Energy Alternative in the U.S." *Progress in Nuclear Energy* 51/2009, pp. 290–296.

Phadke, Roopali, Christie Manning, Samantha Burlager. "Making It Personal: Diversity and Deliberation in Climate Adaptation Planning." *Climate Risk Management* 9/2015, pp. 62–76. https://doi.org/10.1016/j.crm.2015.06.005.

Pickett, Susan E. "Japan's Nuclear Energy Policy: From Firm Commitment to Difficult Dilemma Addressing Growing Stocks of Plutonium, Program Delays, Domestic Opposition and International Pressure." *Energy Policy* 30/2002, pp. 1337–1355.

Pidgeon, Nick F., Irene Lorenzoni, Wouter Poortinga. "Climate Change or Nuclear Power – No Thanks! A Quantitative Study of Public Perceptions and Risk Framing in Britain." *Global Environmental Change* 18/2008, pp. 69–85.

Poetz, Annelise. "What's Your 'Position' on Nuclear Power? An Exploration of Conflict in Stakeholders Participation for Decision-Making about Risky Technologies." *Risk, Hazards & Crisis in Public Policy* 2.2.2/2011, pp. 1–38.

Powell, J. H. "Generating Networks for Strategic Planning by Successive Key Factor Modification." *The Journal of the Operational Research Society* 52.4/2001, pp. 369–382.

Purdy, Jedediah. "The Politics of Nature: Climate Change, Environmental Law, and Democracy." *The Yale Law Journal* 119/2010, pp. 1122–1207.

Ramana, M.V. "Nuclear Power and the Public." *Bulletin of the Atomic Scientists* 67.4/2011, pp. 43–51.

Ramana, M.V. "Shifting Strategies and Precarious Progress: Nuclear Waste Management in Canada." *Energy Policy* 61/2013, pp. 196–206.

Ratchford, J. Thomas, William A. Blanpied. "Paths to the Future for Science and Technology in China, India and the United States." *Technology in Society* 30/2008, pp. 211–233.

Roberts, Mary Roduta, Grace Reid, Meadow Schroeder, and Stephen P. Norris. "Causal or Spurious? The Relationship of Kowledge and Attitudes to Trust in Science and Technology." *Public Understanding of Science* 22.5/2011, pp. 624–641.

Rosa, Eugene A., James Rice. "Public Reaction to Nuclear Power Siting and Disposal." *Encyclopedia of Energy* 5/2004. pp. 181–194.

Rosa, Eugene A. "Long-Term Stewardship and Risk Management: Analytic and Policy Challenges." *Long-Term Management of Contaminated Sites Research in Social Problems and Public Policy* 13/2007, pp. 227–255.

Safarzynska, Karolina, Koen Frenken, Jeroen C.J.M. van den Bergh. "Evolutionary Theorizing and Modelling of Sustainability Transitions." *Research Policy* 41/2012, pp. 1011–1024.

Sauer, Alexandra. "Conflict Pattern Analysis: Preparing the Ground for Participation in Policy Implementation." *Systemic Practice and Action Research (SPAR)* 21/2008, pp. 497–515.

Sbicca, Joshua. "Elite and Marginalised Actors in Toxic Treadmills: Challenging the Power of the State, Military, and Economy." *Environmental Politics* 21.3/2012, pp. 467–485.

Schaffer, Marvin Backer. "Toward a Viable Nuclear Waste Disposal Program." *Energy Policy* 39/2011, pp. 1382–1388.

Sessa, Carlo, Andrea Ricci. "The World in 2050 and the New Welfare Scenario." *Futures* 58/2014, pp. 77–90.

Seyad, Akim, Steven Baeke, Marc de Clercq. "Success Determining Factors for Negotiated Agreements." In: *Co-operative Environmental Governance*, eds. P. Glasbergen, (Kluwer Academic Publishers, 1998), pp. 111–132.

Sheng, Jichuan, Xiao Han. "State Rescaling, Power Reconfiguration and Path Dependence: China's Xin'an River Basin Eco-Compensation Pilot (XRBEP)." *Regional Studies* 2022, DOI:10.1080/00343404.2021.2009454.

Shineha, Ryuma, Masaki Nakamura. "Diversity in STS Communities: A Comparative Analysis of Topics." *East Asian Science, Technology and Society: An International Journal* 7.1/2013, pp. 145–158, DOI: 10.1215/18752160-2075813.

Siegel, M.D., C.R. Bryan. "Environmental Geochemistry of Radioactive Contamination." In: H.D. Holland & K.K. Turekian "*Environmental Geochemistry*," eds. Barbara Sherwood Lollar (2005). (Elsevier Ltd. Sandia National Laboratories, Albuquerque, NM, USA, 2005, chp. 9.06).

Sjoberg, Lennart. "Explaining Individual Risk Perception: The Case of Nuclear Waste." *Risk Management: An International Journal* 6.1/2004, pp. 51–64.

SMEs Energy Technology Market. "KOMIPO KORAD at a Glance." (December 31, 2020). https://www.energytechmarket.or.kr/svc/itd/enOrgIntro.do?org=KORAD

Smith, Adrian, Rob Raven. "What Is Protective Space? Reconsidering Niches in Transitions to Sustainability." *Research Policy* 4/2012, pp. 1025–1036.

Soubbotina, Tatyana, Charles Weiss. "A New Model of Technological Learning for Russia." *Science and Public Policy* 36.4/2009, pp. 271–286.

(van) Staden, Maryke, Francesco Musco. "Three Streams of Local Action: Strategy and Policy; Technology and Measures; People and Lifestyle." In: *Local Governments and Climate Change*, eds. M. van Staden and F. Musco, (Advances in Global Change Research 39/2010, Springer Media B.V), pp. 111–172.

Stoutenborough, James W., Shelbi G. Sturgess, Arnold Vedlitz. "Knowledge, Risk, and Policy Support: Public Perceptions of Nuclear Power." *Energy Policy* 62/2013, pp. 176–184.

Strubelt, Wendelin. *Guiding Principles for Spatial Development in Germany*. (Springer-Verlag Berlin Heidelberg 2009).

Su, Norman Makoto, Hiroko N. Wilensky, and David F. Redmiles. "Doing Business with Theory: Communities of Practice in Knowledge Management." *Computer Supported Cooperative Work* 21/2012, pp. 111–162.

Suksi, Markku. *Sub-State Governance through Territorial Autonomy. A Comparative Study in Constitutional Law of Powers, Procedures and Institutions*. (Springer-Verlag Berlin Heidelberg 2011).

Sundberg, Juanita. "Conservation and Democratization: Constituting Citizenship in the Maya Biosphere Reserve, Guatemala." *Political Geography* 22/2003, pp. 715–740.

Sundqvist, Goran, Mark Elam. "Public Involvement Designed to Circumvent Public Concern? The 'Participatory Turn' in European Nuclear Activities." *Risk, Hazards & Crisis in Public Policy* 1. 4. 8/2010, pp. 203–229.

Swedish Nuclear Fuel and Waste Management Company (Svensk Kärnbränslehantering Aktiebolag, SKB), 2022. "We Take Care of Swedish Radioactive Waste." SKB web portal available at: https://www.skb.com/

Szarka, Joseph. "From Exception to Norm – and Back Again? France, the Nuclear Revival, and the Post-Fukushima Landscape." *Environmental Politics* 22.4/2013, pp. 646–663.

Teravainen, Tuula, Markku Lehtonen, Mari Martiskainen. "Climate Change, Energy Security, and Risk – Debating Nuclear New Build in Finland, France and the UK." *Energy Policy* 39/2011, pp. 3434–3442.

Todtling, Franz, Michaela Trippl. "One Size Fits All? Toward a Differentiated Regional Innovation Policy Approach." *Research Policy* 34/2005, pp. 1203–1219.

Tuler, Seth, Thomas Webler. "Competing Perspectives on a Process for Making Remediation and Stewardship Decisions at the Rocky Flats Environmental Technology Site." *Long-Term Management of Contaminated Sites Research in Social Problems and Public Policy* 13/2007, pp. 49–77. Elsevier Ltd.

Uekoetter, Frank. "Fukushima, Europe, and the Authoritarian Nature of Nuclear Technology." *Environmental History* 17/2012, pp. 277–284.

UK Government, Emergency Preparation, Response and Recovery Section, 2015. "Guidance. Engaging with Nuclear Decommissioning Authority: how to get involved." Nuclear Decommissioning Authority (NDA), 2015. https://www.gov.uk/guidance/engaging-with-nuclear-decommissioning-authority-how-to-get-involved

US Fed News Service, 2007. "U.S. fish and wildlife service establishes Rocky Flats national wildlife refuge." US FED News Service, including US State News (Washington, D.C.), July.

ARF 17491, U.S. Department of Energy, Savannah River Site (SRS), 2011. "Savannah river site federal facility agreement community involvement plan (U)." WSRC-RP-96–120, Revision 7, February 2011.

U.S. Cleanup Sites, Office of Environmental Management, U.S. Department of Energy, accessed on April 2022. https://www.energy.gov/em/cleanup-sites

U.S. Department of Energy, Citizens Advisory Board (CAB), 2022. "GA & SC citizens working together for a better tomorrow at SRS." SRS, CAB page, https://cab.srs.gov/srs-cab.html

U.S. DOE, Office of Environmental Management, 2022. "EM Strategic Vision: 2022–2032." EM Report. Website https://www.energy.gov/sites/defa ult/files/2022-03/DOE-EM-Strategic-Vision-2022-Final-3-8-22.pdf

U.S. Congress, 1985. "Managing the Nation's Commercial High-Level Radioactive Waste." (Washington, DC: U.S. Congress, Office of Technology Assessment, OTA-O-171, March 1985).

Valentine, Scott Victor, Benjamin K. Sovacool. "The Socio-Political Economy of Nuclear Power Development in Japan and South Korea." *Energy Policy* 38/ 2010, pp. 7971–7979.

Vergragt, Philip J., Halina Szejnwald Brown. "Sustainable Mobility: From Technological Innovation to Societal Learning." *Journal of Cleaner Production* 15/2007, pp. 1104–1115.

Waelbers, Katinka. "Doing Good with Technologies. Taking Responsibility for the Social Role of Emerging Technologies." (Springer Science+Business Media B.V. 2011).

Walter, D. Wastl. "Social Movements: Environmental Movements." *International Encyclopedia of the Social & Behavioural Sciences*, Elsevier Science Ltd, 2001, p. 14352–14357. https://doi.org/10.1016/B0-08-043076-7/04191-7

Weber, K. Matthias, Harald Rohracher. "Legitimizing Research, Technology and Innovation Policies for Transformative Change. Combining Insights from Innovation Systems and Multi-Level Perspective in a Comprehensive 'Failures' Framework." *Research Policy* 41/2012, pp. 1037–1047.

Weible, Christopher M. "Political-Administrative Relations in Collaborative Environmental Management." *International Journal of Public Administration* 34/2011, pp. 424–435.

Williams, Gwyndaf. "Metropolitan Governance and Strategic Planning: A Review of Experience in Manchester, Melbourne and Toronto." *Progress in Planning* 52/1999, pp. 1–100.

Williams, Robert W. "Environmental Injustice in America and Its Politics of Scale." *Political Geography* 18/1999, pp. 49–73.

Wilson, John Francis Hamilton, (2000). "Turf Wars in Environmentalism: Competing Discourses in Hydroelectric and Nuclear Power Campaigns in New Zealand." PhD Dissertation. Political Studies, The University of Auckland. New Zealand.

World Nuclear Association, 2021. "Nuclear Power in South Korea." WNA Country Profiles, updated October 2021. https://world-nuclear.org/informat ion-library/country-profiles/countries-o-s/south-korea.aspx

Wolsink, Maarten, Jeroen Devilee. "The Motives for Accepting or Rejecting Waste Infrastructure Facilities. Shifting the Focus from the Planners' Perspective to Fairness and Community Commitment." *Journal of Environmental Planning and Management* 52.2/2009, pp. 217–236.

Woods, Michael. "Deconstructing Rural Protest: The Emergence of a New Social Movement." *Journal of Rural Studies* 19/2003, pp. 309–325.

Woods, Michael. "Social Movements and Rural Politics." Guest Editorial, *Journal of Rural Studies* 24/2008, pp. 129–137.

Xiao, Chenyang. "Public Attitudes Toward Science and Technology and Concern for the Environment: Testing a Model of Indirect Feedback Effects." *Environment and Behaviour* 45.1/2011, pp. 113–137.

Yeung, Henry Wai-chung. "Strategic Governance and Economic Diplomacy in China: The Political Economy of Government-Linked Companies From Singapore." *East Asia* 21.1/2004, pp. 40–64.

Zhou, Yun. "China's Spent Nuclear Fuel Management: Current Practices and Future Strategies." *Energy Policy* 39/2011, pp. 4360–4369.

SUGGESTED LITERATURE

Aaron, James. "Distributive Justice without Sovereign Rule: The Case of Trade," *Social Theory and Practice* 31.4/2005, pp. 533–559.

Abbott, Kenneth W., Duncan J. Snidal. *The Spectrum of International Institutions. An Interdisciplinary Collaboration on Global Governance.* Publisher Routledge, 2021. Taylor & Francis eBooks. ISBN 9780367629731.

Agrawal, Arun, Maria Carmen Lemos. "A Greener Revolution in the Making?: Environmental Governance in the 21st Century," *Environment: Science and Policy for Sustainable Development* 49.5/2007, pp. 36–45.

Aktar, Ismail. "A Comparison of the Effects of the Chernobyl and Three Mile Island Nuclear Accident on the U.S. Electric Utility Industry," *Sosyoekonomi Journal* 2/2005, pp. 13–33.

Albrecht, Steven M. "Forging New Directions in Science and Environmental Politics and Policy: How Can Co-operation, Deliberation and Decision Be Brought Together?" *Environment, Development and Sustainability*, 3/2001, pp. 323–341.

Aldrich, Daniel P. "A Normal Accident or a Sea-Change? Nuclear Host Communities Respond to the 3/11 Disaster," *Japanese Journal of Political Science* 14.2/2013, pp. 261–276.

Alger, Justin, Trevor Findlay. "Strengthening Global Nuclear Governance," *Issues in Science and Technology* 27.1/2010, pp. 73–79.

Amir, Sulfikar. "Nuclear Revival in Post-Suharto Indonesia," *Asian Survey* 50.2/2010, pp. 265–286. https://doi.org/10.1525/as.2010.50.2.265

Anderson, Stuart, Massimo Felici. "Classes of Socio-Technical Hazards: Microscopic and Macroscopic Scales of Risk Analysis," *Risk Management* 11.3–4/2009, pp. 208–240.

Antonelli, Cristiano. "Models of Knowledge and Systems of Governance," *Journal of Institutional Economics* 1.1/2005, pp. 51–73.

Aranguren, Mari Jose, Miren Larrea. "Regional Innovation Policy Processes: Linking Learning to Action," *Journal of the Knowledge Economy* 2.4/2011, pp. 569–585.

Arriaza, Naomi Roht. "Private Voluntary Standard-Setting, the International Organization for Standardization, and International Environmental Lawmaking," *Yearbook of International Environmental Law* 6.1/1996, pp. 107–163.

Asensio, Juan Casado, Reinhard Steurer. "Integrated Strategies on Sustainable Development, Climate Change Mitigation and Adaptation in Western Europe: Communication Rather Than Coordination," *Journal of Public Policy* 34.3/2014, pp. 437–473.

Axelrod, Regina S. "The European Commission and Member States: Conflict Over Nuclear Safety," *Perspectives* 26/2006, pp. 5–22.

B

Bae, Jungah. (2012). "Green Governance Innovation: The Institutional Political Market for Energy Sustainable Communities," PhD Dissertation at the Florida State University, College of Social Sciences and Public Policy, United States. https://www.proquest.com/openview/e840d0c875ac94ec0887c2188bbe0849/1?pq-origsite=gscholar&cbl=18750 (07 Aug. 2022).

Bae, Yooil. "Decentralized Urban Governance and Environmental Collaboration in South Korea: The Case of Hyundai City," *Pacific Affairs* 86.4/2013, pp. 759–783.

Bano, Sayeeda, Sriya Kumarasinghe, Yih Pin Tang. "Comparative Economic Performance and Stock Market Performance: Some Evidence from the Asia-Pacific Region," *Asian Journal of Finance & Accounting* 3.1/2011, pp. 1–22.

Bates, A.Kimberly, E. James Flynn. "Innovation History and Competitive Advantage: A Resource-Based View Analysis of Manufacturing Technology Innovations," *Academy of Management Journal* 1/1995, pp. 235–239.

Beelitz, Annika, Doris M. Merkl-Davies. "Using Discourse to Restore Organizational Legitimacy: 'CEO-Speak' After an Incident in a German Nuclear Power Plant," *Journal of Business Ethics* 108.1/2012, pp. 101–120.

Beken, Tom Vander, Nicholas Dorn, Stijn Van Daele. "Security Risks in Nuclear Waste Management: Exceptionalism, Opaqueness, and Vulnerability," *Journal of Environmental Management* 91.4/2010, pp. 940–948.

Ben, Tai-Ming, Kuan-Fei Wang. "Interaction Analysis among Industrial Parks, Innovation Input, and Urban Production Efficiency," *Asian Social Science* 7.5/2011, pp. 56–71;

Berny, Nathalie. "Europeanization as Organizational Learning: When French ENGOs Play the EU Multilevel Policy Game," *French Politics* 11.3/2013, pp. 217–240.

Birkmann, Jorn, Matthias Garschagen, Frauke Kraas, Nguyen Quang. "Adaptive Urban Governance: New Challenges for the Second Generation of Urban Adaptation Strategies to Climate Change," *Sustainability Science* 5/2010, pp. 185–206.

Bisconti, Ann S., J. Scott Peterson. "Congressional Actions on Nuclear Energy: Do They Follow Public Opinion?" *Natural Gas & Electricity* 20.12/2004, pp. 11–17.

Bixler, Richard Patrick (2014). "Is There an Heir Apparent to the Crown? A More Informed Understanding of Connectivity and Networked Environmental Governance in the Crown of the Continent." PhD Dissertation at the Department of Sociology, Colorado State University, Fort Collins, Colorado, United States. https://www.proquest.com/openview/8a03d2e3fba7c55dbda0b18a5e1a7adc/1?pq-origsite=gscholar&cbl=18750 (08 Aug. 2022).

Boland, Joseph B. (2002). "The Cold War Legacy of Regulatory Risk Analysis: The Atomic Energy Commission and Radiation Safety," PhD Dissertation submitted at the Department of Political Science and the Graduate School of the University of Oregon, United States. https://www.proquest.com/openview/8eb41935504f545dc80e8481c14d2796/1?pq-origsite=gscholar&cbl=18750&diss=y (09 Aug. 2022).

Boonstra, Wiebren J., Florianne W. de Boer. "The Historical Dynamics of Social-Ecological Traps," *Ambio* 43.3/2014, pp. 260–274.

Borenstein, Severin. "The Private and Public Economics of Renewable Electricity Generation," *Journal of Economic Perspectives* 26.1/2012, pp. 67–92.

Bosold, David, Wilfried von Bredow. "Human Security: A Radical or Rhetorical Shift in Canada's Foreign Policy?" *International Journal* 61.4/2006, pp. 829–844.

Bowie, Ryan. "Indigenous Self-Governance and the Deployment of Knowledge in Collaborative Environmental Management in Canada," *Journal of Canadian Studies* 47.1/2013, pp. 91–121.

Bradford, Neil. "Public-Private Partnerships? Shifting Paradigms of Economic Governance in Ontario," *Canadian Journal of Political Science* 36.5/2003, pp. 1005–1033.

Bragdon, Susan H., Marc Pallemaerts, Veerle Heyvaert, Iwona Rummel-Bulska. "International Hazard Management Other than Nuclear," *Yearbook of International Environmental Law* 6.1/1996, pp. 275–292.

Briske, David D. "Translational Science Partnerships: Key to Environmental Stewardship," *BioScience* 62.5/2012, pp. 449–450.

Brouard, Sylvain, Isabelle Guinaudeau. "Policy Beyond Politics? Public Opinion, Party Politics and the French Pro-Nuclear Energy Policy," *Journal of Public Policy* 35.1/2015, pp. 137–170.

Brown, David J., John S. Earle, Scott Gehlbach. "Helping Hand or Grabbing Hand? State Bureaucracy and Privatization Effectiveness," *The American Political Science Review* 103.2/2009, pp. 264–283.

Bruce, Matthews R. "Nuclear Safety: Expect the Unexpected," *Professional Safety* 50.12/2005, pp. 20–27.

Brugge, Doug, Jamie L. deLemos, Cat Bui. "The Sequoyah Corporation Fuels Release and the Church Rock Spill: Unpublicized Nuclear Releases in American Indian Communities," *American Journal of Public Health* 97.9/2007, pp. 1595–1600.

Bruszt, Laszlo, Bela Greskovits. "Transnationalization, Social Integration, and Capitalist Diversity in the East and the South," *Studies in Comparative International Development* 44/2009, pp. 411–434.

Bucheli, Marcelo, Min-Young Kim. "Political Institutional Change, Obsolescing Legitimacy, and Multinational Corporations. The Case of the Central American Banana Industry," *Management International Review* 52.6/2012, pp. 847–877.

Burger, Joanna. "Restoration, Stewardship, Environmental Health, and Policy: Understanding Stakeholders' Perceptions," *Environmental Management* 30.5/2002, pp. 631–640.

Burger, Joanna, James Clarke, Michael Gochfeld. "Information Needs for Siting New, and Evaluating Current, Nuclear Facilities: Ecology, Fate, and Transport, and Human Health," *Environmental Monitoring and Assessment* 172/2011, pp. 121–134.

Busch, Nathan Edward (2001). "Assessing the Optimism-Pessimism Debate: Nuclear Proliferation, Nuclear Risks, and Theories of State Action," PhD Dissertation at the Graduate Department of Political Science, University of Toronto, Canada. https://www.bac-lac.gc.ca/eng/services/theses/Pages/item.aspx?idNumber=1006711395 (07 Aug. 2022).

C

Cable, Sherry. "Political Processes and Institutions – Mobilizing Against Nuclear Energy: A Comparison of Germany and the United States by Christian Joppke," *Contemporary Sociology* 24.1/1995, p. 49.

Caramizaru, Aura, Andreas Uihlein. "Energy Communities: An Overview of Energy and Social Innovation," *JRC Science For Policy Report* 2020, (European Commission, Belgium). https://publications.jrc.ec.europa.eu. (09 Aug. 2022).

Carlisle, Rodney P. "Probabilistic Risk Assessment in Nuclear Reactors: Engineering Success, Public Relations Failure," *Technology and Culture* 38.4/1997, pp. 920–941.

Carruthers, Bruce G. "Contemporary Capitalism: The Embeddedness of Institutions, by J. Rogers Hollingsworth, Robert Boyer," *Contemporary Sociology*, 27.3/1998, pp. 256–257.

Carter, Luther J., Thomas H. Pigford. "Confronting the Paradox in Plutonium Policies," *Issues in Science and Technology* 16.2/2000, pp. 29–36.

Celata, Filippo, Liz Dinnie, and Anne Holsten. "Sustainability Transitions to Low-Carbon Societies: Insights from European Community-Based Initiatives," *Regional Environmental Change* 19/2019, pp. 909–912. https://doi.org/10.1007/s10113-019-01488-6.

Celeste, Richard F. "Strategic Alliances for Innovation: Emerging Models of Technology-Based Twenty-First Century Economic Development," *Economic Development Review* 14.1/1996, pp. 4–8.

Cha, Yong-Jin (1997). "Environmental Risk Analysis: Factors Influencing Nuclear Risk Perception and Policy Implications," PhD Dissertation at Rockefeller College of Public Affairs and Policy, Department of Public Administration and Policy, State University of New York, U.S. https://search.library.albany.edu/ (06 Aug. 2022).

Chen, Hsieh-Sheng. "The Relationship between Technology Industrial Cluster and Innovation in Taiwan," *Asia Pacific Management Review* 16.3/2011, pp. 277–288.

Chattopadhyay, Soumyadip. "Contesting Inclusiveness: Policies, Politics and Processes of Participatory Urban Governance in Indian Cities," *Progress in Development Studies* 15.1/2015, pp. 22–36.

Cho, Sung Ju, (2009). "From Proliferation to Renunciation: Why Some States Give Up Nuclear Ambitions While Others Do Not," PhD Dissertation submitted at the Graduate Faculty of the University of Virginia, The Woodrow Wilson Department of Politics, University of Virginia, United States.

Choi, Seungho, Hoonseok Jung and Dongheup Lee. "The Basic Planning for the Environmental Relationship of Improved Nuclear Power Plant," GENES4/ANP2003, September 15–19, 2003, Kyoto, Japan. Paper 1097. https://www.ipen.br/biblioteca/cd/genes4/2003/papers/1097-final.pdf (10 Aug. 2022).

Choudhury, Upendra. "The Impact of the Fukushima Daiichi Nuclear Crisis on Anti-Nuclear Movements in India," *Social Alternatives* 31.3/2012, pp. 39–44.

Chung, Shan-Shan, Lo Carlos Wing-Hung. "The Roles of Grassroots Local Government in Sustainable Waste Management in China," *International Journal of Sustainable Development and World Ecology* 14.2/2007, pp. 133–144.

Clarke, Beverly, Stocker L., Cannard T., et al. "Enhancing the Knowledge-Governance Interface: Coasts, Climate and Collaboration," *Ocean & Coastal Management* 86.12/ 2013, pp. 88–99. https://doi.org/10.1016/j.ocecoaman.2013.02.009.

Cleff, Thomas, Christoph Grimpe, Christian Rammer. "Demand-Oriented Innovation Strategy in the European Energy Production Sector," *International Journal of Energy Sector Management* 3.2/2009, pp. 108–130.

Conrad, Alexis Jonathan (1999). "Assessing the Adequacy of Intergovernmental Collaboration as an Organizing Principle for Environmental Protection: A Case Study of the Canada-Wide Accord on Environmental Harmonization's Environmental Standards Sub-Agreement," Master of Arts, Thesis submitted at the Department of Political Studies, Queen's University, Kingston, Ontario, Canada.

Corvellec, Hervé, Johan Hultman, "From 'Less Landfilling' to 'Wasting Less': Societal Narratives, Socio-Materiality, and Organizations," *Journal of Organizational Change Management* 25.2/2012, pp. 297–314.

Cosio, Roman Gomez Gonzalez. "Social Constructivism and Capacity Building for Environmental Governance," *International Planning Studies* 3.3/1998, pp. 367–389.

Costa, Carlos A. Bana, Joao Carlos Lourenco, Monica Duarte Oliveira, Joao C. Bana e Costa. "A Socio-Technical Approach For Group Decision Support in Public Strategic Planning: The Pernambuco PPA Case," *Group Decision and Negotiation* 23.1/2014, pp. 5–29.

Craig, Campbell, Jan Ruzicka. "The Nonproliferation Complex," *Ethics & International Affairs* 27.3/2013, pp. 329–348.

Culley, Marci R., Joeph Hughey. "Power and Public Participation in a Hazardous Waste Dispute: A Community Case Study," *American Journal of Community Psychology* 41.1–2/2008, pp. 99–114.

Culley, Marci R., Holly Angelique. "Participation Power, and the Role of Community Psychology in Environmental Disputes: A Tale of Two Nuclear Cities," *American Journal of Community Psychology* 47.3–4/2011, pp. 410–426.

Currie, Wendy L., Matthew W. Guah. "Conflicting Institutional Logics: A National Programme for IT in the Organizational Field of Healthcare," *Journal of Information Technology* 22.3/2007, pp. 235–247.

Curry, Helen Anne. "Industrial Evolution. Mechanical and Biological Innovation at the General Electric Research Laboratory," *Technology and Culture* 54.4/2013, pp. 746–781.

D

Dale, Virginia H., Rebecca A. Efroymson, Keith L. Kline, "The Land Use-Climate Change-Energy Nexus," *Landscape Ecology* 26.6/2011, pp. 755–773.

Davies, Lincoln L. "Beyond Fukushima: Disasters, Nuclear Energy, and Energy Law," *Brigham Young University Law Review* 2011.6/2011, pp. 1937–1989.

Davis, Mary Byrd. "A 'Blue Flash' Hits Tokai-Mura," *Earth Island Journal* 15.1/2000, p. 26.

DeLeon, Peter. "Comparative Technology and Public Policy: The Development of the Nuclear Power Reactor in Six Nations," *Policy Sciences* 11/1980, pp. 285–307.

Delmas, Magali, Bruce Heiman. "Government Credible Commitment to the French and American Nuclear Power Industries," *Journal of Policy Analysis and Management* 20.3/2001, pp. 433–456.

Delmas, Magali A. "The Diffusion of Environmental Management Standards in Europe and in the United States: An Institutional Perspective," *Policy Sciences* 35.1/2002, pp. 91–119.

Delmas, Magali, Michael W. Toffel. "Stakeholders and Environmental Management Practices: An Institutional Framework," *Business Strategy and the Environment* 13. 4/2004, pp. 209–222.

Dent, Christopher M. "Taiwan and the New Regional Political Economy of East Asia," *The China Quarterly* 182/2005, pp. 385–406.

Deyle, Robert E. "Conflict, Uncertainty, and the Role of Planning and Analysis in Public Policy Innovation." *Policy Studies Journal* 22.3/1994, p. 457–473.

Diduck, Alan Paul (2001). "Learning Through Public Involvement in Environmental Assessment: A Transformative Perspective." PhD Dissertation at the University of Waterloo, Department of Geography, Ontario, Canada. https://uwspace.uwaterloo.ca/bitstream/handle/10012/675/NQ65232.pdf? (08 Aug. 2022).

Dierksmeier, Claus. "The Freedom-Responsibility Nexus in Management Philosophy and Business Ethics," *Journal of Business Ethics* 101.2/2011, pp. 263–283.

Dignam, Alan. "Capturing Corporate Governance: The End of the UK Self-Regulating System," *International Journal of Disclosure and Governance* 4/2006, pp. 24–41.

Donohue, Jay (2010). "Resolving Past Liabilities for Future Reduction in Greenhouse Gases; Nuclear Energy and the Outstanding Federal Liability of Spent Nuclear Fuel," Master Thesis at the George Washington University, Law School, United States. https://ui.adsabs.harvard.edu/abs/2010PhDT... ...172D/abstract (07 Aug. 2022).

Dore, Giovanna, Tanvi Nagpal. "Urban Transition in Mongolia: Pursuing Sustainability in a Unique Environment," *Environment: Science and Policy for Sustainable Development* 48.6/2006, pp. 10–24.

Douglas, Sylvester J., Kenneth Wayne Abbott, Gary E. Marchant. "Not Again! Public Perception, Regulation, and Nanotechnology," *SSRN Working Paper Series: Regulation and Governance* 2009. Social Science Research Network.

Driscoll, Charles T., Clarisse M. Hart, *et al.* "Science and Society: The Role of Long-Term Studies in Environmental Stewardship," *BioScience* 62.4/2012, pp. 354–366.

Durant, Robert F. "Sharpening a Knife Cleverly: Organizational Change, Policy Paradox, and the 'Weaponizing' of Administrative Reforms," *Public Administration Review* 68.2/2008, pp. 282–294.

Dusinberre, Martin, Daniel P. Aldrich. "Hatoko Comes Home: Civil Society and Nuclear Power in Japan." *The Journal of Asian Studies* 70.3/2011, pp. 683–705.

Dwyer, Catherine. "The Relationship between Energy Literacy and Environmental Sustainability," *Low Carbon Economy* 2.3/2011, pp. 123–137.

E

Easterling, Douglas, Howard Kunreuther. *The Dilemma of Siting a High Level Nuclear Waste Repository.* Kluwer Academic Publishers, Norwell, Massachusetts, U.S.,1995.

Eberlein, Burkard. "The Making of the European Energy Market: The Interplay of Governance and Government," *Journal of Public Policy* 28.1/2008, pp. 73-92.

Eikeland, Per Ove. "EU Energy Policy Integration – Stakeholders, Institutions and Issue-Linkages," *FNI Report* at Fridtjof Nansen Institute, 13/2012, pp. 139.

Ellison, Brian A. "Intergovernmental Relations and the Advocacy Coalition Framework: The Operation of Federalism in Denver Water Politics." *Publius* 28.4/1998, pp. 35-54.

The-ENPI.ORG, "The Four Concentric Cycles of a Circular Economy," *The European Network of Policy Incubator*, Policy Paper 2/2016, https://the-enpi.org/2016/08/26/project-sustainabilty-paper-1-the-four-concentric-cycles-of-the-circular-economy/ (07 Aug. 2022).

EPA, "What Is the Yucca Mountain Repository," *Unites States Environmental Protection Agency (EPA)*/2021, Radiation Protection Topic. Available at: https://www.epa.gov/radiation/what-yucca-mountain-repository (07 Aug. 2022).

Epstein, Dmitry. "The Making of Institutions of Information Governance: the Case of the Internet Governance Forum," *Journal of Information Technology* 28.2/2013, p. 137-149.

Erawan, I Ketut Putra (2003). "Why Do Regional Actors Comply? Subnational Structure and Collective Action in Indonesia, 1990-2001." PhD Dissertation at Northern Illinois University, Department of Political Science, United States. https://www.proquest.com/openview/3267986d69c1477407b83ffb08247692/1?pq-origsite=gscholar&cbl=18750&diss=y (07 Aug. 2022).

Eusterfeldhaus, Marcel, Barry Barton. "Energy Efficiency: A Comparative Analysis of the New Zealand Legal Framework," *Journal of Energy and Natural Resources Law* 29.4/2011, pp. 431-470.

Evans, Andrew J., Richard Kingston, Steve Carver. "Democratic Input into the Nuclear Waste Disposal Problem: The Influence of Geographical Data on Decision Making Examined through a Web-Based GIS," *Journal of Geographical Systems* 6.2/2004 p. 117-132.

F

Faure, Michael G., Karine Fiore. "The Coverage of Nuclear Risk in Europe: Which Alternative?", *The Geneva Papers on Risk and Insurance. Issues and Practice* 33.2/2008, pp. 288-322.

The Federal Register Report, "Licenses, Certifications, and Approvals for Nuclear Power Plants," Find72, 166, Office of New Reactors, U.S. Nuclear Regulatory Commission, Washington D.C., U.S, 2007.

Feiveson, Harold A. "Faux Renaissance: Global Warming, Radioactive Waste Disposal, and the Nuclear Future," *Arms Control Today* 37.4/2007, pp. 13-17.

Feller, Joseph, Patrick Finnegan, Jeremy Hayes, and Philip O'Reilly. "Institutionalising Information Asymmetry: Governance Structures for Open Innovation," *Information Technology & People* 22.4/2009, pp. 297-316.

Fernandez, Alexander Hertel. "Who Passes Business's 'Model Bills'? Policy Capacity and Corporate Influence in U.S. State Politics," *Perspectives on Politics* 12.3/2014, pp. 582-602.

Fertel, Marvin. "Consortia Confirm Nuclear Resurgence," *Electric Perspectives Magazine* 29.4/2004, pp. 52-53.

Fisher, Susan L. Reynolds (2000). "A Multilevel Theory of Organizational Performance." PhD Dissertation at the Graduate College of the Oklahoma State University, Degree of Doctor of Education, United States. https://www.proquest.com/openview/ae4436edfaee2976f978f69739c751e1/1?pq-origsite=gscholar&cbl=18750&diss=y (07 Aug. 2022).

Flint, Lawrence. "Shaping Nuclear Waste Policy at the Juncture of Federal and State Law," *Boston College Environmental Affairs Law Review* 28.1/2000, pp. 163-190.

Foss, Nicolai J. "Knowledge Governance: Meaning, Nature, Origins, and Implications," in: *Handbook of Economic Organization*, eds. Anna Grandori. Edward Elgar, 2012. SSRN Working Paper Series. (Social Science Research Network).

Foss, Nicolai J., Bo. B. Nielsen. "Researching Multilevel Phenomena: The Case of Collaborative Advantage in Strategic Management," *Journal of CENTRUM Cathedra: The Business and Economics Research Journal* 5.1/2012, pp. 11-23.

G

Gagné, Gilbert. "Policy Diversity, State Autonomy, and the US-Canada Softwood Lumber Dispute: Philosophical and Normative Aspects," *Journal of World Trade* 41.4/2007, pp. 699-730.

Gamal, Ibrahim, Galt Vaughan. "Ethnic Business Development: Toward a Theoretical Synthesis and Policy Framework," *Journal of Economic Issues* 37.4/2003, pp. 1107-1119.

Gao, Anton Ming-Zhi. "Development of a Legal Framework for Climate Change in Taiwan: Lessons from Europe and Germany," *Carbon & Climate Law Review: CCLR* 7.1/2013, pp. 54-70.

Gildart, Keith. "Coal Strikes on the Home Front: Miners' Militancy and Socialist Politics in the Second World War," *Twentieth Century British History* 20.2/2009, pp. 121-151.

Gochfeld, Michael, Sandra Mohr. "Protecting Contract Workers: Case Study of the US Department of Energy's Nuclear and Chemical Waste Management," *American Journal of Public Health (AJPH)* 97.9/2007, pp. 1607-1613.

Gonick, Cy. "Are We Coming to the End of the Growth Era?" *Canadian Dimension* 46.2/2012, pp. 40–43. https://canadiandimension.com/articles/view/are-we-coming-to-the-end-of-the-growth-era (08, Sept. 2022).

Gouldson, Andy. "Risk, Regulation and the Right to Know: Exploring the Impacts of Access to Information on the Governance of Environmental Risk," *Sustainable Development* 12.3/2004, pp. 136–149.

Grafton, Carl, Anne Permaloff. "Public Policy for Business and the Economy: Ideological Dissensus, Change and Consensus," *Policy Sciencess* 34/2001, pp. 403–434.

Gralla, Fabienne, David J. Abson, Anders P. Moller, Daniel J. Lang, Ulli Vilsmaier, Benjamin K. Sovacool, Henrik von Wehrden. "Nuclear Accidents Call for Transdisciplinary Nuclear Energy Research," *Sustainability Science* 10.1/2015, pp. 179–183.

Grandin, Karl, Peter Jagers, Sven Kullander. "Nuclear Energy," *Ambio* 39/2010, suppl.1, Special Report: Energy 2050, pp. 26–30.

Green, Trevor. "Meeting the Rising Energy Demands of a Greener Future," *Cost Engineering* 50.12/2008, pp. 15–16.

Grosjean, Janet Atkinson. "Canadian Science at the Public/Private Divide: The NCE Experiment," *Journal of Canadian Studies* 37.3/2002, pp. 71–91.

(de) Grosbois, John F. P. (2011). "The Impact of Knowledge Management Practices on Nuclear Power Plant Organization Performance," PhD Dissertation submitted at the Faculty of Graduate and Postdoctoral Affairs, Carleton University, Ottawa, Ontario, Canada.

Guess, George M. "Comparative Decentralization Lessons From Pakistan, Indonesia, and the Philippines," *Public Administration Review* 65.2/2005, pp. 217–230.

Gynther, Lea, Irmeli Mikkonen, Antoinet Smits, "Evaluation of European Energy Behavioural Change Programmes," *Energy Efficiency* 5.1/2012, pp. 67–82.

H

Haga, Trond. "Action Research and Innovation in Networks, Dilemmas and Challenges: Two Cases," *AI & Society* 19/2005, pp. 362–383.

Halfacre, Angela C., Albert R. Matheny, Walter A. Rosenbaum. "Regulating Contested Local Hazards: Is Constructive Dialogue Possible Among Participants in Community Risk Management," *Policy Studies Journal* 28.3/2000, pp. 648–667.

Hamdouch, Abdelillah, Marc-Hubert Depret, Jean-Louis Monino, Christian Poncet. "Regional Policies, Key Levers of Regional Innovation Dynamics," *SSRN Working Paper Series* 2009, Copyright Social Science Research Network. http://dx.doi.org/10.2139/ssrn.1366349

Hansen, Teresa. "Positive Image Fuels Nuclear Energy's Resurgence," *Power Engineering* 110.9/2006, pp. 18–30.

Harding, Jim, "Living Behind the Uranium Curtain," *Briar Patch* 37.8/2008, p. 13–15.

Hassler, Bjorn. "Accidental Versus Operational Oil Spills from Shipping in the Baltic Sea: Risk Governance and Management Strategies," *Ambio* 40.2/2011, pp. 170–178.

Heap, Brian. "Man and the Future Environment," *European Review* 12.3/2004, pp. 273–292.

Heath, Joseph, Wayne Norman. "Stakeholder Theory, Corporate Governance and Public Management: What Can the History of State-Run Enterprises Teach Us in the Post-Enron Era?" *Journal of Business Ethics* 53/2004, pp. 247–265.

Heiskanen, Eva, Sirkku Kivisaari, Raimo Lovio, Per Mickwitz. "Designed to Travel? Transition Management Encounters Environmental and Innovation Policy Histories in Finland," *Policy Sciences* 42.4/2009, pp. 409–427.

Hendriks, Carolyn M. "Policy Design without Democracy? Making Democratic Sense of Transition Management," *Policy Sciences* 42.4/2009, pp. 341–368.

Hern, Matt (1997). "Making Space: Radical Democracy in the Megalopolis. An Investigation into Social Ecology and the Creation of Participatory, Self-Designing and Self-Reliant Neighbourhoods within Major Urban Entities." PhD Dissertation at the Union Institute Graduate School, Urban Studies, United States. https://www.proquest.com/openview/18b7a19138c8807baabc4c507a5d959b/1?pq-origsite=gscholar&cbl=18750&diss=y (07 Aug. 2022)

Hickmann, Thomas. "Science-Policy Interaction in International Environmental Politics: An Analysis of the Ozone Regime and the Climate Regime," *Environmental Economics and Policy Studies* 16.1/2014, pp. 21–44.

Hillman, Mick. "Environmental Justice: A Crucial Link Between Environmentalism and Community Development?" *Community Development Journal* 37.4/2002, pp. 349–360.

Hiner, Colleen Crystal (2012). "Changing Landscapes, Shifting Values: A Political Ecology of the Rural-Urban Interface." PhD Dissertation at the Geography Department, Office of Graduate Studies, University of California Davis, United States. https://www.proquest.com/openview/37e3e3d474d61142be3d2fa18d7ba1c2/1?pq-origsite=gscholar&cbl=18750 (07 Aug. 2022).

Hogberg, Lars. "Root Causes and Impacts of Severe Accidents at Large Nuclear Power Plants," *Ambio* 42. 3/2013, pp. 267–284.

Holder, Jane, Antonia Layard. "Drawing Out the Elements of Territorial Cohesion: Re-scaling EU Spatial Governance," *Yearbook of European Law* 30.1/2011, pp. 358–380.

Holger, Rogner H. "Nuclear Power and Sustainable Development," *Journal of International Affairs* 64.1/2010, pp. 137–163.

Holland, Breena (2005). "Environment and Capability: A New Normative Framework for Environmental Policy Analysis." PhD Dissertation at the Department of Political Science, University of Chicago, United States. https://www.elibrary.ru/item.asp?id=9347334 (08 Aug. 2022).

Holloway, Milton L. "Innovation Dynamics and Policy in the Energy Sector," *Academic Press* Copyright @ 2021 Elsevier Inc. All rights reserved. https://doi.org/10.1016/C2020-0-01005-4. (08 Aug. 2022).

Hostovsky, Charles (2002). "Integrating Planning Theory and Waste Management: A Critical Analysis of Current EIA Practice in Ontario," PhD Dissertation, at the University of Waterloo, Ontario, Canada. https://www.bac-lac.gc.ca/eng/services/theses/Pages/item.aspx?idNumber=56920538 (07 Aug. 2022).

Howe, Christopher. "Taiwan in the 20th Century: Model or Victim? Development Problems in a Small Asian Economy," *The China Quarterly* 165/2001, pp. 37–60.

Huang, Gillian Chi-Lun, Tim Gray, Derek Bell. "Environmental Justice of Nuclear Waste Policy in Taiwan: Taipower, Government, and Local Community," *Environment, Development, and Sustainability* 15/2013, pp. 1555–1571.

Hudson, Ray. "Cultural Political Economy Meets Global Production Networks: A Productive Meeting?" *Journal of Economic Geography* 8.3/2008, pp. 421–440.

Hurlbert, Margot. "Evaluating Public Consultation in Nuclear Energy: The Importance of Problem Structuring and Scale," *International Journal of Energy Sector Management* 8.1/2014, pp. 56–75.

I

International Atomic Energy Agency, IAEA, Technical Reports Series No. 464. "Managing the Socioeconomic Impact of the Decommissioning of Nuclear Facilities." IAEA, Vienna, 2008.

Iglesias, Ana, Sonia Quiroga, Marta Moneo, Luis Garrote. "From Climate Change Impacts to the Development of Adaptation Strategies: Challenges for Agriculture in Europe," *Climatic Change* 112.1/2012, pp. 143–168.

Interfax: Ukraine Business Weekly. "Economic Policy; Energy Ministry Developing Nuclear Fuel of Ukraine State Program for 2014–2019," *Interfax-America, Inc.* 11/2011.

J

Jackson, Gregory, Richard Deeg. "Comparing Capitalisms: Understanding Institutional Diversity and Its Implications for International Business,"

Journal of International Business Studies, suppl. Part Special Issue: Institutions and International Business 39.4/2008, pp. 540–561.

Jenkin, Thomas. "Exploitation or 'Wise Use' of the Coongie Lakes, South Australia: Issues Arising from a Petroleum Exploration Proposal," *Australian Geographer*, 30.3/1999, pp. 355–371.

Jerneck, Anne, Lennart Olsson, Barry Ness, Stefan Anderberg, Matthias Baier, Eric Clark, Thomas Hickler, Alf Hornborg, Annica Kronsell, Eva Lovbrand, Johannes Persson. "Structuring Sustainability Science," *Sustainability Science* 6.1/2011, pp. 69–82.

Jessop, Bob. "Revisiting the Regulation Approach: Critical Reflection on the Contradictions, Dilemmas, Fixes and Crisis Dynamics of Growth Regimes," *Capital & Class* 37.1/2013, pp. 5–24.

Johannessen, Jon-Arild, Bjorn Olsen. "Systemic Knowledge Processes, Innovation and Sustainable Competitive Advantages," *Kybernetes* 38.3–4/2009, pp. 559–580. Emerald Group Publishing Limited.

John, Peter, Alistair Cole. "When Do Institutions, Policy Sectors, and Cities Matter? Comparing Networks of Local Policy Makers in Britain and France," *Comparative Political Studies* 33.2/2000, pp. 248–268.

Johnson, Daniel K.N, Kristina M. Lybecker. "Challenges to Technology Transfer: A Literature Review of the Constraints on Environmental Technology Dissemination," *Colorado College Working Paper* No. 2009-07 – SSRN Working Paper Series, Aug. 2010.

Johnson, Erik W., Jon Agnone, John D. McCarthy. "Movement Organizations, Synergistic Tactics and Environmental Public Policy," *Social Forces* 88.5/2010, pp. 2267–2292.

Johnson, Genevieve Fuji. "The Discourse of Democracy in Canadian Nuclear Waste Management Policy," *Policy Sciences* 40.2/2007, pp. 79–99.

Jones, Alistar Colin, and Sudhir Rama Murthy. "Mutuality and Concepts of Responsible Business." In: *Putting Purpose into Practice: The Economics of Mutuality*. Oxford University Press, chapter 6. 2021. At: https://purposeintopractice.org/mutuality-and-concepts-of-responsible-business (08 Aug. 2022).

Jones, Peter. "Local Environment Agency Plans: New Environmental Management Initiatives in England and Wales," *Management Research News* 22.10/1999, pp. 12–16.

Jones, Peter Vincent. "Contractual Governance: Institutional and Organizational Analysis," *Oxford Journal of Legal Studies* 20.3/2000, pp. 317–351.

Joskow, Paul L., John E. Parsons. "The Economic Future of Nuclear Power," *Daedalus* 138.4/2009, pp. 45–59.

Ju, Chang Bum (2008). "Institutional Contestation, Network Legitimacy and Organizational Heterogeneity: Interactions between Government and Environmental Nonprofits in South Korea." PhD Dissertation at the Graduate School University of Southern California, Public Administration, United States. https://www.proquest.com/openview/75d835d87fc93c9cb16273cb2 7404f4d/1?pq-origsite=gscholar&cbl=18750 (09 Aug. 2022).

K

Kakabadse, Andrew, Nada K. Kakabadse, Alexander Kouzmin, "Reinventing the Democratic Governance Project through Information Technology? A Growing Agenda for Debate," *Public Administration Review* 63.1/2003, pp. 44–60.

Kalfagianni, Agni. "Addressing the Global Sustainability Challenge: The Potential and Pitfalls of Private Governance from the Perspective of Human Capabilities," *Journal of Business Ethics* 122.2/2014, pp. 307–320.

Kalin, Ivanov. "Legitimate Conditionality? The European Union and Nuclear Power Safety in Central and Eastern Europe," *International Politics* 45.2/2008, pp. 146–167.

Kane, Chen, Stephanie C. Lieggi, Miles A. Pomper. "Time for Leadership: South Korea and Nuclear Nonproliferation," *Arms Control Today* 41.2/2011, pp. 22–28.

Kristian Kallenberg. "Operational Risk Management in Swedish Industry: Emergence of a New Risk Paradigm," *Risk Management* 11.2/2009, pp. 90–110.

Krupar, Shiloh R. "Transnatural Ethics: Revisiting the Nuclear Cleanup of Rocky Flats, CO, through the Queer Ecology of Nuclia Waste," *Cultural Geographies* 19.3/2012, pp. 303–327.

Kao, Shu-Fen (2002). "Risk Perceptions and Environmental Mobilization. Tracking the Transformation of Collective Actions in a Radiation Contamination Incident in Taiwan," PhD Dissertation at Michigan State University, Department of Sociology, U.S. https://www.proquest.com/openv iew/a9c79560b348609af1fc7dca4ec7b29a/1?pq-origsite=gscholar&cbl= 18750&diss=y (07 Aug. 2022).

Karlsson, Sylvia I. "Institutionalized Knowledge Challenges in Pesticide Governance: The End of Knowledge and Beginning of Values in Governing Globalized Environmental Issues." *International Environmental Agreements: Politics, Law and Economics* 4/2004, pp. 195–213.

Katz, Jonathan L. "A Web of Interests: Stalemate on the Disposal of Spent Nuclear Fuel," *Policy Studies Journal* 29.3/2001, pp. 456–477.

Keast, Robyn, Myrna Mandell, Kerry Brown. "Mixing State, Market and Network Governance Modes: The Role of Government in 'Crowded' Policy Domains," *International Journal of Organization Theory and Behavior* 9.1/2006, pp. 27–50.

Kemp, René, Jan Rotmans. "Transitioning Policy: Co-Production of a New Strategic Framework for Energy Innovation Policy in the Netherlands," *Policy Sciences* 42/2009, pp. 303–322.

Kennedy, Harold. "A Big Job: Cleaning Up Nation's Nuclear Waste," *National Defense* 83.546/1999, pp. 24–27.

Kilmurray, Heather C. McLeod, Gavin Smith. "Unsustainable Development in Canada: Environmental Assessment, Cost-Benefit Analysis, and Environmental Justice in the Tar Sands," *Journal of Environmental Law and Practice* 21/2010, pp. 65–105.

Kim, Seong-Jai. "Korean Waste Management and Eco-Efficient Symbiosis – a Case Study of Kwangmyong City," *Clean Technologies and Environmental Policy* 3/2002, pp. 371–382.

Kim, Sunhyuk. "Collaborative Governance in South Korea: Citizen Participation in Policy Making and Welfare Service Provision," *Asian Perspective* 34.3/2010, pp. 165–190.

Kingston, Jeff. "Nuclear Power Politics in Japan, 2011–2013," *Asian Perspective* 37.4/2013, pp. 501–521.

Koehler, Dinah A. "The Effectiveness of Voluntary Environmental Programs – A Policy at a Crossroads?" *Policy Studies Journal* 35.4/2007, pp. 689–722.

Komori, Yasumasa. "Evaluating Regional Environmental Governance in Northeast Asia," *Asian Affairs: An American Review* 37.1/2010, pp. 1–25.

Kralj, Davorin. "Dialectal System Approach Supporting Environmental Innovation for Sustainable Development," *Kybernetes* 37.9–10/2008, pp. 1542–1560.

Krutli, Pius, Michael Stauffacher, Dario Pedolin, Corinne Moser, Roland W. Scholz. "The Process Matters: Fairness in Repository Siting for Nuclear Waste," *Social Justice Research* 25.1/2012, pp. 79–101.

Kuhn, Richard G. "Social and Political Issues in Siting a Nuclear-Fuel Waste Disposal Facility in Ontario, Canada," *The Canadian Geographer* 42.1/1998, pp. 14–28.

Kulas, Darko Pavo (2004). "Contemporary Political Risk Analysis and Management: Civil Society and Sustainable Development," Master's Thesis at the University of Calgary, Department of Management, Alberta, Canada. https://prism.ucalgary.ca/handle/1880/41691 (07 Aug.2022).

L

Lai, Po-Hsin, Michael G. Sorice, Sanjay K. Nepal, Chia-Kuen Cheng. "Integrating Social Marketing into Sustainable Resource Management at Padre Island National Seashore: An Attitude-Based Segmentation Approach," *Environmental Management* 43.6/2009, pp. 985–998.

Lane, R., E. Dagher, J. Burtt, P. A. Thompson. "Radiation Exposure and Cancer Incidence (1990 to 2008) Around Nuclear Power Plants in Ontario, Canada," *Journal of Environmental Protection* 4.9/2013, pp. 888–913.

Laurian, Lucie. "Public Participation in Environmental Decision Making: Findings from Communities Facing Toxic Waste Cleanup," *Journal of the American Planning Association* 70.1/2004.

Layzer, Judith A. "Citizen Participation and Government Choice in Local Environmental Controversies," *Policy Studies Journal* 30.2/2002, pp. 193–207.

Lazar, Marc. "Testing Italian Democracy," *Comparative European Politics* 11.3/2013, pp. 317–336.

Lecours, André, Daniel Béland. "The Institutional Politics of Territorial Redistribution: Federalism and Equalization Policy in Australia and Canada," *Canadian Journal of Political Science* 46.1/2013, pp. 93–113.

Lee, Soonsil (2003). "Successful Implementation Strategies for Environmental Management Systems in Public Organizations." PhD Dissertation at the Department of Work Environment, University of Massachusetts Lowell, United States. https://www.proquest.com/openview/b2253a488bb5fa2f5732e3014fcbc11b/1?pq-origsite=gscholar&cbl=18750&diss=y (08 Aug. 2022).

Lee, Sung Roe (1997). "A Policy Deadlock in South Korea: The Case Study of Siting Nuclear Facilities," PhD Dissertation at the University of Tennessee, Knoxville, United States. https://www.proquest.com/openview/d7673fd43278a7c801cf41b5556e0b91/1?pq-origsite=gscholar&cbl=18750&diss=y (07 Aug. 2022).

Leibovitz, Yosseph (2001). "Associative Governance? The Political Economy of Institutional Change in Two Ontario City-Regions." PhD Dissertation at the Graduate Department of Geography, University of Toronto, Canada. https://tspace.library.utoronto.ca/handle/1807/16454 (07 Aug. 2022).

Letourneau, Carmel (2010). "Adaptive Management of Complex Environmental Problems – Comparison of National Nuclear Waste Management Policies," PhD Dissertation at Graduate College University of Nevada, Las Vegas, United States. https://digitalscholarship.unlv.edu/thesesdissertations/841/ (06 Aug. 2022).

Leuenberger, Deniz Zeynep, Michele Wakin. "Sustainable Development in Public Administration Planning: An Exploration of Social Justice, Equity, and Citizen Inclusion," *Administrative Theory & Praxis* 29.3/2007, p. 394–411.

Li, Dan, Lorraine Eden, Michael A. Hitt, R. Duane Ireland, Robert P. Garrett. "Governance in Multilateral R&D Alliances" *Organization Science* 23.4/2012, pp. 1191–1210.

Li, Xiaofei (Sarah). "Chinese Political Transition Makes Change Possible in Energy Approach," *Oil & Gas Journal* 111.2/2013, pp. 46–52.

(de) Lima, Ismar Borges, Leszek Buszynski. "Local Environmental Governance, Public Policies and Deforestation in Amazonia," *Management of Environmental Quality: An International Journal* 22.3/2011, pp. 292–316. Emerald Group Publishing Limited.

Lin, Chih-Pin, Hsin-Mei Lin. "Maker-Buyer Strategic Alliances: An Integrated Framework," *Journal of Business & Industrial Marketing* 25.1/2010, pp. 43–56.

Lock, Gwendolyn E. (2010). "Who Shares? Managerial Knowledge Transfer Practices in British Columbia's Ministry of Health Services." PhD Dissertation at the College of Management and Technology, Walden University, United States. https://www.proquest.com/openview/e385896922369033c67e01742e29c0d4/1?pq-origsite=gscholar&cbl=18750 (08 Aug. 2022).

Lofstedt, Ernst Ragnar (1993). "The Swedish Energy Policy Dilemma and the Role of Behavioral Energy Conservation." PhD Dissertation submitted at Clark University, Worcester Massachusetts. The Graduate School of Geography, United States. https://www.proquest.com/openview/22dba31a14e07e16c36f32380e79bc0b/1?pq-origsite=gscholar&cbl=18750&diss=y (10 Aug. 2022.)

Loh, Christine. "Hong Kong-Mainland Innovations in Environmental Protection since 1980," *Asian Survey* 51.4/2011, pp. 610–632.

Lowry, Robert C. "All Hazardous Waste Politics Is Local: Grass-roots Advocacy and Public Participation in Siting and Cleanup Decisions," *Policy Studies Journal* 26.4/1998, pp. 748–759.

Ludwiszewski, Raymond B. "The Interdependence of Economic Development and Environmental Protection," *Economic Development Review* 10.3/1992, pp. 5.

Lukasiewicz, Anna, Geoffrey J. Syme, Kathleen H. Bowmer, Penny Davidson, "Is the Environment Getting Its Fair Share? An Analysis of the Australian Water Reform Process Using a Social Justice Framework," *Social Justice Research* 26.3/2013, pp. 231–252.

Luna-Reyes, Luis F., Jing Zhang, J. Ramòn Gil-Garcìa, Anthony M. Cresswell. "Information Systems Development as Emergent Socio-Technical Change: A Practice Approach," *European Journal of Information Systems* 14/2005, pp. 93–105.

Lundberg, Kristina. "A Systems Thinking Approach to Environmental Follow-Up in a Swedish Central Public Authority: Hindrances and Possibilities for Learning from Experience," *Environmental Management* 48.1/2011, pp. 123–133.

M

Macaneiro, Marlete Beatriz, Sieglinde Kindl da Cunha. "Theoretical Analysis Model of the Adoption of Reactive and Proactive Eco-Innovation Strategies: The Influence of Contextual Factors Internal and External to Organizations," *Brazilian Business Review* 11.5/2014, pp. 1–23. ISSN 1808-2386.

Malayang, Ben S. III Ma, Zita B. Toribio, (2002). "Networking, Environmental Management and Governance," *Produced by the Department of Environment and Natural Resources-United States Agency for International Development's (DENR-USAID)*, Philippine Environmental Governance (EcoGov) project through the assistance of the USAID under USAID PCE-1-00-99-00002-00. https://faspselib.denr.gov.ph/sites/default/files/Publication%20Files/Networking%20Environmental%20Mgmt%20and%20Governance.pdf?msclkid=c834552da91011ecb7eb95ff5f7d574f (10 Aug. 2022).

Malgosia, Fitzmaurice. "Public Participation in the North American Agreement on Environmental Cooperation," *The International & Comparative Law Quarterly* 52.2/2003, pp. 333–368.

Malik, J. Mohan. "China and the Nuclear Non-Proliferation Regime," *Contemporary Southeast Asia* 22.3/2000, pp. 445–478.

Mallavarapu, Bravishwar (2013). "Regional Pathways to Technological Upgrading: The Impact of Agglomeration Economies and Its Regional Covariates on Upgrading in Post-Reforms India's Manufacturing Sector." PhD Dissertation at the University of California, Los Angeles, Department of Urban Planning, United States. https://www.proquest.com/openview/605ff9d13fbc12931391e2fe153d2763/1?pq-origsite=gscholar&cbl=18750 (07 Aug. 2022).

Mansfield, Carol, Georgie Van Houtven, Joel Huber. "The Efficiency of Political Mechanisms for Siting Nuisance Facilities: Are Opponents More Likely to Participate than Supporters?" *The Journal of Real Estate Finance and Economics* 22.2.3/2001, pp. 141–161.

Masco, Joseph. "Survival Is Your Business: Engineering Ruins and Affect in Nuclear America," *Cultural Anthropology* 23.2/2008, p. 361–398.

Mastroeni, Michele (2008). "How Small Economies Finance High-Tech Industries: Seeking Variance in Innovation Commercialization." PhD Dissertation at the Department of Political Science, University of Toronto, Canada. https://www.collectionscanada.gc.ca/obj/thesescanada/vol2/002/NR39909.PDF?oclc_number=649886906 (08 Aug. 2022).

Matsui, Kenichi. "Global Demand Growth of Power Generation, Input Choices and Supply Security," *The Energy Journal* 19.2/1998, pp. 93–107.

McCauley, Stephen M., Jennie C. Stephens. "Green Energy Clusters and Socio-Technical Transitions: Analysis of a Sustainable Energy Cluster for Regional Economic Development in Central Massachusetts, USA," *Sustainability Science* 7.2/2012, pp. 213–225.

McCorquodale, Robert, Raul Pangalangan. "Pushing Back the Limitations of Territorial Boundaries," *European Journal of International Law (EJIL)* 12.5/2001, pp. 867–888.

Mccreary, Tyler, Vanessa Lamb. "A Political Ecology of Sovereignty in Practice and on the Map: The Technicalities of Law, Participatory Mapping, and Environmental Governance," *Leiden Journal of International Law* 27.3/2014, pp. 595–619.

McDonald, Nicole C, Joshua M. Pearce. "Renewable Energy Policies and Programs in Nunavut: Perspectives from the Federal and Territorial Governments," *Arctic* 65.4/2012, pp. 465–475.

McGlynn, Grace, Gregg Butler, Alan Pearman. "Stakeholder Preference Mapping—Seeking a Way Forward for the Processing of Spent Nuclear Fuel," *The Journal of the Operational Research Society* 66.2/2015, pp. 219–230.

McNichol, Jason Hall (2002). "Contesting Governance in the Global Marketplace: A Sociological Assessment of Business-NGO Partnership to Build Markets for Certified Wood," PhD Dissertation at Sociology Graduate Division of the University of California, Berkeley, U.S. https://www.econbiz.de/Record/contesting-governance-the-global-marketplace-sociological-assessment-business-ngo-partnerships-build-markets-for-certified-wood-mcnichol-jason/10003629885 (06 Aug.2022).

Mele, Valentina, Donald H. Schepers. "E Pluribus Unum? Legitimacy Issues and Multi-Stakeholder Codes of Conduct," *Journal of Business Ethics* 118.3/2013, pp. 561–576.

Mercer, Doug, Thomas Leschine, Christina H. Drew, William Griffith, Timothy Nyerges. "Public Agencies and Environmental Risk: Organizing Knowledge in a Democratic Context," *Journal of Knowledge Management* 9.2/2005, pp. 129–147.

Miller, Marvin, Lawrence Scheinman. "Israel, India, and Pakistan: Engaging the Non-NPT States in the Nonproliferation Regime," *Arms Control Today* 33.10/2003, pp. 15–20.

Miodownik, Dan, Britt Cartrite. "Does Political Decentralization Exacerbate or Ameliorate Ethnopolitical Mobilization? A Test of Contesting Propositions," *Political Research Quarterly* 63.4/2010, pp. 731–746.

Molis, Arunas, Justina Gliebute. "Prospects for the Development of Nuclear Energy in the Baltic Region," *Lithuanian Annual Strategic Review* 10.1/2012, pp. 121–151.

Mondejar, Reuben, Hongxin Zhao. "Antecedents to Government Relationship Building and the Institutional Contingencies in a Transition Economy," *Management International Review* 53.4/2013, pp. 579–605.

Morgan, Jonathan Quentin (2004). "The Role of Regional Industry Clusters in Urban Economic Development: An Analysis of Process and Performance." PhD Dissertation at the Graduate Faculty of North Carolina State University, Public Administration, United States. https://www.proquest.com/openview/5b8067601e6b862139de1aa160080e3a/1?pq-origsite=gscholar&cbl=18750&diss=y (07 Aug. 2022).

Morgan, Jonathan Q. "Governance, Policy Innovation, and Local Economic Development in North Carolina," *Policy Studies Journal* 38.4/2010, pp. 679–702.

Murakami, Tomoko, Venkatachalam Anbumozhi. "Public Acceptance of Nuclear Power Plants in Hosting Communities: A Multilevel System Analysis," *Economic Research Institute for ASEAN and East Asia (ERIA), ERIA*, Research Project 2018, No.18, published in December 2019. https://www.eria.org/uploads/media/Public-Acceptance-Nuclear-Power-Plants/RPR_FY2018_18.pdf (06 Aug.2022).

N

Neuwirth, Rostam J. "'Novel Food' for Thought' on Law and Policymaking in the Global Creative Economy," European Journal of Law and Economics 37.1/2014, pp. 13–50.

Newell, Peter. "The Political Economy of Global Environmental Governance," *Review of International Studies* 34.3/2008, pp. 507–529.

Nierenberg, Danielle. "U.S. Environmental Policy: Where Is It Headed?" *World Watch* 14.4/2001, pp. 12–21.

NPR, "Profile: How The Residents and Activists Surrounding Three Mile island Feel Regarding their Community 20 Years After the Nuclear Accident," *Weekend Edition. Saturday* 1/1999. Washington, D.C.: National Public Radio.

Nohrstedt, Daniel. "The Politics of Crisis Policymaking: Chernobyl and Swedish Nuclear Energy Policy," *Policy Studies Journal* 36.2/2008, pp. 257–278.

O

OECD report. "National Reports on Uranium Exploration, Resources, Production, Demand, and the Environment," Chapter 3. *Uranium 2011: Resources, Production and Demand*, ISBN 978-92-64-17803-8, OECD 2012.

Omi, Koji. "Alternative Energy for Transportation," *Issues in Science and Technology* 25.4/2009, pp. 31–34.

Omoto, Akira. "Where Was the Weakness in Application of Defence-in-Depth Concept and Why?," in: *Reflection on the Fukushima Daiichi Nuclear Accident*, eds. J. Ahn et al. (chp. 8, 2015) DOI 10.1007/978-3-319-12090-4.

P

Park, Seung Ho, Shaomin Li, David K. Tse. "Market Liberalization and Firm Performance during China's Economic Transition," *Journal of International Business Studies* 37.1/2006, pp. 127–147.

Parr, Joy. "A Working Knowledge of the Insensible? Radiation Protection in Nuclear Generating Stations, 1962–1992," *Comparative Studies in Society and History* 48.4/2006, pp. 820–851.

Parson, Edward A., Karen Fisher-Vanden. "Integrated Assessment Models of Global Climate Change," *Annual Review of Energy and the Environment* 22/1997, pp. 589–628.

Parthasarathy, Shobita. "Whose Knowledge? What Values? The Comparative Politics of Patenting Life Forms in the United States and Europe," *Policy Sciences* 44.3/2011, pp. 267–288.

Paul, T.V. "The US-India Nuclear Accord: Implications for the Nonproliferation Regime," *International Journal: Canada's Journal of Global Policy Analysis* 62.4/2007, pp. 845–861.

Peace, Ade. "Anatomy of a Blockade. Towards an Ethnography of Environmental Dispute (Part 2), Rural New South Wales 1996," *The Australian Journal of Anthropology* 10.2/1999, pp. 144–162.

Pellerin-Carlin, Thomas, Jean-Arnold Vinois, Eulalia Rubio, and Sofia Fernandes. "Making the Energy Transition a European Success," *Notre Europe*, Jacques Delors Institute, Studies & Reports 114/2017.

Peters, Katherine McIntire. "Power player," *Government Executive* 39.12/2007, pp. 40–46.

(delle) Piane, Sebastian, AvellaNeda. "Good Governance, Institutions and Economic Development: Beyond the Conventional Wisdom," *British Journal of Political Science* 40.1/2010, pp. 195–224.

Pidgeon, Nick. "Public Understanding of, and Attitudes to, Climate Change: UK and International Perspectives and Policy," *Climate Policy* 12/2012, pp. S85–S106.

Plater, Zygmunt J.B. "Law, Media, & Environmental Policy: A Fundamental Linkage in Sustainable Democratic Governance," *Boston College Environmental Affairs Law Review* 33/2006, pp. 511–550.

Potoski, Matthew, Aseem Prakash. "The Regulation Dilemma: Cooperation and Conflict in Environmental Governance," *Public Administration Review* 64.2/2004, pp. 152–163.

Powell, Lydia, Akhilesh Sati, and Vinod Kumar Tomar. "Nuclear Energy in India: Small May Not Be Beautiful." Observer Research Foundation (ORF), War Fare, Feb. 03 2022. https://www.orfonline.org/expert-speak/nuclear-energy-in-india/ (10 Aug. 2022).

Powell, Maria C., Martin P. A. Griffin, Stephanie Tai. "Bottom-Up Risk Regulation? How Nanotechnology Risk Knowledge Gaps Challenge Federal and State Environmental Agencies" *Environmental Management* 42.3/2008, pp. 426–443.

PR Report, "Groups: Nuclear Energy Institute Is Pushing NRC to Shortchange Environmental Review of Long-Term Radioactive Waste Storage," *PR Newswire Association LLC*, New York, U.S, 2013.

Price, Richard. "Review: Transnational Civil Society and Advocacy in World Politics," *World Politics* 55.4/2003, pp. 579–606.

R

Raco, Mike. "Competition, Collaboration and the New Industrial Districts: Examining the Institutional Turn in Local Economic Development," *Urban studies* 36.5/6/1999, pp. 951–968.

Ratliff, Jeanne Nelson. "The Politics of Nuclear Waste: An Analysis of a Public Hearing on the Proposed Yucca Mountain Nuclear Waste Repository," *Communication Studies* 48.4/1997, pp. 359–380.

Rego, Josep Vives, Serge Caschetto, Jordi Faraudo, Diego Prior, "Management Options for the Increasing Demand of Energy and Water: Is the Problem Soluble in Technosciences Only?" *Ambio* 37.2/2008, pp. 134–136.

Richardson, Benjamin J. "Keeping Ethical Investment Ethical: Regulatory Issues for Investing for Sustainability," *Journal of Business Ethics* 87/2009, pp. 555–572.

Rietig, Katharina. "Reinforcement of Multilevel Governance Dynamics: Creating Momentum for Increasing Ambitions in International Climate Negotiations," *International Environmental Agreements: Politics, Law and Economics* 14.4/2014, pp. 371–389.

(del) Rio, Pablo, Xavier Labandeira. "Barriers to the Introduction of Market-Based Instruments in Climate Policies: An Integrated Theoretical Framework," *Environmental Economics and Policy Studies* 10/2009, pp. 41–68.

Robertson, Philip G., Janet C. Broome, Elisabeth A. Chornesky, Jane R. Frankenberger; et al. "Rethinking the Vision for Environmental Research in US Agriculture," *BioScience* 54.1/2004, pp. 61–65.

Robyn, Linda. "Indigenous Knowledge and Technology: Creating Environmental Justice in the Twenty-First Century," *American Indian Quarterly* 26.2/2002, pp. 198–220.

Rochon, Gilbert L., Dev Niyogi, Souleymane Fall, Joseph E. Quansah, Larry Biehl, Bereket Araya, Chetan Maringanti, Angel Torres Valcarcel, Lova Rakotomalala, Hildred S. Rochon, Bertin Hilaire Mbongo, Thierno Thiam. "Best Management Practices for Corporate, Academic and Governmental Transfer of Sustainable Technologies to Developing Countries," *Clean Technologies and Environmental Policy* 12.1/2009, pp. 19–30.

Roesler, Axel. "Lessons from Three Mile Island: The Design of Interactions in a High-Stakes Environment," *Visible Language* 43.2/3/2009, pp. 170–195.

Rogner, Holger H. "Nuclear Power and Sustainable Development." *Journal of International Affairs* 64.1/2010, pp. 137–163.

Rojavin, Yuri, Mark J. Seamon, Ravi S. Tripathi, Thomas J. Papadimos, Sagar Galwankar, et al. "Civilian Nuclear Incidents: An Overview of Historical, Medical, and Scientific Aspects," *Journal of Emergencies, Trauma, and Shock* 4.2/2011, pp. 260–272.

Romero, Bazan J. (2000). "Sovereignty or Environmental Inequity: The Complexities Surrounding the Proposed Controversial Nuclear Waste Facility at Mescalero," PhD Dissertation submitted at the Graduate College of Bowling Green State University, Ohio, United States. https://www.proquest.com/openview/32642b884aed795158335676288b420b/1?pq-origsite=Gscholar&cbl=18750&diss=yhttps://www.proquest.com/openview/32642b884aed795158335676288b420b/1?pq-origsite=gscholar&cbl=18750&diss=y (09 Aug. 2022).

Rootes, Christopher. "Climate Change, Environmental Activism, and Community Action in Britain," *Social Alternatives* 31.1/2012, pp. 24–28.

Rose, Gregory, Ben Milligan. "Law for the Management of Antarctic Marine Living Resources: From Normative Conflicts towards Integrated Governance?" *Yearbook of International Environmental Law* 20.1/2010, pp. 41–87.

Roy, Jeffrey (1999). "Government and Governance in High-Technology Localities: Ottawa-Carleton and Canada's Technology Triangle." PhD Dissertation at Carleton University, Ottawa, Ontario, Canada. https://curve.carleton.ca/system/files/etd/f1ca23c1-6b5d-45f0-9d75-e51c5a8b55ab/etd_pdf/0bdac7d0591fe18d607874f7100de528/roy-governmentandgovernanceinhightechnologylocalities.pdf (08 Aug. 2022).

Rubinson, Harold Paul (2008). "Containing Science: The U.S. National Security State and Scientists' Challenge to Nuclear Weapons During the Cold War," PhD Dissertation at the University of Texas, Austin, United States. https://repositories.lib.utexas.edu/handle/2152/17997 (07 Aug. 2022).

Rumiel, Lisa. "Getting to the Heart of Science: Rosalie Bertell's Eco-Feminist Approach to Science and Anti-Nuclear Activism," *Journal of Women's History* 26.2/2014, pp. 135–159.

Russell, Laura Jeanne (2002). "Collective Reciprocal Causation: A Model of the Relationships Among Group Behavior, Group Environment, Group Cognition, and Group Effectiveness," PhD Dissertation at the Graduate Faculty of Auburn University, Alabama, United States. https://psycnet.apa.org/record/2002-95015-005 (06 Aug. 2022).

Ryland, Diane. "The Future of Nuclear Energy in Europe: Questions, Problems, and Perceptions," *Managerial Law* 44.4/2002, pp. 91–111.

S

Sachse, V. Edward (1989). "Hegemonic Stability Theory: An Examination," PhD Dissertation at the Louisiana State University and Agricultural and Mechanical College, United States. https://www.proquest.com/openview/78cfc7eecb0d38d4b66eb1f20857f695/1?pq-origsite=gscholar&cbl=18750&diss=y (07 Aug. 2022).

Sacko, David H.(2003). "Hegemonic Governance and the Process of Conflict," PhD Dissertation at the Pennsylvania State University, Department of Political Science, United States. https://www.proquest.com/openview/d182e7e18856d19bf8a4bc50bdad1ab2/1?pq-origsite=gscholar&cbl=18750&diss=y (07 Aug. 2022).

(de) Sadeleer, Nicolas. "Environmental Governance and the Legal Bases Conundrum," *Yearbook of European Law* 31.1/2012, pp. 373–401.

Salvati, Luca, Marco Zitti. "Territorial Disparities, Natural Resource Distribution, and Land Degradation: A Case Study in Southern Europe," *GeoJournal* 70.2-3/2007, pp. 185–194.

Scarpellini, Sabina, Alfonso Aranda, Juan Aranda, Eva Llera, Miguel Marco. "R&D and Eco-Innovation: Opportunities for Closer Collaboration Between Universities and Companies through Technology Centers," *Clean Technologies and Environmental Policy* 14.6/2012, pp. 1047–1058.

Scholz, John T. "American Politics – Hostages of Each Other: The Transformation of the Nuclear Power Industry After Three Mile Island by Joseph V. Rees," *The American Political Science Review* 89.2/1995, p. 503–504.

Schienke, Erich William, (2006). "Greening the Dragon: Environmental Imaginaries in the Science, Technology, and Governance of Contemporary China," PhD Dissertation submitted at the Graduate Faculty of Rensselaer Polytechnic Institute, New York, United States, 2006. https://dspace.rpi.edu/bitstream/handle/20.500.13015/3798/6061_schienke_abs.pdf?sequence=3 (09 Aug. 2022).

Schreurs, Miranda A. "Federalism and the Climate: Canada and the European Union," *International Journal: Canada's Journal of Global Policy Analysis* 66.1/2011, pp. 91–108.

Seabrook, Charles. "Nuclear Power Industry Tries to Improve Image with Public," *Journal Record [Oklahoma City, Okla]* 04/1991.

SEI report, 2021. "The Local 2030 Coalition for the Decade of Action." *Stockholm Environment Institute*, DOI: 10.51414/sei2021.035. At: https://cdn.sei.org/wp-content/uploads/2021/11/local2030-coalition-for-the-decade-of-action.pdf (08 Aug. 2022).

Sengenberger, Werner. "Local Development and International Economic Competition," *International Labour Review* 132.3/1993, pp. 313–330.

Sheingate, Adam D. "Political Entrepreneurship, Institutional Change, and American Political Development," *Studies in American Political Development* 17.2/2003, pp. 185–203.

Shiroyama, Hideaki, Masaru Yarime, Makiko Matsuo, Heike Schroeder, Roland Scholz, Andrea E. Ulrich. "Governance for Sustainability: Knowledge Integration and Multi-Actor Dimensions in Risk Management." *Sustainability Science*, supplement 7.1/2012, pp. 45–55.

Simmons, Erica. "Grievances Do Matter in Mobilization," *Theory and Society* 43.5/2014, pp. 513–546. published by: Springer.

Simpson, Adam (2009). "Transnational Energy Projects and Green Politics in Thailand and Burma. A Critical Approach to Activism and Security," PhD Dissertation, School of History and Politics. The University of Adelaide. https://digital.library.adelaide.edu.au/dspace/handle/2440/58974 (10 Aug. 2022).

Sjoberg, Lennart. "Risk Perception, Emotion and Policy: The Case of Nuclear Technology," *European Review* 11.1/2003, pp. 109–128.

Sjoberg, Lennart. "Antagonism, Trust and Perceived Risk," *Risk Management* 10.1/2008, pp. 32–55. Palgrave Macmillan Ltd.

Skelcher, Chris, Erik-Hans Klijn, Daniel Kubler, Eva Sorensen, Helen Sullivan. "Explaining the Democratic Anchorage of Governance Networks. Evidence from Four European Countries," *Administrative Theory & Praxis* 33.1/2011, pp. 7–38.

Skjaerseth, Jon Birger. "Governance by EU Emissions Trading: Resistance or Innovation in the Oil Industry?" *International Environmental Agreements: Politics, Law and Economics* 13.1/2013, pp. 31–48.

Smith, Adrian, Al Rainnie, Mick Dunford, Jane Hardy, Ray Hudson, David Sadler. "Networks of Value, Commodities and Regions: Reworking Divisions of Labour in Macro-Regional Economies," *Progress in Human Geography* 26.1/2002, pp. 41–63.

Soberon, Jorge M., Jose K. Sarukhan. "A New Mechanism for Science-Policy Transfer and Biodiversity Governance?" *Environmental Conservation* 36.4/ 2009, pp. 265–267.

Solingen, Etel. "Macropolitical Consensus and Lateral Autonomy in Industrial Policy: The Nuclear Sector in Brazil and Argentina," *International Organization* 47.2/1993, p. 263–298.

Sovacool, Benjamin K., Scott Victor Valentine. "The Socio-Political Economy of Nuclear Energy in China and India," *Energy* 35.9/ 2010, pp. 3803–3813, https://doi.org/10.1016/j.energy.2010.05.033 (6 Aug.2022).

Spence, David B. "Regulation, 'Republican Moments,' and Energy-Policy Reform," *Brigham Young University Law Review* 5/2011, pp. 1561–1623.

Spranger, Michael Scott (1999). "Citizen Participation in Environmental Issues: A Comparative Study of Citizen Involvement Programs in Water Quality and Nuclear Waste Management," PhD Dissertation at Portland State University, Urban Studies, U.S. https://www.proquest.com/openview/37229 bc367b00524d541e522176fff7b/1?pq-origsite=gscholar&cbl=18750&diss=y (07 Aug. 2022).

Stokke, Olav Schram. "Regime Interplay in Arctic Shipping Governance; Explaining Regional Niche Selection," *International Environmental Agreements: Politics, Law and Economics* 13.1/2013, pp. 65–85.

Strachan, Neil, Stephen Pye, Nicholas Hughes. "The Role of International Drivers on UK Scenarios of a Low-Carbon Society," *Climate Policy* 8/2008, pp. S125–S139.

Summers, Craig, Donald W. Hine. "Nuclear Waste Goes On the Road: Risk Perceptions and Compensatory Tradeoffs in Single-Industry Communities," *Canadian Journal of Behavioural Science* 29.3/1997, p. 211–223.

Sutherland, Siobhan. (2002). "First Nations and Nuclear Fuel Waste Management: An Analysis of Stakeholder Position," Master of Arts Thesis submitted at the Faculty of Graduate Studies of the University of Guelph, Canada.

Suzuki, Atsuyuki. "Toward a Robust Nuclear Management System," *Daedalus* 139.1/2010, pp. 82–92.

Symon, Andrew. "Southeast Asia's Nuclear Power Thrust: Putting ASEAN's Effectiveness to the Test?" *Contemporary Southeast Asia* 30.1/2008, pp. 118–139.

T

Tang, Shui-Yan, Carlos Wing-Hung Lo. "The Political Economy of Service Organization Reform in China: An Institutional Choice Analysis," *Journal of Public Administration Research and Theory* 19.4/2009, pp. 731–767.

Tannenwald, Nina. "The Nuclear Taboo: The United States and the Normative Basis of Nuclear Non-Use," *International Organization* 53.3/1999, pp. 433–468.

Taubes, Gary. "Whose Nuclear Waste?," *Technology Review* 105.1/2002, pp. 60–67.

Taylor, Bryan C., Brian Freer. "Containing the Nuclear Past: The Politics of History and Heritage at the Hanford Plutonium Works," *Journal of Organizational Change Management* 15.6/2002, p. 563–588.

Teng, Bing-Sheng, T.K. Das. "Governance Structure Choice in Strategic Alliances. The Roles of Alliance Objectives, Alliance Management Experience, and International Partners," *Management Decision* 46.5/2008, pp. 725–742.

Terjesen, Siri, Jolanda Hessels. "Varieties of Export-Oriented Entrepreneurship in Asia," *Asia Pacific Journal of Management* 26/2009, pp. 537–561.

Teubal, Morris. "What Is the Systems Perspective to Innovation and Technology Policy (ITP) and How Can We Apply It to Developing and Newly Industrialized Economies?" *Journal of Evolutionary Economics* 12.1–2/2002, pp. 233–257.

Tilleman, Suzanne Gladys (2009). "Aligning Institutional Logics to Enhance Regional Cluster Emergence: Evidence from the Wind and Solar Energy Industries." PhD Dissertation at the Department of Management, the Graduate School of the University of Oregon, United States. https://www.proquest.com/openview/0f20d3f47fa2a1fa703a90ddb819f811/1?pq-origsite=gscholar&cbl=18750 (07 Aug. 2022).

Titoff, Dmitri A., Caitlin A. Buckley, Dmitry Novak, Richard Weitz, "Assessing Kazakhstan's Proposal to Host a Nuclear Fuel Bank," *UNISCI Discussion Papers* 28/2012, pp. 99–125.

Tod, Dave. "Ontario Power Generation to Assume Armed Security Duties at Nuclear Plants," *SNL Energy Power Week Canada* 08/2007.

Tollefson, Jeff. "Battle of Yucca Mountain Rages On," *Nature* 473.7347/2011, pp. 266–267.

Truffer, Bernhard, and Lars Coenen. "Environmental Innovation and Sustainability Transitions in Regional Studies," *Regional Studies* 46.1/2012, https://doi.org/10.1080/00343404.2012.646164.

Turnock, David. "Environmental Problems and Policies in East Central Europe: A Changing Agenda," *GeoJournal* 55/2001, pp. 485–505.

U

US Fed News Service. "NRC Advisory Committee on Nuclear Waste and Materials Issues Report on Engagement with International Commission on Radiological Protection," *US State News [Washington, D.C.]* 9/2007, Publisher HT Media Ltd. U.S.

US Fed News Service, Including US State News, "NRC Chairman Klein Addresses to American Nuclear Society at Raleigh," *HT Media Ltd.* 2008, Washington, D.C., U.S.

V

Vail, Benjamin. "Ecological Modernization at Work? Environmental Policy Reform in Sweden at the Turn of the Century," *Scandinavian Studies* 80.1/2008, p. 85–108.

Valberg, Anna Helena. "Brazil's Role in Environmental Governance," *Report for the Norwegian Ministry of the Environment*, (FNI) Fridtjof Nansen Institute, Report 8/2011, pp. 1–49.

Vasi, Ion Bogdan, Brayden G. King. "Social Movements, Risk Perceptions, and Economic Outcomes: The Effect of Primary and Secondary Stakeholder Activism on Firms' Perceived Environmental Risk and Financial Performance," *American Sociological Review* 77.4/2012, pp. 573–596.

Vigoda, Eran. "From Responsiveness to Collaboration: Governance, Citizens, and the Next Generation of Public Administration," *Public Administration Review* 62.5/2002, pp. 527–540.

W

Wachsmuth, David (2014). "Post-City Politics: US Urban Governance and Competitive Multi-City Regionalism." PhD Dissertation at the New York University, Department of Sociology, UMI Dissertation Publishing, United States. https://www.mcgill.ca/urbanplanning/people-0/wachsmuth (07 Aug. 2022).

Wagner, Antonin. "Good Governance: A Radical and Normative Approach to Nonprofit Management," *Voluntas: International Journal of Voluntary and Nonprofit Organizations* 25.3/2014, pp. 797–817.

Walcott, Susan M. "Chinese Science and Technology Industrial Parks." Aldershot, Hants: Ashgate, England. 2003.

Wallenburg, Carl Marcus, Thorsten Schaffler. "The Interplay of Relational Governance and Formal Control in Horizontal Alliances: A Social Contract Perspective," *Journal of Supply Chain Management* 50.2/2014, pp. 41–58.

Wang, Chen-Yu (2006). "The Impact of Regional Economic Integration under the GATT/WTO Regime toward the Peace Process: the Case of Conflict Resolution between Taiwan and Mainland China." PhD Dissertation at the Faculty of the Washington College of Law, American University, United States. https://digitalcommons.wcl.american.edu/stu_sjd_abstracts/7/ (07 Aug. 2022).

Wang, Hongli, Zhenlong Peng, Feng Gu. "The Emerging Knowledge Governance Approach Within Open Innovation: Its Antecedent Factors and Interior Mechanism," *International Journal of Business and Management*, 6.8/2011, p. 94–104.

Watkins, Claire A. "Nuclear Power Rate Regulation after Eastern Enterprises: Are Ratepayers Being Taken for a Ride?," *Boston College Environmental Affairs Law Review* 28.1/2000, pp. 191–228.

Weber, Edward P. "A Wish list for 21st Century Environmental Policy: Decentralization, Integration, Cooperation, Flexibility, and Enhanced Participation by Citizens and Local Governments," *Policy Studies Journal* 26.1/1998, pp. 185–195.

White, Robert. "Implementing a Strategic Approach for Energy Efficiency Regulations," *Report on WTO Workshop on Nontariff Barriers Trade in Information Technology Products*, 2015. https://www.wto.org/english/tratop_e/inftec_e/workshopmay15_e/white.pdf (07 Aug. 2022).

Whitford, Josh, Cuz Potter, "The State of the Art. Regional Economies, Open Networks and the Spatial Fragmentation of Production," *Socio-Economic Review* 5/2007, pp. 497–526.

Wiek, Arnim, Barry Ness, Petra Schweizer-Ries, Fridolin S. Brand, Francesca Farioli. "From Complex Systems Analysis to Transformational Change: A Comparative Appraisal of Sustainability Science Projects," *Sustainability Science*, 7/2012, (Supplement 1), pp. 5–24.

Wiliarty, Sarah Elise. "Nuclear Power in Germany and France," *Polity* 45.2/2013, pp. 281–296.

Wilson, David. "Exploring the Limits of Public Participation in Local Government," *Parliamentary Affairs* 52.2/1999, pp. 246–259.

Winchester, Brian N. "Emerging Global Environmental Governance," *Indiana Journal of Global Legal Studies* 16.1/2009, pp. 7–23.

Wittman, Hannah, Charles Geisler. "Negotiating Locality: Decentralization and Communal Forest Management in the Guatemalan Highlands," *Human Organization* 64.1/2005, pp. 62–74.

WISE Uranium Project. "Issues at Canon City Uranium Mill (Colorado)." *World Information Service on Energy Uranium Project (WISE)*/2014, Available at: https://www.wise-uranium.org/umopcc.html (07 Aug 2022).

Wolff, Alan W.M. "China's Drive Toward Innovation," *Issues in Science and Technology* 23.3/2007, pp. 54–62.

Wolfgang, Sofka. "Globalizing Domestic Absorptive Capacities," *MIR: Management International Review* 48.6/2008, pp. 769–792.

Wong, Lai Tim, Gerald E. Fryxell. "Stakeholder Influences on Environmental Management Practices: A Study of Fleet Operations in Hong Kong (SAR), China," *Transportation Journal* 43.4/2004, pp. 22-35.

World Bank, *PPPLRC Public-Private-Partnership Legal Resource Center*, "Energy and Power Ppps." 2019. https://ppp.worldbank.org/public-private-partnership/sector/energy (10 Aug. 2022).

Wurzel, Rudiger K.W., Anthony Zito R., Andrew Jordan J. "From Government Towards Governance? Exploring the Role of Soft Policy Instruments," *German Policy Studies* 9.2/2013, pp. 21-48.

X

Xu, Yi-Chong. "IAEA: Facilitating Multilateral Cooperation on Nuclear Fuel Services," *Social Alternatives* 28.2/2009, pp. 36-41.

Y

Yee, Wai-Hang, Carlos Wing-Hung Lo, Shui-Yan Tang. "Assessing Ecological Modernization in China: Stakeholder Demands and Corporate Environmental Management Practices in Guangdong Province." *The China Quarterly* 213.3/2013, pp. 101-129.

Yi, Yaqun, Yi Liu, Hong He, Yuan Li. "Environment, Governance, Controls, and Radical Innovation during Institutional Transitions," *Asia Pacific Journal of Management* 29.3/2012, pp. 689-708.

Yoon, Jungwon (2011). "Exploring Regional Innovation Capacities of PR China: Toward the Study of Knowledge Divide." PhD Dissertation at the Georgia Institute of Technology, United States. https://www.proquest.com/openview/0f20d3f47fa2a1fa703a90ddb819f811/1?pq-origsite=gscholar&cbl=18750 (07 Aug. 2022).

Yoon, Woojin, Eunjung Hyun, "Economic, Social and Institutional Conditions of Network Governance," *Management Decision* 48.8/2010, pp. 1212-1229.

Z

Zhu, Kevin, Shutao Dong, Sean Xin Xu, Kenneth L. Kraemer. "Innovation Diffusion in Global Contexts: Determinants of Post-Adoption Digital Transformation of European Companies," *European Journal of Information System* 15/2006, pp. 601-616.

Index of Names

A
Achillas, Ch. 66
Adamantiades, A. 119
Al-Rodhan, N. R. F. 184
(d') Aquino, P. 157
Aronson, J. 30
Arts, B. 87, 235
Asheim, B. T. 228
Ates, D. 248

B
Barben, D. 268, 269
Beken, T. V. 44, 110, 167, 168
Bhadra, M. 262, 263
Bierly, D. 206, 207
Bjorklund, A. E. 224
Blowers, A. 33
Bluth, C. 181
Boschma, R. A. 234, 237
Bozeman, B. 177
Breuste, J. 234
Bridge, G. 254, 255
Brownell, E. 271
Burchell, J. 16, 17, 147, 148
Burger, J. 15, 43, 205, 206, 208
Butler, C. 152, 181
(van) Buuren, A. 66, 67, 70–72

C
Chan, G. 145
Chang, H.- C. 247
Chen, L.- C. 74, 75
Chung, J. B. 41, 42
Ciarli, T. 62
Clark, S. M. 19, 22
Coenen, L. 45, 51, 54, 55, 57, 228
Conde, M. 258
Cooke, P. 47–49, 226, 230, 231

Cortes, M. I. C. 269

D
Davies, A. R. 150, 151
Davoudi, S. 150, 232, 233
Dawson, J. I. 153, 154, 241
Dennis, M. A. 175
Deutz, P. 150
Doherty, B. 271
Doloreux, D. 215–217
Durant, D. 100, 103, 163
Dusinberre, M. 152, 191

E
Edelenbos, J. 76, 77
Edwards, A. 91–93
Eizaguirre, S. 97
Endres, D. 246
Escobar, A. 259

F
Falk, R. 270
Farla, J. 51, 54, 55
Feldman, D. L. 224, 225
Finnveden, G. 218, 224
Fox, M. K. 243, 244
Frey, B.S. 85

G
Gee, G.W. 138
Gibbs, D. 150, 210
Gibson, K. 17
Gimholt, E. 284
Gouldson, A. 28, 105–107
Grant, D. 209, 210
Greenberg, M. 108, 109
Grossmann, M. 154
Grote, J. R. 237

H
(van der) Heijden, J. 267, 268
Heijungs, R. 219
Hess, D. J. 264
Hisschemoller, M. 28, 29
Hollingsworth, R. J. 91, 94
Holmes, J. C. 243, 245
Hogselius, P. 118
Honneland, G. 139, 140
Hulbert, M. 125–128, 171, 172
Hung, H. - C. 44

I
Iles, A. 264

J
Jasanoff, S. 265
Jenkins-Smith, H. C. 44
Jennings, A. B. 206
Jessop, B. 91, 95, 96, 232
Johnson, G. F. 18, 19, 204
Jones, C.F. 265
Jorgensen, U. 51, 59

K
Kennedy, E. H. 143, 144
Kessides, I. N. 196, 197
Khoo, S.-M. 272
Kitschelt, H. P. 198–200
Knop, K. C. 250
Kostelnik, K. M. 203, 204
Krupar, S. R. 183, 202
Kunreuther, H. 40, 43
Kurath, M. 192
Kurtz, H. E. 239, 240

L
Lagendijk, A. 87, 88, 90
Laird, F. N. 265
Larsen, R. K. 251, 252
Lawhon, M. 271
Leong, C. C. 160, 161

Lesbirel, S. H. 141
Leschine, T. M. 43, 206, 207
Lin, S. S. 247, 248
Linkov, I. 22, 23
Loveland, P. J. 137

M
Markard, J. 45, 47, 54, 55
Marshall, A. 40
Maythorne, L. 249
McIntyre, J. 156
Messer, C. M. 41
Miller, C. A. 178
Mochizuki, M. M. 244
Morrissey, A. J. 218, 219
Moulaert, F. 237
Mulvaney, D. 265

N
Nichols, R. W. 165, 166
Nielsen, H. 179, 180
Nykvist, B. 227

O
Ottinger, G. 266

P
Paraskevopolou, E. 64, 66
Parthasarathy, S. 148, 179
Pennington, C. W. 169, 170
Phadke, R. 36
Pickett, S. E. 194, 195
Pidgeon, N. F. 129, 130, 181
Poetz, A. 36–38
Powell, J. H. 201
Purdy, J. 270

R
Ramana, M.V. 38, 39, 132, 133
Ratchford, J. T. 164
Roberts, M. R. 192
Rosa, E. A. 133, 207

Index of Names

S
Safarzynska, K. 62
Sauer, A. 157, 158
Sbicca, J. 267
Schaffer, M. B. 120
Sessa, C. 253
Seyad, A. 32
Sheng, J. 51
Shineha, R. 35
Siegel, M.D. 134–136
Sjoberg, L. 38, 40, 153
Smith, A. 59, 61, 127
Soubbotina, T. 179
(van) Staden, M. 29
Stoutenborough, J. W. 133
Strubelt, W. 236
Su, N. M. 185
Suksi, M. 236
Sundberg, J. 241, 242
Sundqvist, G. 36
Szarka, J. 149

T
Teravainen, T. 121–123
Todtling, F. 229
Tuler, S. 202, 203

U
Uekoetter, F. 182

V
Valentine, S. V. 18, 171, 172
Vergragt, P. J. 231, 232

W
Waelbers, K. 187, 188
Walter, D. W. 261, 262
Weber, K. M. 63, 64
Weible, C. M. 208, 209
Williams, G. 51
Williams, R. W. 242, 243
Wilson, J. F. H. 260, 261
Wolsink, M. 151, 152
Woods, M. 255, 257

X
Xiao, C. 189

Y
Yeung, H. W. - C. 81, 82

Z
Zhou, Y. 131

Studies in Politics, Security and Society

Edited by Stanisław Sulowski

Vol. 1   Robert Wiszniowski (ed.): Challenges to Representative Democracy. A European Perspective. 2015.
Vol. 2   Jarosław Szymanek: Theory of Political Representation. 2015.
Vol. 3   Alojzy Z. Nowak (ed.): Global Financial Turbulence in the Euro Area. Polish Perspective. 2015.
Vol. 4   Jolanta Itrich-Drabarek: The Civil Service in Poland. Theory and Experience. 2015.
Vol. 5   Agnieszka Rothert: Power of Imagination. Education, Innovations and Democracy. 2016.
Vol. 6   Zbysław Dobrowolski: Combating Corruption and Other Organizational Pathologies. 2017.
Vol. 7   Vito Breda: The Objectivity of Judicial Decisions. A Comparative Analysis of Nine Jurisdictions. 2017.
Vol. 8   Anna Sroka: Accountability and democracy in Poland and Spain. 2017.
Vol. 9   Anna Sroka / Fanny Castro-Rial Garrone / Rubén Darío Torres Kumbrián (eds.): Radicalism and Terrorism in the 21st Century. Implications for Security. 2017.
Vol. 10 Filip Pierzchalski: Political Leadership in Morphogenetic Perspective. 2017.
Vol. 11 Alina Petra Marinescu: The Discursive Dimension of Employee Engagement and Disengagement. Accounts of keeping and leaving jobs in present-day Bucharest organizations. 2017.
Vol. 12 Jacek Giedrojć: Competition, Coordination, Social Order. Responsible Business, Civil Society, and Government in an Open Society. 2017.
Vol. 13 Filip Ilkowski: Capitalist Imperialism in Contemporary Theoretical Frameworks. 2017.
Vol. 14 Leszek Leszczyński / Adam Szot (eds.): Discretionary Power of Public Administration. Its Scope and Control. 2017.
Vol. 15 Tadeusz Klementewicz: Understanding Politics. Theory, Procedures, Narratives. 2017.
Vol. 16 Tomasz Bichta: Political Systems of the Former Yugoslavia. 2018
Vol. 17 Miroslav Palárik / Alena Mikulášová / Martin Hetényi / Róbert Arpáš: The City and Region Against the Backdrop of Totalitarianism. 2018
Vol. 18 Jolanta Itrich-Drabarek / Stanisław Mazur / Justyna Wiśniewska-Grzelak (eds.): The Transformations of the Civil Service in Poland in Comparison with International Experience. 2018..
Vol. 19 Jerzy Juchnowski / R. Jan Sielezin / Ewa Maj: The Image of "White" and "Red" Russia in the Polish Political Thought of the 19th and 20th Century. Analogies and Parallels. 2018.
Vol. 20 Roman Kuźniar: Europe in the International Order. 2018.
Vol. 21 Piotr Jaroszyński: Europe - the Clash of Civilisations. 2018.
Vol. 22 Stanisław Filipowicz: Truth and the Will to Illusion. 2018.
Vol. 23 Andrzej Szeptycki: Contemporary Relations between Poland and Ukraine. The "Strategic Partnership" and the Limits Thereof. 2019.

Vol. 24 Sylwester Gardocki / Rafał Ożarowski / Rafał Ulatowski (eds.): The Islamic World in International Relations. 2019.

Vol. 25 Jacek Zaleśny (ed.): Constitutional Courts in Post-Soviet States. Between the Model of a State of Law and Its Local Application. 2019.

Vol. 26 Andrzej Antoszewski / Przemysław Żukiewicz / Mateusz Zieliński / Katarzyna Domagała: Formation of Government Coalition in Westminster Democracies. Towards a Network Approach. 2020.

Vol. 27 Joanna Osiejewicz: Global Governance of Oil and Gas Resources in the International Legal Perspective. 2020.

Vol. 28 Anita Oberda-Monkiewicz: Poland-Mexico towards a Strategic Partnership. 2020.

Vol. 29 Bartosz Czepil / Wojciech Opioła: Ethnic diversity and local governance quality. The case of Opole Province in Poland. 2020.

Vol. 30 Adam Szymański / Jakub Wódka / Wojciech Ufel / Amanda Dziubińska: Between Fair and Rigged. Elections as a Key Determinant of the 'Borderline Political Regime' - Turkey in Comparative Perspective. 2020.

Vol. 31 Tomasz Grzegorz Grosse (ed.): Fuel for Dominance. On the Economic Bases of Geopolitical Supremacy. 2020.

Vol. 32 Dariusz Jarosz / Maria Pasztor: From Subjection to Independence. Post-World War II Polish-Italian Relations. 2020.

Vol. 33 Paweł Sekuła: Chernobyl Liquidators. The Unknown Story. With the Testimony of the President of Latvia. 2020.

Vol. 34 Kamil Glinka (ed.): Urban Policy System in Strategic Perspective: From V4 to Ukraine. 2020.

Vol. 35 Paweł Lesiński: At the Origins of German Liberalism: the State in the Thought of Robert von Mohl. 2020.

Vol. 36 Jana Popovicsová (ed.): Reflexions about a Cultural and Social Phenomenon: Identity. 2020.

Vol. 37 Agnes Bernek: Geopolitics of Central and Eastern Europe in the 21st Century. From the Buffer Zone to the Gateway Zone. 2021.

Vol. 38 Marek Antoni Musioł: The European Union as a Post-Lisbon Regional Security Complex. 2021.

Vol. 39 Alina Petra Marinescu: Manipulation in the Disclosure of the *Securitate* Files. The Case of Mona Muscă. 2021

Vol. 40 Jiří Zákravský: Cycling Diplomacy. Undemocratic Regimes and Professional-Road Cycling Teams Sponsorship. 2021.

Vol. 41 Mirosław Karwat / Filip Pierzchalski / Marcin Tobiasz (eds.): Constituents of Political Theory. Selected Articles of the Warsaw School of Political Theory. 2021.

Vol. 42 Angieszka Bieńczyk-Missala: Preventing Mass Human-Rights Violations and Atrocity Crimes. 2021.

Vol. 43 Jan Garlicki: Political Participation Capital. 2021.

Vol. 44 Mirosław Karwat: Theory of Provocation. In Light of Political Science. 2022.

Vol. 45 Anna Sroka / Piotr Potejko / Rubén Darío Torres Kumbrián (eds.): Migration and Border Security. Global Perspectives. 2021.

Vol.	46	Stanisław Sulowski / Tomasz Słomka (eds.): The Political System of Poland. Tradition and Contemporaneity. 2022.
Vol.	47	Ewelina Kancik-Kołtun / Josef Smolík (eds.): New Political Parties in the Party Systems of the Czech Republic. 2022.
Vol.	48	Ewa Feder-Sempach: Economic Populism in British and American Political Discourse. A Comparative Analysis of Boris Johnson's and Donald Trump's Speeches. 2022.
Vol.	49	Katarzyna Kołodziejczyk (ed.): The EU Towards the Global South During the COVID-19 Pandemic. 2022.
Vol.	50	Silvia Amato: Public Undertakings of Nuclear Waste Storage. The Role of the Government in the Dissemination of Public Knowledge. 2022.

www.peterlang.com

www.ingramcontent.com/pod-product-compliance
Ingram Content Group UK Ltd.
Pitfield, Milton Keynes, MK11 3LW, UK
UKHW041912140426
5217IPUK00002B/10